Cancer:
Basic Science and
Clinical Aspects

Craig A. Almeida and Sheila A. Barry

Department of Biology, Stonehill College, Easton,
Massachusetts, USA

D1275407

⟨W⟩WILEY-BLACKWELL

A John Wiley & Sons, Ltd., Publication

This edition first published 2010, © 2010 by Craig A. Almeida and Sheila A. Barry

Wiley-Blackwell is an imprint of John Wiley & Sons, formed by the merger of Wiley's global Scientific, Technical and Medical business with Blackwell Publishing.

Registered office: John Wiley & Sons Ltd, The Atrium, Southern Gate, Chichester, West Sussex, PO19 8SQ, UK

Other editorial offices:
9600 Garsington Road, Oxford, OX4 2DQ, UK
111 River Street, Hoboken, NJ 07030-5774, USA

For details of our global editorial offices, for customer services and for information about how to apply for permission to reuse the copyright material in this book please see our website at www.wiley.com/wiley-blackwell

Library of Congress Cataloguing-in-Publication Data
Almeida, Craig A.
  Cancer : basic science and clinical aspects / Craig A. Almeida and Sheila A. Barry.
      p. ; cm.
  Includes bibliographical references and index.
  ISBN: 978-1-4051-5606-6 (pbk. : alk. paper)   1. Cancer–Textbooks.   I. Barry,
Sheila A.   II. Title.
  [DNLM: 1. Neoplasms. QZ 200 A447c 2010]
  RC262.A378 2010
  616.99′4–dc22

                                                                              2009015237

A catalogue record for this book is available from the British Library.

Set in 10/12.5pt Meridien
by Graphicraft Limited, Hong Kong

1   2010

Contents

Companion website www.wiley.com/go/almeida/cancer

Preface

We have authored a textbook on cancer that is unique in its coverage in a number of respects. This book stands out from others because it is written for both nonscience and science majors. The coverage spans the spectrum from the molecular, cellular, and genetic through to the applied aspects of the disease. The book has been structured so that it will be an appropriate text for use by an instructor regardless of the depth to which he/she desires to cover any of the material. The amount of material is manageable within a single semester, and individual chapters can be excerpted for study on each of the major cancers.

We believe this book is appropriate for cancer courses offered to either science or nonscience majors at any level. A target audience with such a variant science background is accommodated by a series of introductory chapters that provide the molecular, cellular, and genetic information needed to comprehend the material of the subsequent chapters. A reader without a science background could study the chapter on breast cancer and learn the risk factors, symptoms, diagnostic testing, and treatment methods without being overwhelmed. If after reading about the risks associated with the *BRCA1* and *2* genes, a student wants a better understanding of what a gene is, he or she could then refer back to the appropriate section of one of the introductory chapters. This cross-referencing ability is what we feel is the basis for the success of the text from the perspectives of both student and instructor. The introductory chapters can be used by lower or upper class science majors to review foundational information.

The chapters of the book are grouped into two sections. The first seven chapters contain introductory information that will be most helpful to the nonscientist while serving as a review for the scientist in training. The second section contains nine chapters, each focusing on a specific form of cancer in areas such as risk factors, diagnostic and treatment methods, and relevant current research. Each of the chapters includes review questions as marginal insertions at points through the text, key words/terms in bold in the text, boxed articles highlighting stories of an individual's experience, and complex questions in the section "Expand your knowledge" for the student to answer with some additional reading.

We have taught an undergraduate biology of cancer course open to all majors since the fall of 2005. The organization of the book reflects the format that we have used successfully when teaching the course. Since this text is intended for use in either a nonscience or science course, it addresses a wide range of issues associated with cancer. Depending on each course design, it could be either an elective or satisfy a requirement within a general education program or a natural science or allied health major. One of the major strengths of the book is that it can be used in any level undergraduate course. There are no specific prerequisites assumed; the information in the introductory chapters is sufficient to bring the non-scientist to the level needed to read and understand the later chapters. The ultimate intent of the book is to have appeal to students who are either at the beginning or intermediate stages of scientific inquiry into the study of cancer.

Acknowledgments

We would like to thank the many editors from Blackwell and John Wiley and Sons that have worked with us during the development of the text for their thoughtful and creative contributions: Nancy Whilton, Elizabeth Frank, Humbert Haze, Steve Weaver, Karen Chambers, Kelvin Matthews, and Pat Croucher. We are appreciative of the talents of Ali McNeill, the illustrator of many of the text's figures. A special thanks is extended to Sandy Parker for her enthusiastic Blackwell introduction. We are indebted to those who critically read and provided valuable feedback on draft chapters: anonymous faculty reviewers; our colleagues, Greg Maniero, Jane De Luca, David Gilmore, Lucy Dillon, and Brenda Sweeney; and our students, Jacqueline Tenaglia, Kristen McCarthy, Lauren Bennett, Samantha April, and Tyler Herbert. We are so appreciative of the valuable assistance provided by our administrative assistant Romelle Berry. Two additional students are due a tremendous thank you: Melissa Martin for her thoughtful insight and superior editorial skills, and Sarah Wilson for her creative draft illustrations. It is quite likely that our fellow faculty and administrators at Stonehill College are unaware of how truly grateful we are for their constant encouragement and generous support throughout the development and writing of this book.

A special appreciation goes to all of the students who have taken our biology of cancer course, for their subtle and unknowing influences can be found in the organization, content and pedagogy of this book.

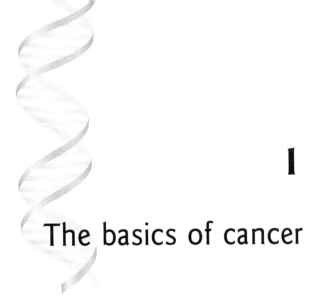

The basics of cancer

> If the three worst words are, 'You have cancer', then the four worst are 'Your cancer is back'.
>
> Katie Couric, American newscaster and journalist

Cancer: Basic Science and Clinical Aspects, 1st edition. By C. A. Almeida and S. A. Barry. Published 2010 by Blackwell Publishing, ISBN 978-1-4051-5606-6.

Very little strikes more fear into peoples' hearts than being told they have cancer. Such a diagnosis can turn a person's world upside down and conjure up thoughts of what lies ahead: pain, disfigurement, disability, nausea, hair loss, or even death. Recent years, however, have seen extraordinary advances in basic cancer research and in the development of more effective methods for the detection, diagnosis, and treatment of cancer. Consequently, while the phrase "You have cancer," *may* be life-altering, it is not necessarily the devastating, life-threatening diagnosis of generations past.

CANCER IS A COMPLEX ENTITY

In the most basic sense, cancer is the abnormal, uncontrolled growth of previously normal cells. The transformation of a cell results from alterations to its DNA that accumulate over time. The change in the genetic information causes a cell to no longer carry out its functions properly. A

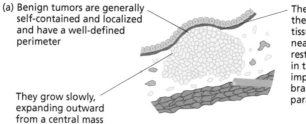

(a) Benign tumors are generally self-contained and localized and have a well-defined perimeter

They are dangerous when they compress surrounding tissues. A benign tumor near a blood vessel could restrict the flow of blood; in the abdomen it could impair digestion; in the brain it could cause paralysis

They grow slowly, expanding outward from a central mass

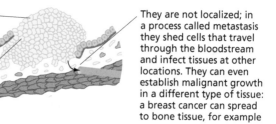

(b) Malignant tumors are not self-contained, and usually do not compress surrounding tissues. Their growth is an irregular invasion of adjacent cells

They are not localized; in a process called metastasis they shed cells that travel through the bloodstream and infect tissues at other locations. They can even establish malignant growth in a different type of tissue: a breast cancer can spread to bone tissue, for example

Although they may grow slowly, they are also capable of very rapid growth

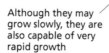

Figure 1.1 Benign vs. malignant cancers. (a) A benign tumor is a mass of cells that remains within the tissue in which it originally developed. (b) The invasion of cancer cells into surrounding tissues is the hallmark of a malignant tumor. Malignant cells may break free from the tumor and travel to other locations in the body through the process of metastasis. Source: http://health.stateuniversity.com/pages/1580/Tumor-Removal.html

primary characteristic of cancer cells is their ability to rapidly divide, and the resulting accumulation of cancer cells is termed a tumor. As the tumor grows and if it does not invade the surrounding tissues, it is referred to as being benign (Figure 1.1a). If, however, the tumor has spread to nearby or distant tissues then it is classified as malignant (Figure 1.1b). Metastasis is the breaking free of cancer cells from the original primary tumor and their migration to either local or distant locations in the body where they will divide and form secondary tumors.

> **Do benign or malignant cells form metastatic tumors?**

There are many types of cancer

Cancer is not a single disease; there are over 100 identified types, all with different causes and symptoms. To distinguish one form from another the cancers are named according to the part of the body in which they originate. Some tumors are identified to reflect the type of tissues they arise from, with the suffix *-oma*, meaning tumor, added on. For example, *myelos-* is a Greek term for marrow. Thus, myeloma is a tumor of the bone marrow, whereas hepatoma is liver cancer (*hepato-* = liver), and melanoma is a cancer of melanocytes, cells found primarily in the skin that produce the pigment melanin. (Table 1.1)

There are four predominant types of cancer

The four major types of cancer are carcinomas, sarcomas, leukemias, and lymphomas. Approximately 90% of human cancers are carcinomas, which arise in the skin or epithelium (outer lining of cells) of the internal organs, glands, and body cavities. Tissues that commonly give rise to carcinomas are breast, colorectal, lung, prostate, and skin. Sarcomas are less common than carcinomas and involve the transformation of cells in connective tissue such as cartilage, bone, muscle, or fat. There are a variety of sarcoma subtypes and they can develop in any part of the body, but most often arise in the arms or legs. Liposarcoma is a malignant tumor of fat tissue (lipo- = fat) whereas a sarcoma that originates in the bone is called osteosarcoma (osteo- = bone).

> **What is the difference between the terms hepatoma and hepatocarcinoma?**

Certain forms of cancer do not form solid tumors. For example, leukemias are cancers of the bone marrow, which leads to the over-production and early release of immature leukocytes (white blood cells). Lymphomas are cancers of the lymphatic system. This system, which is a component of the body's immune defense, consisting of lymph, lymph vessels, and lymph nodes, serves as a filtering system for the blood and tissues.

Each cancer is unique

While there are certain commonalities shared by cancers of a particular type, each may be unique to a single individual. This is because of different cellular mutations that are possible, and can depend on whether the disease is detected at an early or advanced stage. As a result, two women diagnosed with breast cancer may or may not receive the same treatment. The impact of the disease on the individual, as well as the final outcome of the disease, is unique in every case. Still, several types of cancers can have a similar set of symptoms, which may be shared with several other conditions, making screening, detection, and diagnosis a complex problem.

A tumor can impact the function of the tissue in which it resides or those in the surrounding areas. Tumors provide no useful function themselves and may be considered "parasites," with every step of their advance being at the expense of healthy tissue (Figure 1.2). While most types of cancers form tumors, many do not form discrete masses. As previously stated, leukemia is a cancer of the blood that does not produce a tumor, but rather rapidly produces abnormal blood cells in the bone marrow at the expense of normal blood cells.

Table 1.1 Tumor terminology

Prefix	Cell type	Benign tumor	Malignant tumor	Tissue affected
Tumors of epithelial cells:				
Adeno-	Gland	Adenoma	Adenocarcinoma	Breast, colon/rectum, lung, ovary, pancreas, prostate
Basal cell	Basal cell	Basal cell adenoma	Basal cell carcinoma	Skin
Squamous cell	Squamous cell	Keratoacanthoma	Squamous cell carcinoma	Esophagus, larynx, lung, oral cavity, pharynx, skin, cervix
Melano-	Pigmented cell	Mole	Melanoma	Skin
Tumors of supporting tissue origin:				
Hemangio-	Blood vessels	Hemangioma	Hemangiosarcoma	Blood vessels
Lipo-	Fat	Lipoma	Liposarcoma	Fat cells
Meningio-	Meninges	Meningioma	Meningiosarcoma	Brain
Myo-	Muscle	Myoma	Myosarcoma	Muscle
Osteo-	Bone	Osteoma	Osteosarcoma	Bone
			Ewing's sarcoma	
Cancers of blood and lymphatic origin:				
Lympho-	Lymphocyte		Lymphoma	Lymphocytes
			Lymphocytic leukemia	
Myelo-	Bone marrow		Myeloma,	Granulocytes
			Myelogenous leukemia	

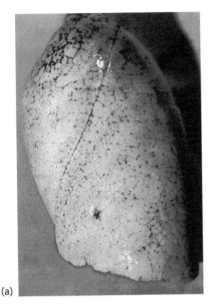

(a)   (b)

Figure 1.2 (a) Healthy lung and (b) cancerous lung. Reprinted with permission © American Lung Association. For more information about the American Lung Association or to support the work it does, call 1-800-LUNG-USA (1-800-586-4872 or log on to www.lungusa.org)

The development of tumors

All tumors begin with mutations (changes) that accumulate in the DNA (genetic information) of a single cell causing it and its offspring to function abnormally. DNA alterations can be sporadic or inherited. Sporadic mutations occur spontaneously during the lifespan of a cell for a number of reasons: a consequence of a mistake made when a cell copies its DNA prior to dividing, the incorrect repair of a damaged DNA molecule, or chemical modification of the DNA, each of which interferes with expression of the genetic information. Inherited mutations are present in the DNA contributed by the sperm and/or egg at the moment of conception. To date, 90–95% of diagnosed cancers appear to be sporadic in nature and thus have no heredity basis. Whether the mutations that result in a cancer are sporadic or inherited, certain genes are altered that negatively affect the function of the cells.

Genetic influence on tumors

A link between a particular genetic mutation and one or more types of cancers is made by analyzing and comparing the DNA of malignant tissue samples obtained from patients and members of families with a high incidence of a particular cancer and comparing it to the DNA from healthy individuals. For example, a study could be conducted in which the DNA isolated from tumor cells obtained from liver cancer patients is analyzed and determined to possess certain versions of genes whereas different versions of those same genes are present in the DNA of liver cells of healthy persons. An association could then be drawn between the "bad" versions of those genes and liver cancer.

This type of analysis has been crucial in identifying certain versions of genes associated with a predisposition for the development of particular forms of cancer. For example, studies have demonstrated that there is an elevated risk of breast or ovarian cancer associated with certain versions of the BRCA1 and/or BRCA2 genes (Chapter 8). Another example is retinoblastoma, a rare tumor of the eye typically found in infants and young children, which is associated with alterations within the Rb gene (Chapter 4).

CANCER THROUGH THE AGES

Although not specifically identified as such, cancer has been known for many centuries. In fact, there is evidence of tumors in the bones of five thousand year old mummies from Egypt and Peru. The disease itself was not very common, nor explored or understood, because in ancient times fatal infectious diseases resulted in shorter lifespans. Given that the vast majority of cancers are sporadic, there was less opportunity for the accumulation of the mutations necessary to transform normal cells into cancerous ones.

The word "cancer" was first introduced by Hippocrates (460–370 BC), the Greek physician and "father of medicine" (Figure 1.3). He coined the term carcinoma, from the Greek word *karcinos*, meaning "crab," when describing tumors. This is because tumors often have a central cell mass with extensions radiating outward that mimic the shape of the shellfish (Figure 1.4).

Hippocrates believed that the body contained four "humors" or body fluids, and that each fluid was associated with a specific personality or temperament characteristic. "Blood" persons had a sanguine or optimistic personality with a passionate, joyous disposition. Someone with a dull or sluggish temperament would have "Phlegm." Possessing "Yellow Bile" meant that one was quick to anger, while having "Black Bile" indicated a person was melancholic or depressed. In medieval times, it was believed that disease was a result of an imbalance of any of the four humors and physicians could restore health or harmony by purging, starving, vomiting, or bloodletting. In particular, an excess of black bile was thought to be the primary cause of cancer. This theory was accepted and taught for over 1300 years through the Middle Ages and championed by Claudius Galen (131–201 AD), a Greek physician and writer who described many diseases using this hypothesis (Figure 1.5). Both Hippocrates and Galen defined disease as a natural process, a theory that remained for centuries.

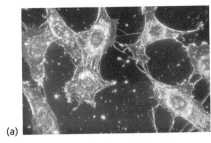
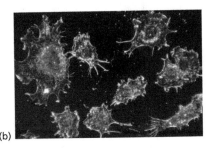

(a)   (b)

Figure 1.8 (a) Note the abundance of the thin, sheet-like extensions from the cell bodies of the healthy cells. (b) Note the rounded appearance of the cancer cells. Source: National Cancer Institute Visuals Online; http://visualsonline.cancer.gov/

in both appearance and behavior from normal cells within the same tissue or organ (Figure 1.8). Early twentieth century accomplishments in the development of cell culture (an *in vitro* technique for studying the activity of cells in an organism under simulated physiological conditions[1.1]), new and improved diagnostic techniques, the discovery of chemical carcinogens, and the use of chemotherapy (powerful anticancer drugs) all had significant impacts upon the understanding and treatment of cancer.

> [1.1] The term *in vitro* is Latin for "in glass" meaning that an experiment is being conducted in an artificial environment, such as a glass test tube. The complementary Latin term *in vivo* means "in life," implying that something is occurring in a living organism.

MODERN DAY CANCER RESEARCH AND TREATMENT

The radioactive element radium, isolated by Marie and Pierre Curie in 1898, was found to be effective in the treatment of tumors in 1903. While both healthy and cancerous cells are susceptible to the damage caused by X-rays, cancer cells are inherently less able to repair the damage and recover. Once safe dosage levels were determined, radiation therapy became a standard form of treatment for many cancers.

Public awareness of cancer

As the access to scientific knowledge increased, so too did the public's fear and misconceptions. In the early 1900s the word "cancer" was rarely spoken in public and it was omitted from obituaries, similar to how the use of the word AIDS was avoided in the 1980s and 1990s. Surprisingly, the first widely published discussion of cancer was a 1913 *Ladies Home Journal* magazine article that asked "What Can We Do About Cancer?" and listed the disease's principal warning signs. With its large circulation,

the magazine brought the word "cancer" to the forefront of public awareness and discussion. That same year in New York City, a group of ten
physicians and five laypeople established the American Society for the
Control of Cancer (ASCC). The group raised $10,000 that year, and pub-
lished a pamphlet titled *The Facts About Cancer*. In 1945 the ASCC changed
its name and to this day is known as the American Cancer Society (ACS).
The following year, volunteer Mary Lasker and her colleagues raised over
$4 million, an extraordinary amount at that time and even today, for the
Society. Of that sum, $1 million was allocated to establish the Society's
cancer research program. In 1947 the ACS initiated its public education
campaign concentrating on "The seven signs and symptoms of cancer,"
utilizing the word "Caution" as a way to remember the first letter of each
symptom:

- Change in bowel or bladder habits.
- A sore that does not heal.
- Unusual bleeding or discharge.
- Thickening or a lump in the breast or other parts of the body.
- Indigestion that is chronic or difficulty in swallowing.
- Obvious changes in a wart or mole.
- Nagging cough or hoarseness.

Although currently outdated due to its lack of specificity and discontinued
in the 1980s, the campaign served to provide enormous awareness among
people who, for the first time, were encouraged to be aware of and on
the lookout for possible signs and symptoms of cancer.

Unexpected discoveries result in some cures

The first clinical use of a dramatic new form of cancer treatment –
chemotherapy – occurred in the 1940s. Soldiers in World War I who
were exposed to the chemical warfare agent known as mustard gas
would develop painful blisters on their skin and eyes. If the gas was inhaled,
their ability to breathe was severely compromised because the gas would
destroy the mucous linings of their lungs. It was noted that those who
survived an attack developed low white
blood cell counts. This led researchers to
question whether mustard gas or a derivative of it could be used to treat certain cancers that affect white blood cells.

1.2 The use of chemicals as weapons
was considered so horrific and inhu-
mane that it was outlawed by an international chemical weapons convention
treaty in 1997. What is significant is
that the evil use of chemical warfare
against people gave rise to the beneficial practice of chemotherapy in the
war against cancer.

Shortly after World War II, a clinical
study determined that the injection (rather
than inhalation) of the principal component of mustard gas caused the temporary
remission of lymphoma, a type of white
blood cell cancer.[1.2] As is commonly done
when a molecule demonstrates potential

clinical effectiveness, a large number of chemical variants were synthesized in an attempt to lessen the highly reactive nature of the original mustard gas compound while retaining, and eventually improving upon, its desired function.

The synthesis and testing of literally hundreds of molecules in animal and human trials have led to several of them becoming conventional cancer treatment regimens. Serendipitous findings similar to those associated with mustard gas tend to occur often in scientific discoveries and are examples of some good coming from an unfortunate event.

Another milestone came in 1947 when Dr Sidney Farber of the Harvard Medical School treated children with acute leukemia, another form of white blood cell cancer, with the drug aminopterin. The drug blocked the synthesis of nucleotides, the building blocks of DNA. This effectively inhibited the division of the cancer cells because they were unable to replicate their DNA and the children entered into a cancer-free state of remission. Even though the remission period was only temporary, Dr Farber is credited with the first successful use of chemotherapy.

Government funding was/is necessary

The National Cancer Institute Act was passed by the US Congress in 1937, creating the National Cancer Institute (NCI), with the goal of conducting

---

Box 1.2

The Legacy of Henrietta Lacks

Henrietta Lacks died of cervical cancer in 1951 at the age of 31 at Johns Hopkins University Hospital in Baltimore, Maryland. She might have long been forgotten by all except her family and friends. Instead, she lives on in the form of highly malignant cells taken from the quarter-sized tumor that was removed from her cervix and that eventually invaded almost every organ of her body. Laboratories and research centers worldwide have been using these cells for many years. Given the code name HeLa, for the first two letters of her first and last names, these were the first human cells known to thrive and multiply outside of the body. Still alive and rapidly dividing to this day, HeLa cells have been used in the development of the polio vaccine, the search for causes of cancer and a cure for leukemia, the study of the growth of viruses, the mechanisms that control the expression of our genetic information, and the effects of drugs and radiation on cellular functions. Henrietta's cells grow aggressively, producing an entirely new generation of cells every 24 hours. Unfortunately, at the time of her death and for years after, there was no system of "informed consent" in medical research, so it took many years for her family to discover the impact that she has had on science.

and promoting cancer research. In 1971, President Richard M. Nixon signed the National Cancer Act, legislation that provided federal funds for cancer research to fight the disease. This Act infused money and authority into the NCI in order to more effectively carry out the national effort to understand and fight cancer. Thus the phrase "War on Cancer" was born. After more than three decades the war remains ongoing, although great strides occur almost daily.

The field of oncology was born

One primary outcome of the research and focus on cancer has been the establishment of the field of oncology, the medical subspecialty dealing with the study and treatment of cancer. Previously, primarily family physicians treated people diagnosed with cancer and followed their care throughout their illnesses. If a woman developed the disease she might have been treated by her gynecologist, a doctor who specializes in female medical issues and who did not have extensive cancer training. The body of knowledge physicians had to work with was limited, to say the least. Only 51 physicians attended the first meeting of the American Society of Clinical Oncology in 1964. Today the organization comprises more than 25,000 members representing the principal oncology disciplines (medical, radiological, and surgical) as well as several subspecialties (geriatric oncology, pediatric oncology, gastrointestinal oncology, etc.). Government funding and private research spurred by charitable giving and the tireless support of groups such as the American Cancer Society, St Jude Children's Research Hospital, and the Boston-based Jimmy Fund have helped to raise cancer awareness. Unfortunately, as aging baby boomers become increasingly cancer-prone, medical schools cannot train enough new oncologists to satisfy the need for them. A 2007 report by the Association of American Medical Colleges estimates that by 2020, visits to oncologists will increase by 48% while the projected number of oncologists is estimated to grow by only 14% over the same time period. It is hoped that medical schools will acknowledge this shortage and address the issue by taking an active role in recruiting a greater number of future oncologists.

Currently, there are a wide variety of methods to diagnose and treat cancer. The ones used are dependent upon the type of cancer and its current stage and location within the body. Genetic links to the disease are being discovered at a rapid pace, enabling physicians to identify those patients who have a predisposition for the development of specific cancers or will respond best to certain forms of treatment. It is hoped that this knowledge will encourage people to be more diligent in maintaining healthy lifestyles and be more consistent in following recommendations for regular screening tests. Some treatments will be personalized based on one's genetic make-up and/or the specific genetic make-up of the tumor.

While not all treatment options offer a "cure," they enable a large number of individuals to manage their disease and live fairly normal and full lives.

The field of oncology is evolving and expanding, with research articles being published almost weekly announcing links between certain versions of genes and the risk for the development of a particular form of cancer, methods to either reduce certain cancer risks or detect the disease at an earlier stage, and treatment options that are more effective and increase the chance for survival. Media reports are often based on a single study, and they are designed so as to have a "wow!" factor that will garner maximum public attention. As a result, one must be able to differentiate between a reputable study and a report that is not scientifically valid, and learn to distinguish hype from reality. All too frequently, today's established protocols are challenged by tomorrow's newest discoveries. For this reason, the integrity and accuracy of new research findings must withstand the test of time and be constantly questioned and reviewed.

PREVALENCE AND MORTALITY VARIES WITH EACH CANCER

Unfortunately, most of us know or will know someone who has been diagnosed with cancer. In 2008 an estimated 1.4 million Americans were newly diagnosed with cancer and approximately 565,000 died from the disease (Table 1.2). This ACS estimate does not include two forms of cancer. The first are those classified as carcinoma *in situ* (in the original site), which are more commonly known as benign tumors. They are confined and have not invaded surrounding tissues. An exception to this rule is carcinoma *in situ* of the urinary bladder which is reported because of its tendency to aggressively grow and progress to a malignant state. The second form consists of nonmelanomas or basal and squamous cell skin cancers, which are the most common forms of skin cancer with over one million new cases per year. Because these skin cancers rarely spread and are not life threatening, they are routinely removed and treated on an outpatient basis in nonhospital settings and, as such, are rarely reported to cancer registries.

Cancer rates are different for men and women

For men, the five most prevalent cancers, in decreasing order, are prostate, lung and bronchus, colorectal, urinary bladder, and skin melanoma. The most commonly diagnosed cancers in women are breast, lung and bronchus, colorectal, non-Hodgkin's lymphoma, and skin melanoma. Lung and bronchus, colorectal, and pancreatic cancers are among the top five most fatal forms of the disease in both men and women. Ironically, the number one cancer killer for both men and women, lung and bronchus,

Table 1.2 Estimated new cancer cases and deaths by sex, USA, 2008*

	Estimated new cases			Estimated deaths		
	Both sexes	Male	Female	Both sexes	Male	Female
All sites	1,437,180	745,180	692,000	565,650	294,120	271,530
Oral cavity and pharynx	35,310	25,310	10,000	7,590	5,210	2,380
Tongue	10,140	7,280	2,860	1,880	1,210	670
Mouth	10,820	6,590	4,230	1,840	1,120	720
Pharynx	12,410	10,060	2,350	2,200	1,620	580
Other oral cavity	1,940	1,380	560	1,670	1,260	410
Digestive system	271,290	148,560	122,730	135,130	74,850	60,280
Esophagus	16,470	12,970	3,500	14,280	11,250	3,030
Stomach	21,500	13,190	8,310	10,880	6,450	4,430
Small intestine	6,110	3,200	2,910	1,110	580	530
Colon†	108,070	53,760	54,310	49,960	24,260	25,700
Rectum	40,740	23,490	17,250			
Anus, anal canal, and anorectum	5,070	2,020	3,050	680	250	430
Liver and intrahepatic bile duct	21,370	15,190	6,180	18,410	12,570	5,840
Gallbladder and other biliary	9,520	4,500	5,020	3,340	1,250	2,090
Pancreas	37,680	18,770	18,910	34,290	17,500	16,790
Other digestive organs	4,760	1,470	3,290	2,180	740	1,440
Respiratory system	232,270	127,880	104,390	166,280	94,210	72,070
Larynx	12,250	9,680	2,570	3,670	2,910	760
Lung and bronchus	215,020	114,690	100,330	161,840	90,810	71,030
Other respiratory organs	5,000	3,510	1,490	770	490	280
Bones and joints	2,380	1,270	1,110	1,470	820	650
Soft tissue (including heart)	10,390	5,720	4,670	3,680	1,880	1,800
Skin (excluding basal and squamous)	67,720	38,150	29,570	11,200	7,360	3,840
Melanoma	62,480	34,950	27,530	8,420	5,400	3,020
Other non-epithelial skin	5,240	3,200	2,040	2,780	1,960	820
Breast	184,450	1,990	182,460	40,930	450	40,480
Genital system	274,150	195,660	78,490	57,820	29,330	28,490
Uterine cervix	11,070		11,070	3,870		3,870

	Estimated New Cases			Estimated Deaths		
	Total	Male	Female	Total	Male	Female
Uterine corpus	40,100		40,100	7,470		7,470
Ovary	21,650		21,650	15,520		15,520
Vulva	3,460		3,460	870		870
Vagina and other genital, female	2,210		2,210	760		760
Prostate	186,320	186,320		28,660	28,660	
Testis	8,090	8,090		380	380	
Penis and other genital, male	1,250	1,250		290	290	
Urinary system	125,490	85,870	39,620	27,810	18,430	9,380
Urinary bladder	68,810	51,230	17,580	14,100	9,950	4,150
Kidney and renal pelvis	54,390	33,130	21,260	13,010	8,100	4,910
Ureter and other urinary organs	2,290	1,510	780	700	380	320
Eye and orbit	2,390	1,340	1,050	240	130	110
Brain and other nervous system	21,810	11,780	10,030	13,070	7,420	5,650
Endocrine system	39,510	10,030	29,480	2,430	1,110	1,320
Thyroid	37,340	8,930	28,410	1,590	680	910
Other endocrine	2,170	1,100	1,070	840	430	410
Lymphoma	74,340	39,850	34,490	20,510	10,490	10,020
Hodgkin lymphoma	8,220	4,400	3,820	1,350	700	650
Non-Hodgkin lymphoma	66,120	35,450	30,670	19,160	9,790	9,370
Myeloma	19,920	11,190	8,730	10,690	5,640	5,050
Leukemia	44,270	25,180	19,090	21,710	12,460	9,250
Acute lymphocytic leukemia	5,430	3,220	2,210	1,460	800	660
Chronic lymphocytic leukemia	15,110	8,750	6,360	4,390	2,600	1,790
Acute lymphocytic leukemia	13,290	7,200	6,090	8,820	5,100	3,720
Chronic lymphocytic leukemia	4,830	2,800	2,030	450	200	250
Other leukemia[†]	5,610	3,210	2,400	6,590	3,760	2,830
Other and unspecified primary sites[‡]	31,490	15,400	16,090	45,090	24,330	20,760

*Rounded to the nearest 10; estimated new cases exclude basal and squamous cell skin cancers and in situ carcinomas except urinary bladder. About 67,770 female carcinoma in situ of the breast and 54,020 melanoma in situ will be newly diagnosed in 2008.

†Estimated deaths for colon and rectum cancers are combined.

‡More deaths than cases suggests lack of specificity in recording underlying causes of death on death certificates.

Source: Estimated new cases are based on 1995–2004 incidence rates from 41 states and the District of Columbia as reported by the North American Association of Central Cancer Registries (NAACCR), representing about 85% of the US population. Estimated deaths are based on data from US Mortality Data, 1969 to 2005, National Center for Health Statistics, Centers for Disease Control and Prevention, 2008. Reproduced with permission from: American Cancer Society. *Cancer Facts and Figures 2008*. Atlanta: American Cancer Society, Inc.

is the most preventable as well since smoking, the greatest risk factor for this disease, is a lifestyle choice.

Americans are living longer; in 1900 men and women lived to an average age of 48 and 51 years, respectively, whereas today they can expect to live on average to 74.1 and 79.5 years. Over the same period there have been tremendous advances in the prevention and treatment of infectious diseases, the disposal and recycling of sewage and solid waste, and the quality and safety of drinking water. As a result, there has been a shift in the causes of death from infectious to chronic, noninfectious diseases. Today, cancer is second only to cardiovascular (heart) disease as a leading cause of death. According to the most recent information available from the Centers for Disease Control and Prevention (CDC), more than 630,000 Americans died of heart disease and approximately 560,000 died from cancer in 2006; together, these noninfectious diseases account for more than 60% of all deaths.

> **What is the difference between infectious and noninfectious diseases?**
>
> **Name two infectious diseases and their causes.**
>
> **Why is cancer more prevalent in older rather than younger people?**

The mortality rate of cancer has decreased

One of the most common means of determining the overall health of a population is to assess the number of deaths caused by disease. In a stunning 2006 announcement, the American Cancer Society reported that the actual number of Americans who died of cancer between 2002 and 2003 dropped for the first time since 1930. This decline in cancer deaths was unexpected given the increases in US population and lifespans. Recent statistics have also shown decreases in mortality rates for specific cancers such as breast and colorectal. Without question, an increase in public awareness, changes in diet and exercise, the number of people undergoing screening and diagnostic tests, the availability and improved sensitivity and accuracy of tests that can detect the disease at earlier, more treatable stages, and the development of more effective and tolerable treatment methods, are all responsible for this downturn.

> **How does an increase in the number of people undergoing screening and diagnostic tests for cancer play a role in decreasing the number of people who die from the disease?**

RISK FACTORS HAVE BEEN IDENTIFIED

There are a wide variety of factors that determine a person's risk of developing a particular disease and cancer is no different. Nearly half of men and two-thirds of women live their lives *without* developing or having been diagnosed with cancer. Why are they cancer-free and not others? Conversely, what makes the others susceptible? There are no absolute answers to these questions but there are certain factors that have been shown to contribute to a person's risk of developing all, many, or only certain forms of cancer, and they include a person's age, gender, weight, family history, sociological and economic status, lifetime exposure to sex hormones, lifestyle choices (whether or not a person eats a healthy diet, smokes, drinks alcohol, is physically active or inactive, etc.), and environmental and occupational exposures to carcinogens.

Aging and genetics contribute to the highest incidence of cancer cases

Disease does not discriminate – young, middle-aged, the elderly, men, and women are affected. As one grows older, however, the risk of developing virtually all serious conditions and illnesses increases. This is a corollary of life as susceptibility to all diseases increases as the body's natural defense mechanisms begin to erode. There are a number of reasons why the rate of cancer incidence increases with age (Table 1.3). The longer a person lives, the more cell divisions take place within the body. Prior to when a cell divides, the entire genome (all 46 chromosomes) is replicated so that each daughter cell will obtain the same genetic information.[1.3] The process of DNA synthesis is not perfect though and errors do get incorporated into the chromosomes. Problems arise when the alterations negatively affect genes that promote or inhibit the ability of a cell to divide. Since mutations will accumulate over time, the longer a person lives, the greater the chance that an individual will acquire a collection of mutations. If a set of key regulatory genes is affected, it can cause a cell to go down the path of uncontrolled growth.

[1.3] The use of the term "daughter cells" does not refer to the gender of the cells, but rather simply to indicate that they are the descendants of a parental cell.

Other risk factors are also important

At the present time, an American man has a 50% chance of being diagnosed with cancer in his lifetime, while an American woman has a 33% chance. This gender difference is likely the result of a number of factors, such as differences in occupations, lifestyle choices, and in exposure to and levels of certain hormones. For example, both men and women develop breast cancer but less than one man develops the disease for every 100 women who are diagnosed. It has been determined that a large factor in this gender discrepancy is the amount of estrogen a woman is exposed to during her lifetime.

Ethnicity and race are also risk factors for the development of certain forms of the disease (Table 1.4). Obstacles faced by minorities include being

Table 1.3 Probability of developing invasive cancers over selected age intervals by sex, US, 2002–2004*

		Birth to 39 (%)	40 to 59 (%)	60 to 69 (%)	70 and Older (%)	Birth to Death (%)
All sites[†]	Male	1.42 (1 in 70)	8.58 (1 in 12)	16.25 (1 in 6)	38.96 (1 in 3)	44.94 (1 in 2)
	Female	2.04 (1 in 49)	8.97 (1 in 11)	10.36 (1 in 10)	26.31 (1 in 4)	37.52 (1 in 3)
Urinary bladder[‡]	Male	0.02 (1 in 4,477)	0.41 (1 in 244)	0.96 (1 in 104)	3.50 (1 in 29)	3.70 (1 in 27)
	Female	0.01 (1 in 9,462)	0.13 (1 in 790)	0.26 (1 in 384)	0.99 (1 in 101)	1.17 (1 in 85)
Breast	Female	0.48 (1 in 210)	3.86 (1 in 26)	3.51 (1 in 28)	6.95 (1 in 15)	12.28 (1 in 8)
Colon and rectum	Male	0.08 (1 in 1,329)	0.92 (1 in 109)	1.60 (1 in 63)	4.78 (1 in 21)	5.65 (1 in 18)
	Female	0.07 (1 in 1,394)	0.72 (1 in 138)	1.12 (1 in 89)	4.30 (1 in 23)	5.23 (1 in 19)
Leukemia	Male	0.16 (1 in 624)	0.21 (1 in 468)	0.35 (1 in 288)	1.18 (1 in 85)	1.50 (1 in 67)
	Female	0.12 (1 in 837)	0.14 (1 in 705)	0.20 (1 in 496)	0.76 (1 in 131)	1.06 (1 in 95)
Lung and bronchus	Male	0.03 (1 in 3,357)	1.03 (1 in 97)	2.52 (1 in 40)	6.74 (1 in 15)	7.91 (1 in 13)
	Female	0.03 (1 in 2,964)	0.82 (1 in 121)	1.81 (1 in 55)	4.61 (1 in 22)	6.18 (1 in 16)
Melanoma of the skin	Male	0.15 (1 in 656)	0.61 (1 in 164)	0.66 (1 in 151)	1.56 (1 in 64)	2.42 (1 in 41)
	Female	0.26 (1 in 389)	0.50 (1 in 200)	0.34 (1 in 297)	0.71 (1 in 140)	1.63 (1 in 61)
Non-Hodgkin lymphoma	Male	0.13 (1 in 760)	0.45 (1 in 222)	0.57 (1 in 174)	1.61 (1 in 62)	2.19 (1 in 46)
	Female	0.08 (1 in 1,212)	0.32 (1 in 312)	0.45 (1 in 221)	1.33 (1 in 75)	1.87 (1 in 53)
Prostate	Male	0.01 (1 in 10,553)	2.54 (1 in 39)	6.83 (1 in 15)	13.36 (1 in 7)	16.72 (1 in 6)
Uterine cervix	Female	0.16 (1 in 638)	0.28 (1 in 359)	0.13 (1 in 750)	0.19 (1 in 523)	0.70 (1 in 142)
Uterine corpus	Female	0.06 (1 in 1,569)	0.71 (1 in 142)	0.79 (1 in 126)	1.23 (1 in 81)	2.45 (1 in 41)

*For people free of cancer at beginning of age interval.
†All sites exclude basal and squamous cell skin cancers and in situ cancers except urinary bladder.
‡Includes invasive and in situ cancer cases.

Source: DevCan: Probability of Developing or Dying of Cancer Software. Version 6.2.1. Statistical Research and Applications Branch, National Cancer Institute, 2007. www.srab.cancer.gov/devcan.
Reproduced with permission from: American Cancer Society. *Cancer Facts and Figures 2008.* Atlanta: American Cancer Society, Inc.

Table 1.4 Cancer incidence and mortality rates* by site, race, and ethnicity, USA, 2000–2004

	White	African American	Asian American and Pacific Islander	American Indian and Alaska Native[†]	Hispanic/ Latino[‡§¶]
Incidence					
All sites					
Males	556.7	663.7	359.9	321.2	421.3
Females	423.9	396.9	285.8	282.4	314.2
Breast (female)	132.5	118.3	89.0	69.8	89.3
Colon and rectum					
Males	60.4	72.6	49.7	42.1	47.5
Females	44.0	55.0	35.3	39.6	32.9
Kidney and renal pelvis					
Males	18.3	20.4	8.9	18.5	16.5
Females	9.1	9.7	4.3	11.5	9.1
Liver and bile duct					
Males	7.9	12.7	21.3	14.8	14.4
Females	2.9	3.8	7.9	5.5	5.7
Lung and bronchus					
Males	81.0	110.6	55.1	53.7	44.7
Females	54.6	53.7	27.7	36.7	25.2
Prostate	161.4	255.5	96.5	68.2	140.8
Stomach					
Males	10.2	17.5	18.9	16.3	16.0
Females	4.7	9.1	10.8	7.9	9.6
Uterine cervix	8.5	11.4	8.0	6.6	13.8
Mortality					
All sites					
Males	234.7	321.8	141.7	187.9	162.2
Females	161.4	189.3	96.7	141.2	106.7
Breast (female)	25.0	33.8	12.6	16.1	16.1
Colon and rectum					
Males	22.9	32.7	15.0	20.6	17.0
Females	15.9	22.9	10.3	14.3	11.1
Kidney and renal pelvis					
Males	6.2	6.1	2.4	9.3	5.4
Females	2.8	2.8	1.1	4.3	2.3
Liver and bile duct					
Males	6.5	10.0	15.5	10.7	10.8
Females	2.8	3.9	6.7	6.4	5.0
Lung and bronchus					
Males	72.6	95.8	38.3	49.6	36.0
Females	42.1	39.8	18.5	32.7	14.6
Prostate	25.6	62.3	11.3	21.5	21.2
Stomach					
Males	5.2	11.9	10.5	9.6	9.1
Females	2.6	5.8	6.2	5.5	5.1
Uterine cervix	2.3	4.9	2.4	4.0	3.3

*Per 100,000, age adjusted to the 2000 US standard population.
[†]Data based on Contract Health Service Delivery Areas (CHSDA), 624 counties comprising 54% of the US American Indian/Alaska Native population; for more information, please see: Espey DK, Wu XC, Swan J, et al. Annual report to the nation on the status of cancer, 1975–2004, featuring cancer in American Indians and Alaska Natives.
[‡]Persons of Hispanic/Latino origin may be of any race.
[§]Incidence data unavailable from the Alaska Native Registry and Kentucky.
[¶]Matality data unavailable from Minnesota, New Hampshire, and North Dakota.

Source: Ries LAG, Melbert D, Krapcho M, et al. (eds.). *SEER Cancer Statistics Review, 1975–2004*, National Cancer Institute, Bethesda, MD, www.seer.cancer.gov/csr/1975_2004/, 2007.
Reproduced with permission from: American Cancer Society. *Cancer Facts and Figures 2008*. Atlanta: American Cancer Society, Inc.

stereotyped, communicating with difficulty due to language barriers, distrusting the medical establishment, and not having health insurance, all of which can hinder the prevention, early detection, and successful treatment. These factors, however, do not always explain the disparities in cancer risks and prevalence rates. Cultural and genetic differences that exist may also increase or reduce the risk of developing cancer and these will be examined in subsequent chapters.

Environment and lifestyle choices affect the risk of cancer development

Many cases of cancer have been linked to the environment, either by a direct result of lifestyle or from exposure to carcinogens. For example, among the thousands of chemical compounds in cigarette smoke, more than 40 are known carcinogens. It is well-established that the use of tobacco products or exposure to second-hand smoke in the environment drastically increases a person's risk of developing lung and oral cancer. Smoking has also been demonstrated to contribute to kidney, pancreatic, cervical, and stomach cancers as well as acute myeloid leukemia. Another example is the risk of skin cancer from exposure to the sun's ultraviolet (UV) light.[1.4] UV radiation damages and mutates DNA, which can significantly compromise the workings of a cell. The correlation between carcinogen exposure and cancer development implies that many cancers are preventable. It is the goal that with improved education, changes in lifestyles, the use of proper precautions, and implementation of environmental regulations that there will be a decrease in the number of new cancer cases.

> **Name three environmental causes of cancer and what could be done to reverse the risk in each case.**

> [1.4] Generations ago, people believed extensive sunlight exposure was an indicator of low social class, since having a tan was an indication of being a day laborer who worked outside. The dawn of commercial air travel made it easier for the wealthy to visit beach resorts for extended winter vacations and eventually the "healthy glow" of a tan was associated with wealth.

Health disparities exist for many diseases

Unfortunately, those most frequently burdened by cancer and other catastrophic illnesses are oftentimes those who are least capable of coping with them. People with low incomes often lack health insurance and traditionally have less access to quality medical care. As a consequence, the poor and uninsured are less likely to receive quality treatment and services, are more

> **Why would a person be more likely to die from cancer if it was detected at a later stage?**

likely to be treated for cancer at a later stage of the disease, and are thus more likely to die from the disease.

WILL CANCER BE CONQUERED WITHIN OUR LIFETIME?

The number of cancer survivors increases with each passing year; some are deemed cancer-free while others continue to undergo treatment for their disease. The five-year relative survival rate for all cancers diagnosed between 1991 and 2001 was 66%, which is significantly higher than the documented rate of 50% from 1974 to 1976 (Table 1.5). This rate increase is certainly an encouraging sign. Nevertheless, not all cancers are the same and each type has its own mortality and survival statistics. The prognosis for recovery for particular cancers will be discussed in later chapters.

Cancer is not an automatic death sentence

It is important to stress the fact that many people not only survive their battle with cancer but thrive and live normal lives. The NCI has set an ambitious goal for the elimination of suffering and death due to cancer by the year 2015, hoping that cancer will be considered a chronic disease by then. Former NCI director Dr Andrew C. von Eschenbach said that the establishment of this goal did not mean that cancer would be "cured," but that many cancers could be eliminated and others controlled at acceptable levels. He referred to this as the 3-D approach to cancer research:
Discovery – the process of generating new information at the genetic, molecular, cellular, individual, and population levels.
Development – the improvement of cancer detection, diagnosis, predictions, treatments, and prevention.
Delivery – the method of distributing cancer interventions to everyone through means of research, communication, education and training, and technical assistance.
If this goal becomes a reality, the effects on millions of people and their families would be stunning. The three words "You have cancer" will have a far less dramatic impact, and the fear and ignorance associated with the disease will be significantly diminished. Without question, this goal is an extremely ambitious one, but it is filled with much promise and hope and its achievement would be an extraordinary milestone in the course of medical science.

EXPAND YOUR KNOWLEDGE

1 A myosarcoma would originate in what type of tissue?
2 A study that was done concluded that men possessing certain versions of two genes have a predisposition for the development of testicular

Table 1.5 Trends in 5-year relative survival rates* (%) by race and year of diagnosis, US, 1975–2003

Site	White			African American			All races		
	1975–77	1984–86	1996–2003	1975–77	1984–86	1996–2003	1975–77	1984–86	1996–2003
All sites	51	55	67†	40	41	57†	50	54	66†
Brain	23	28	34†	27	33	37†	24	29	35†
Breast (female)	76	80	90†	62	65	78†	75	79	89†
Colon	52	60	66†	46	50	55†	51	59	65†
Esophagus	6	11	18†	3	8	11†	5	10	16†
Hodgkin lymphoma	74	80	87†	71	75	81†	74	79	86†
Kidney	51	56	66†	50	54	66†	51	56	66†
Larynx	67	68	66	59	53	50	67	66	64
Leukemia	36	43	51†	34	34	40	35	42	50†
Liver#	4	6	10†	2	5	7†	4	6	11†
Lung and bronchus	13	14	16†	12	11	13†	13	13	16†
Melanoma of the skin	82	87	92†	60†	70§	77	82	87	92†
Myeloma	25	27	34†	31	32	32	26	29	34†
Non-Hodgkin lymphoma	48	54	65†	49	48	56	48	53	64†
Oral cavity	55	57	62†	36	36	41	53	55	60†
Ovary	37	39	45†	43	41	38	37	40	45†
Pancreas	3	3	5†	2	5	5†	2	3	5†
Prostate	70	77	99†	61	66	95†	69	76	99†
Rectum	49	58	66†	45	46	58†	49	57	66†
Stomach	15	18	22†	16	20	24†	16	18	24†
Testis	83	93	96†	82‡	87†	88	83	93	96†
Thyroid	93	94	97†	91	90	94	93	94	97†
Urinary bladder	75	79	81†	51	61	65†	74	78	81†
Uterine cervix	71	70	74†	65	58	66	70	68	73†
Uterine corpus	89	85	86†	61	58	61	88	84	84†

*Survival is adjusted for normal life expectancy and based on cases diagnosed in the SEER 9 areas from 1975–1977, 1984–1986, and 1996–2003, and followed through 2004.
†The difference in rates between 1975–1977 and 1996–2003 is statistically significant ($N < 0.05$).
‡The standard error of the survival rate is between 5 and 10 percentage points.
§The standard error of the survival rate is greater than 10 percentage points.
#Includes intrahepatic bile duct.

Source: Ries LAG, Melbert D, Krapcho M, et al (eds). *SEER Cancer Statistics Review, 1975–2004*. National Cancer Institute, Bethesda, MD, www.seer.cancer.gov/csr/1975_2004/, 2007.
Reproduced with permission from: American Cancer Society. *Cancer Facts and Figures 2008*. Atlanta: American Cancer Society, Inc.

cancer that is higher than that for the general population. What would you want to know about the study that would help determine how valid the conclusions are?

3 Provide three reasons why the probabilities of developing cancer for men and women are different.

4 Give some examples of what can be done about disparities in the health-care system today.

5 Do you think irradiated food (i.e., food exposed to radiation to sterilize it) is a potential cancer risk? Why or why not?

ADDITIONAL READINGS

American Cancer Society, Surveillance Research, 2007 & 2008. http://www.cancer.org/docroot/PRO_1_1_Cancer_Statistics_2008_Presentation.asp

Cancer and the Environment: What You Need to Know. What You Can Do. National Cancer Institute (NCI) and the National Institute of Environmental Health Sciences (NIEHS) NIH publication no. 03-2039, August 2003.

Closing In On Cancer, Solving a 5000-Year-Old Mystery. A Publication of the National Cancer Institute, US Department of Health and Human Services, Public Health Service, National Institutes of Health.

Erikson, C., Salsberg, E., Forte, G., Bruinooge, S., and Goldstein, M. (2007) Future supply and demand for oncologists. *Journal of Oncology Practice,* **3**: 79–86.

Farber, S., Diamond, L. K., Mercer, R. D., Sylvester, Jr, R. R., and Wolff, J. A. (1948) Temporary remissions in acute leukemia in children produced by folic acid antagonist . . . aminopterin. *New England Journal of Medicine,* **238**: 787–793.

National Institute for Occupational Safety and Health (NIOSH): www.cdc.gov/niosh/

2

Cells: the fundamental unit of life

The uniformity of Earth's life, more astonishing than its diversity, is accountable by the high probability that we derived, originally, from some single cell, fertilized in a bolt of lightning as the Earth cooled.

Lewis Thomas, physician, researcher, educator, and essayist

CHAPTER CONTENTS

- Seven hierarchal levels of organization
- Four types of macromolecular polymers
- Cell structure and function
- Relationship between structure and function is important
- Expand your knowledge
- Additional readings

When our bodies work properly we have the tendency to take their complex structure and functions for granted. It is important to realize that the more we know about healthy body function, the better the position we will be in to fix what is wrong when we are ill. This and the following two chapters will provide a basic understanding of the way cells function normally and how an attack on the body by rogue cells that divide uncontrollably and function abnormally can result in cancer. This chapter will take a stepwise approach to gradually build a working knowledge of subcellular components in order to understand how they work together as a single entity – the cell. The whole is more than the sum of its parts, and all components of the cell must work together seamlessly to carry out the processes that give rise to what we know as life.

Cancer: Basic Science and Clinical Aspects, 1st edition.
By C. A. Almeida and S. A. Barry. Published 2010
by Blackwell Publishing, ISBN 978-1-4051-5606-6.

SEVEN HIERARCHAL LEVELS OF ORGANIZATION

Using the stepwise approach to understand how a living organism as complex as a human is put together first requires some knowledge of the seven levels of biological organization (Figure 2.1), and there is no better place to start than at the beginning, at the level of the atom.

Atoms are the building blocks of all molecules

Everything around us is composed of atoms – the building blocks of matter (Figure 2.1a). An atom is composed of negatively charged particles called electrons that spin in orbitals around a central nucleus possessing positively charged protons and uncharged neutrons. A chemical or covalent bond can be created between two atoms when their orbitals overlap, allowing their electrons to be shared. A molecule is formed by such bonding of two or more atoms, and when molecules bond with other molecules they can serve as building blocks for the formation of macromolecules such as proteins, carbohydrates, fats, and DNA (Figure 2.1b). A cell is a collection of atoms, molecules, macromolecules, and macromolecular structures (Figure 2.1c). The coordinated aggregation of countless molecules and macromolecules results in complex cellular structures such as the cell membrane, nucleus, and mitochondrion, to name a few.

2.1 The existence of cells was not known until the mid seventeenth century when the English scientist Robert Hooke viewed thin sections of cork with the newly invented compound microscope. Cork consists of densely packed dead cells found in the inner bark of woody plants. The symmetrically arranged hollow boxes Hooke saw reminded him of the small, empty rooms in which monks lived and so he used the descriptive term *cellulae*, which is Latin for "tiny rooms," to describe the structures. Over time, the derivative term *cells* has been used.

Cells are the simplest units of life

Life, in its simplest sense, is the collective reactions that occur between the atoms and molecules within cells. With a few rare exceptions, cells are unable to be seen with the unaided eye (Table 2.1). A compound microscope, an instrument that uses two lenses, one in front of the other, is needed to magnify cells in order to see them in sufficient detail.[2.1] All cells possess an outer plasma membrane that acts as a semipermeable barrier between the external environment and the internal contents, enabling them to maintain a steady state of operation

What distinguishes living organisms from nonliving things?

Why is it important that a membrane be semipermeable? What are some characteristics of a plasma membrane that allow for this property?

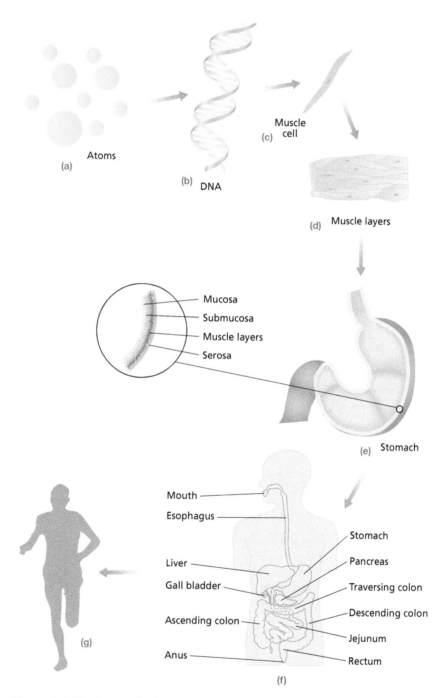

Figure 2.1 The human body consists of seven hierarchal levels of organization. (a) Atoms combine to form (b) molecules, which combine to form (c) cells that work together to form (d) tissues, that work together to form (e) organs, that work together to form (f) organ systems, that work together to form an (g) organism.

Table 2.1 A range of cells and their sizes

Category	Cell type	Size
Smallest cell	Microplasma (bacterium)	0.1 micrometer (µm) or 0.000004 inch
Smallest human cell	Sperm	3 µm or 0.0001 inch
Largest human cell	Egg	35 µm in diameter or 0.0014 inch in diameter
Longest cell	Nerves	Over 3 meters long! or over 9.84 feet long!

known as homeostasis. Without the integrity that the plasma membrane provides, the contents of a cell would not be maintained as a single unit, nor would their activities be coordinated. As a result, life processes would not occur.

There is a wide variety of cells in the body and each type has its own distinct shape, chemical composition, and internal structures that enable it to carry out its unique function. For example, red blood cells (RBCs) are small and packed with hemoglobin, a protein that has an ability to bind oxygen (Figure 2.2a).[2.2] The small size of red blood cells (RBCs) enables them to easily pass through the smallest of blood vessels while their unique round, biconcave (i.e., dimpled on both sides), disc shape provides a large outer surface area that dramatically increases the efficiency of both oxygen pick-up in the lungs and oxygen delivery to the body's tissues. Cells in the outer layer of our skin, the epidermis, are flattened and tightly packed together so as to create a barrier between the environment and the internal tissues of the body (Figure 2.2b). Nerve cells have many long extensions called dendrites and axons that extend outward from the central cell body. These allow a single cell to make contact, and thereby communicate, with large numbers of neighboring cells (Figure 2.2c).

2.2 Even though RBCs are small (approximately 8 µm (0.0003 in) in diameter, 2 µm (0.00008 in) thick at the outside edge, and 1 µm (0.00004 in) thick in the center), a single one contains approximately 250 million hemoglobin molecules, and about 5 million RBCs are in a single cubic millimeter (1 mm^3, 1/25 in^3) of blood (the reference volume clinicians use). RBCs are replaced at a rate of 2–3 million per second. This means that your body must synthesize 500–750 trillion hemoglobin molecules every second!

The shape of a red blood cell provides greater efficiency by creating a larger surface area. Explain the importance of a cell's surface area to volume ratio and how you think it may increase the efficiency of cellular processes.

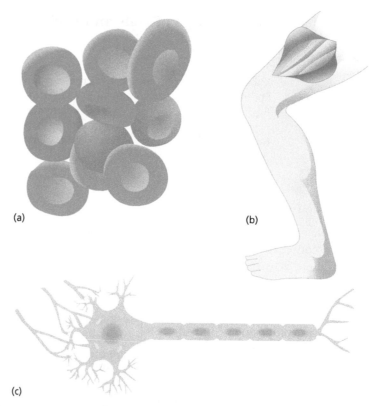

(a)

(b)

(c)

Figure 2.2 Representative types of cells. (a) Red blood cells. (b) Outer layer of skin cells. (c) Nerve cell.

Tissues are composed of cells, which determine their functions

A tissue is a collection of cells working together to perform a particular function (Figure 2.1d). Cells within a tissue interact with one another as well as with the extracellular matrix, a network of fibrous proteins and sugars found just outside the cell. The extracellular matrix is similar to the framing of a house in that it acts as a support to which the flooring and the drywall, used to create the walls and ceilings, are attached. If each room is thought of as a cell, with the drywall and flooring being the plasma membrane, then the membrane of each cell is bound to the extracellular matrix (the cellular framing). The shape and stability of the rooms depend in part on the structure and organization of the extracellular matrix.

In order for the cells of a tissue to work together as a cohesive unit, it is essential that they communicate with one another. They accomplish this through the exchange of chemical and physical signals. An example of chemical signaling is when cells release signaling (messenger), molecules

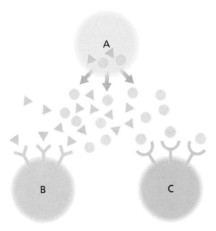

Figure 2.3 Cell A releases two types of signaling molecules and cell B possesses receptors that are only capable of binding the signaling molecules, whereas cell C is capable of only receiving signals because of the types of receptors it produces.

that other target cells detect through means of a receptor and then respond to in an appropriate way. This form of cell signaling is controlled in part by the fact that only those cells that produce the appropriate receptor for a messenger molecule have the ability to respond to it (Figure 2.3).

Physical signals, on the other hand, can be exchanged between adjacent cells through direct cell-to-cell contacts or along the extracellular matrix materials around them. This works in a manner that is similar to how a spider can detect the struggles of a trapped moth on the opposite side of its web through the vibrations of the silk fibers. Cells are therefore able to work together to perform the particular function of a tissue as a result of how they are structurally organized and by their ability to communicate with one another.

Organs are composed of more than one tissue

The next level of organization is represented by the organ – two or more tissues working in a coordinated fashion to carry out one or more specific functions (Figure 2.1d). In this way, an organ can perform a function that neither of the individual tissues can achieve on its own. For example, the heart is made up of muscle, nervous, and connective tissues that work together to pump blood throughout the body in an intricate series of branching and interconnected blood vessels. The blood provides cells with nutrients and removes toxic wastes that are byproducts of cellular metabolism. The eye also has connective, muscle, and nervous tissues, but their structures are different from those same types of tissues that make up the heart and therefore function differently. The tissues of the

> Name three additional organs in the human body and their specific functions.

eye collect, focus, and convert light into a chemical signal that is sent to the brain where it is interpreted as an image. These examples illustrate the importance of cells and tissues having cellular differences in order to accomplish specialized functions.

Organisms are collections of organ systems working in a coordinated fashion

The ultimate in cooperation and complex dependency within living beings is the organ system, which is composed of multiple organs working together to perform a particular function (Figure 2.1e). For example, the digestive system is a series of organs – the stomach, small intestine, large intestine, pancreas, and gall bladder – that enables a person to digest and absorb food, extract the energy and nutrients that the body needs, and eliminate the remaining unusable waste material.

Multiple organ systems, when present, cooperate with a complex dependency upon one another for the proper functioning of the organism as a whole (Figure 2.1f). Therefore, the proper functioning of the body is dependent upon the ability of a virtually uncountable number of cells possessing a variety of shapes and functions to work as a cohesive unit. Perhaps even more remarkable is how a single "rogue" cell arises, divides uncontrollably, wreaks havoc on healthy cells and tissues, and brings about serious harm or even death.

> Name two additional organ systems in the human body and which organs comprise them.
>
> What are the hierarchal levels of structure present in a human?

FOUR TYPES OF MACROMOLECULAR POLYMERS

Many of the molecules that carry out the processes vital to life are in the form of macromolecular polymers – large molecules formed from the linking of smaller molecules called monomers. The four major macromolecules in living systems are nucleic acids, proteins, carbohydrates, and lipids (fats). The roles these molecules play in the functioning of our cells, tissues, organs, and organ systems will be discussed in the following sections of this chapter and periodically throughout subsequent chapters. An understanding of the structure–function relationship of each macromolecule and the relationships that exist between them is key to understanding how an organism as complex as a human functions. In a diseased state, it is the functioning of these macromolecules that is negatively affected.

Nucleic acid molecules are information molecules

Genetic information is the resource material a cell uses to perform all of its functions. Nucleic acids are the macromolecules that store and transmit genetic information. There are two types of nucleic acids in cells: DNA (deoxyribonucleic acid) and RNA (ribonucleic acid). Both of these molecules are built from monomer molecules called nucleotides. Nucleotides have three components: a phosphate group, a sugar, and a nitrogenous base (Figure 2.4).

The terms DNA and RNA are based in part on the type of sugar each molecule possesses. DNA molecules are built with nucleotides that contain the sugar deoxyribose whereas RNA nucleotides possess the sugar ribose (Figure 2.4). Single capital letters are used to represent individual nucleotides and depend on which of the five different nitrogenous bases the nucleotide possesses: adenine (A), thymine (T), cytosine (C), guanine (G), and uracil (U) (Figure 2.4). Three out of the five, A, G, and C, are found in both DNA and RNA molecules whereas T is unique to DNA and U is found only in RNA.

A final and distinct difference between these two types of nucleic acid molecules is the number of chains, or strands, of nucleotides they contain. A single DNA molecule consists of two strands of nucleotides bonded to each other, whereas RNA molecules consist of only a single chain of nucleotides (Figure 2.5a,b). The nitrogenous bases on one strand of a DNA molecule are bonded with their complementary nitrogenous bases on the second strand: A bonds with T and G bonds with C; thus the term complementary base pairing (Figure 2.5a). The structure of a DNA molecule resembles a rope ladder – the phosphate-sugar "backbones" of the two individual strands are held together by the rungs, which are the nitrogenous bases bonded to one another. Twisting the rope ladder along its longitudinal axis will yield a structure that resembles the true helical structure of a DNA molecule. It is important to note that all DNA molecules possess the same phosphate-sugar backbone. The true uniqueness of each DNA molecule is found within its complementary nitrogenous base-pair sequence.

> **Provide three ways that DNA and RNA differ in their structures.**
>
> **Draw three different RNA and three different DNA molecules. For simplicity, represent phosphate groups with a "P", deoxyribose and ribose sugars with "DR" and "R", respectively, and use the appropriate single letter abbreviations for each of the nitrogenous bases.**
>
> **How does one strand of a DNA molecule differ from the other strand?**
>
> **Given the sequence of nitrogenous bases along one strand of a DNA molecule can you predict the sequence of nitrogenous bases on the other strand? Explain.**

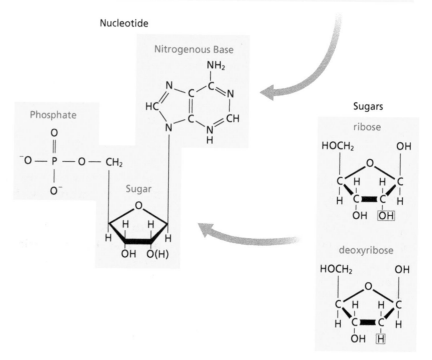

Figure 2.4 The structure of a nucleotide. Nucleotides contain a phosphate group bound to a sugar (ribose or deoxyribose) attached to one of five nitrogenous bases (A, T, C, G, or U). DNA nucleotides contain deoxyribose and A, T, C, or G, whereas RNA nucleotides contain ribose and A, U, C, or G.

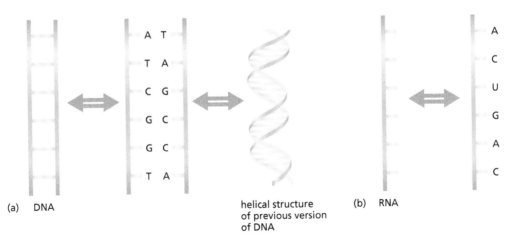

(a)   DNA   helical structure   (b)   RNA
                                           of previous version
                                           of DNA

Figure 2.5 Comparative structures of DNA and RNA. (a) DNA is double stranded, contains the sugar deoxyribose, and the nitrogenous bases A, T, C, and G. DNA is twisted to form a helix. (b) RNA is single stranded, possesses the sugar ribose, and the nitrogenous bases A, U, C, and G.

The relationship between DNA, chromosomes, and genes

Individual DNA molecules within cells are known as chromosomes. Human cells have 46 chromosomes. The information that cells use to perform their functions is stored in the sequences of As, Ts, Cs, and Gs within each of the chromosomes and is therefore referred to as genetic information. This is analogous to how a software program enables a computer to perform specific functions based on its binary code, which is a string of 0s and 1s. Consider what would happen if a computer, which has no information on its hard drive, was turned on . . . nothing would occur because there is an absence of information that would "instruct" the computer what to do. Likewise, a cell that had its chromosomes removed would be lacking a set of "directions" and would soon die.

Continuing with the computer analogy . . . all of the information a computer uses to perform tasks is organized into programs, which are individual strings of binary code. Each program enables a computer to carry out a particular task. For example, word processing is performed using Microsoft Word®, spread sheets are analyzed with Microsoft Excel®, and presentations are created using Microsoft PowerPoint®. Genetic information is organized in a similar way. A specific stretch of As, Ts, Cs, and Gs on a chromosome is a gene (Figure 2.6). Each gene contains the information

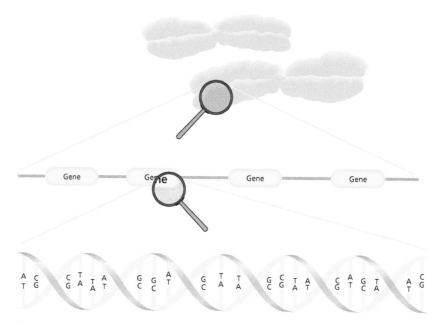

Figure 2.6 A gene is a coding sequence on a chromosome. A stretch of As, Ts, Cs, and Gs on a chromosome is a unit of information called a gene and there are multiple genes on a single chromosome.

a cell needs to make a specific molecule that will perform a particular function. It is currently estimated that there are 20,000–25,000 genes present on the chromosomes in a single human cell. Since each cell in a person's body has the same genetic information, then upwards of 25,000 pieces of information are needed to "build" a human. A person's complete set of genetic information is referred to as his or her genome.

The importance of genes lies not in their composition of As, Ts, Cs, and Gs, but rather in the exact *order* or *sequence* of these nucleotides. This can be illustrated simply by considering how words can be radically altered with only a subtle change in some of their letters. For example, "lamp" can become "lamb" or "medical" can become "medieval".

The central dogma: converting genetic information into functional proteins

Genetic information is *stored* in the form of DNA but to be useful to the cell it must be *expressed*, most often in the form of protein. Proteins are polymers of amino acids, which will be discussed later in this section. The expression of genetic information is a fundamental activity critical to the function and, therefore, survival of every living cell on the planet.

To indicate this fact, the expression of genetic information is referred to as the **Central Dogma of Biology**. This term clearly stresses the importance of this dogma to all of life's processes.[2.3]

> 2.3 A dogma is a law of nature that has been accepted as an essentially irrefutable fact.

The Central Dogma of Biology is composed of essentially two processes (Figure 2.7). During the first step, transcription, the DNA nucleotide sequence of one of the two strands of a gene is used as a template to

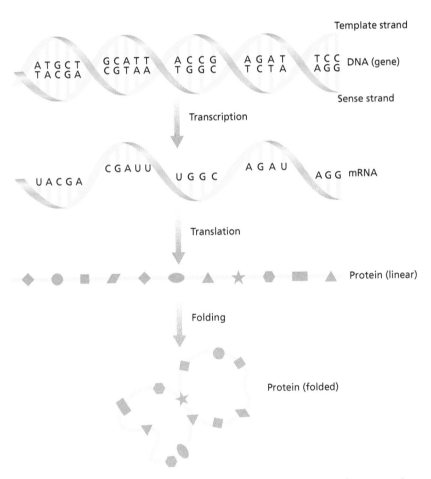

Figure 2.7 The Central Dogma of Biology. (a) The first step in the expression of genetic information is the process of transcription, the synthesis of an RNA molecule with a sequence that is complementary to a gene's sequence. (b) The second step in the process is translation, the synthesis of a protein with a sequence of amino acids that is dictated by the sequence of an RNA molecule. (c) The amino acids interact with one another and fold into a three-dimensional shape that is associated with its function.

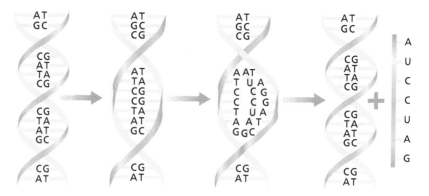

Figure 2.8 Transcription is the synthesis of RNA. An RNA molecule is produced by linking together nucleotides using one of the two DNA strands of a gene as a template. The nucleotide sequence of an RNA molecule will be complementary to the template strand of DNA.

> **What would be the sequence of an mRNA molecule made from a template DNA strand with the sequence: TACCAAGATTAACAGAAA? Hint: Keep in mind the differences between RNA and DNA.**
>
> **How does the sequence of an mRNA molecule compare to the DNA strand of the gene that is complementary to the template strand?**

make an mRNA (*m* = messenger) molecule. The process begins by having the two strands of a gene separate so that the sequence of one of the two strands, the template strand, dictates the order in which the RNA nucleotides are linked during synthesis of the mRNA (Figure 2.8). As a result, the nucleotide sequence of the mRNA is complementary to the DNA template strand and in this way, the gene's code is now in the mRNA.

The message in the mRNA molecule is then used to synthesize a specific protein during the process of translation. The translational machinery of a cell uses the order of nucleotides in an mRNA as a code to determine the order in which amino acids, the building blocks of proteins, will be linked together in a linear chain. Each of the 20 different amino acids used to build proteins has its own unique chemical characteristic. Rarely does a linear strand of amino acids remain as a straight chain. The chemical characteristics of the amino acids result in their interacting and bonding with one another. For example, a positively charged amino acid will bond with a negatively charged one. This is a very dynamic process that ultimately results in a linear chain of amino acids folding into an often times elaborate three-dimensional structure (Figure 2.7). The sequence of amino acids in a protein is thus a strong determinant of its final folded structure, and its structure is critical to its particular function. The relationship between the genetic information stored in a cell and the functionality of that cell can be summarized in the following way:

a gene's DNA nucleotide sequence
↓ dictates
the mRNA nucleotide sequence
↓ dictates
the sequence of amino acids of the protein
↓ influences
the folding (interaction of the amino acids) of the protein
↓ influences
the structure of the protein
↓ influences
the function of the protein

Therefore, within a cell, the expression of a wide range of genes produces the proteins that give it its unique structure, composition, and ability to carry out life processes.

Carbohydrates function as energy and signaling molecules

A third class of macromolecules important to the function of cells is carbohydrates, which are composed of carbon, hydrogen, and oxygen. Commonly known carbohydrates are glucose (blood sugar), fructose (fruit sugar), sucrose (table sugar), lactose (milk sugar), cellulose and starch (plant carbohydrates), and glycogen (animal carbohydrate). Monosaccharides (*mono-* = one; *-saccharide* = sugar) are often referred to as simple sugars and are the building blocks of more complex carbohydrates (Figure 2.9c–e). The general chemical formula for a monosaccharide is $(CH_2O)_n$ where n is the number of times this grouping of atoms is repeated. From the chemical formula it can be seen that for every carbon there is a molecule of water; in other words, the carbon is hydrated, therefore the term carbohydrate. Carbohydrates can have the same chemical formula but a different spatial orientation of the atoms (Figure 2.9a). Such differences affect their chemical and biological properties. This is another example of a structure–function relationship similar to that exhibited by proteins.

Linking two monosaccharides together forms a disaccharide (*di-* = two) (Figure 2.9c,d). An oligosaccharide (*oligo-* = few) is formed from the linkage of three to ten monosaccharides. Any carbohydrate containing

> **Would two amino acids, one with a positive charge and the other with a negative charge, interact if brought close together?**
>
> **Would two amino acids, both with a positive charge or both with a negative charge, interact if brought close together?**
>
> **Do you suppose that all cells constantly express all genes all the time? Why or why not? If not, what do you think determines which genes are constantly expressed and which are not?**

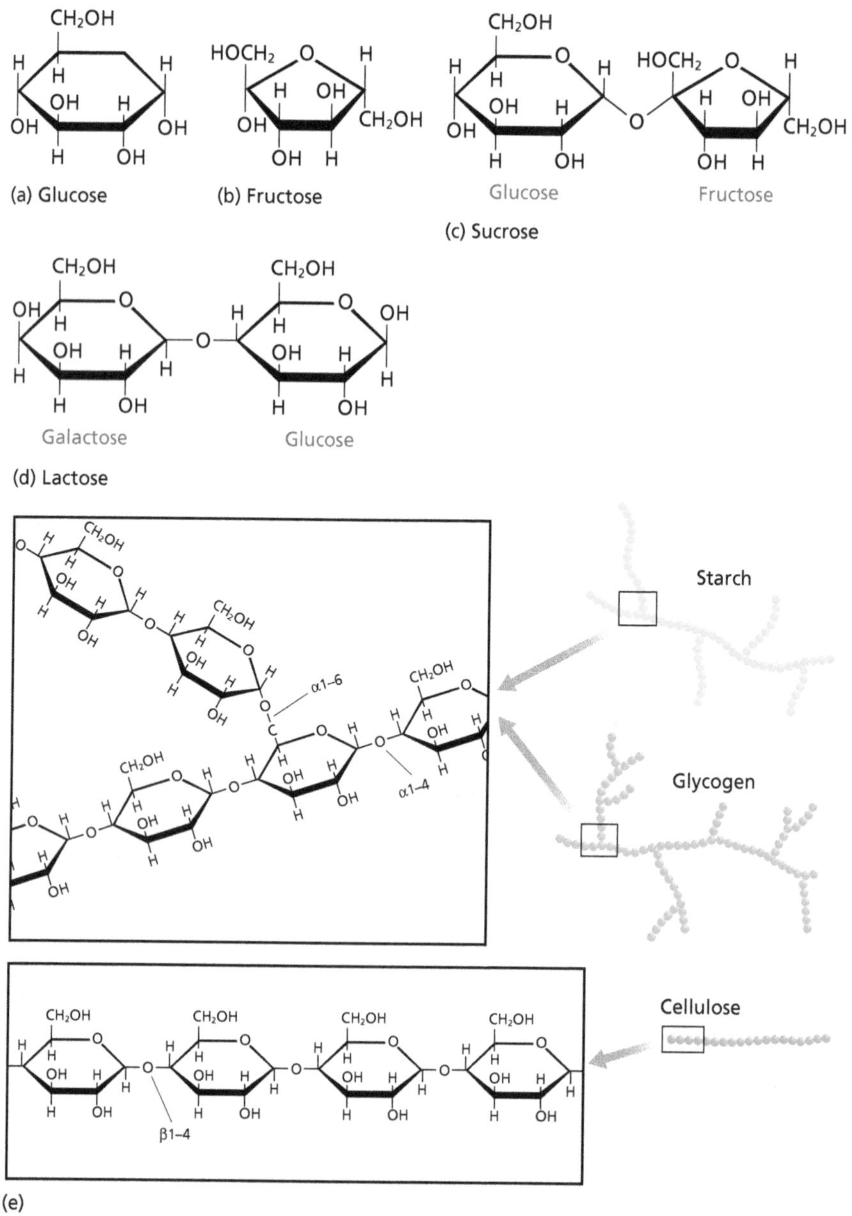

Figure 2.9 Structures of representative carbohydrates. (a,b) Glucose and fructose are both monosaccharides; (c) sucrose is a disaccharide of glucose and fructose; (d) lactose is a disaccharide of galactose and glucose; (e) starch, glycogen, and cellulose are polysaccharides of glucose.

more than 10 monosaccharides is called a **polysaccharide** (*poly-* = many) (Figure 2.9e). While nucleotides and amino acids are linked together to form only straight chains, monosaccharides can be linked to form either straight or branched chains. For example, cellulose, a component of plant cell walls that is indigestible by humans, is a straight chain of many glucose molecules, whereas starch and glycogen, the storage forms of glucose in plants and animals, respectively, are highly branched polymers.

> **What are the chemical formulas for glucose and fructose?**
>
> **If glucose and fructose have the same chemical formula then can you tell based on Figure 2.9 why they are different molecules?**

Roles carbohydrates play in living systems

Most people are familiar with carbohydrates as food – a source of energy. Certain mono- and disaccharides, commonly known as sugars (glucose, fructose, and sucrose), are broken down by cells for energy. Both starch and glycogen are referred to as complex carbohydrates because they are large polymers that, when broken down, release the simple sugar glucose.

Oligosaccharides are commonly bound to the proteins and lipids (fats) that make up the outer membrane of each cell and are referred to as glycoproteins (*glyco-* = sugar) and glycolipids, respectively. They function as receptors for signaling molecules and as cell surface markers. In this way, the types of carbohydrates present on the surface of a cell act as an address of sorts, providing it with a particular identity. The designations given for the ABO blood types are a classic example of the use of carbohydrates as cell identifiers. The three principal oligosaccharides bound to lipids in the outer membrane of red blood cells (RBCs) and used in their identification are designated as A, B, and O.[2.4] A person with only glycolipid A or both A and O on his RBCs would have blood type A, whereas someone with blood type AB would have glycolipids A and B on his RBCs.

> [2.4] The A, B, and O oligosaccharides are also referred to as blood group antigens. A fourth antigen, D, is a protein and is often referred to as the Rh factor. Red blood cells are typically designated as positive if antigen D is present in their outer membranes and negative if it is absent (e.g., AB+ or AB−).

> **What blood group antigen(s) could a person with blood type B have on the surface of his or her RBCs?**
>
> **What blood group antigen(s) could a person with blood type O have on the surface of his or her RBCs?**

The role of cellular lipids in membrane structure and function

Lipids are more commonly known as fats and their distinguishing feature is that they do not dissolve in water, so they are referred to as hydrophobic (*hydro-* = water; *-phobic* = fearing). A common everyday example of this is the well known fact that oil, a liquid fat, does not mix with water. The simple reason for this is that water is a polar (charged)

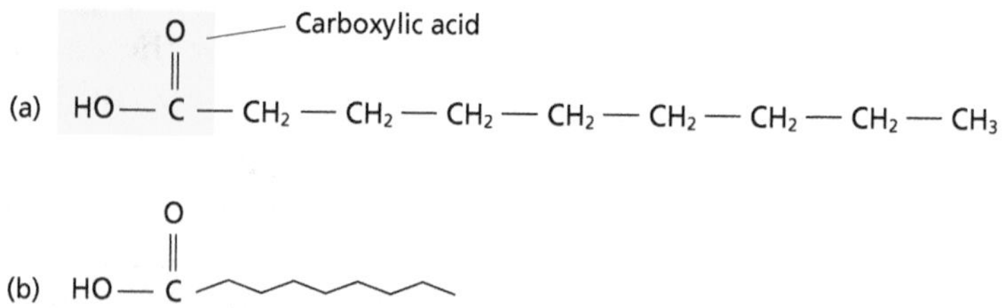

O — Carboxylic acid
‖
(a)   HO— C — CH$_2$ — CH$_2$ — CH$_2$ — CH$_2$ — CH$_2$ — CH$_2$ — CH$_2$ — CH$_3$

O
‖
(b)   HO— C ∿∿∿∿

Figure 2.10 Structures of a fatty acid. (a) A fatty acid contains a long stretch of carbons and hydrogens (a hydrocarbon chain) with a carboxylic acid chemical group at one end. (b) The hydrocarbon chain is often represented by a jagged line with each bend representing a carbon with hydrogen atoms bound to it.

molecule while lipids are nonpolar (neutral or uncharged); polar molecules can interact or bond with other polar molecules but not with nonpolar molecules.

The majority of cellular lipids contain one or more fatty acids. A fatty acid consists of a long chain of carbons and hydrogens (hydrocarbon chain; the "fatty" or hydrophobic portion) with a carboxylic *acid* at one end (Figure 2.10). It is traditional for the hydrocarbon chain to be represented as a jagged line. The oxygen in the acid portion is capable of bonding with water, thereby making that portion of the molecule hydrophilic (*-philic* = loving). The hydrophilic and hydrophobic portions are referred to as the polar head and nonpolar tails, respectively.

The principal component of cellular membranes is phospholipids, which possess two fatty acids bound to two of the three carbons of a glycerol molecule (Figure 2.11). Bound to the third carbon is a phosphate group (PO$_4^{2-}$), similar to that found in nucleotides. Most often there is an additional hydrophilic chemical group attached to the phosphate that adds diversity and functionality to the lipid. It is common when illustrating phospholipids to simply represent the polar head as a circle. Cell membranes consist of phospholipid bilayers, with the molecules in each layer oriented such that their polar heads are on the outside surfaces and their nonpolar tails are positioned inward towards the core of the membrane (Figure 2.11). The arrangement of the bilayer with the polar heads on the two surfaces enables them to interact with the water on either side of the membrane, while the hydrophobic tails remain sequestered in the interior of the membrane away from the water.

Given what was discussed about hydrophobic (nonpolar) and hydrophilic (polar) molecules, how would you expect amino acids with these same characteristics to affect the folding of a protein?

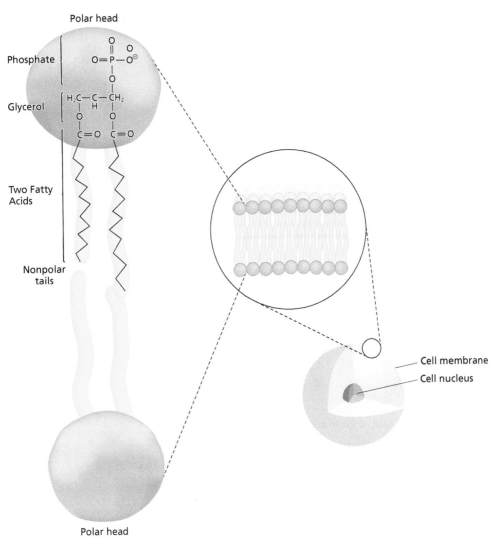

Figure 2.11 Structures of a phospholipid and phospholipid bilayer. A phospholipid consists of two hydrocarbon chains and a phosphate group bonded to the three carbons of glycerol. A phospholipid bilayer is an aggregation of phospholipids into two layers with the nonpolar tails facing inward toward each other and the polar heads facing the aqueous environment on either side of the membrane.

The bilayer structure of membranes provides a selectively permeable barrier; only small, uncharged molecules are able to move across a membrane by passing between the closely associated phospholipids (Figure 2.12). This selectivity occurs because the large nonpolar region within the membrane readily repels charged molecules, preventing their

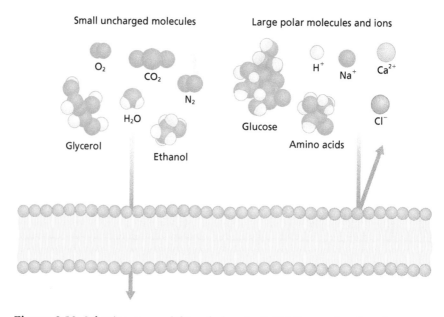

Figure 2.12 Selective permeability of phospholipid bilayers. Small and uncharged molecules can diffuse freely across the membrane. Large, polar molecules cannot diffuse across the membrane.

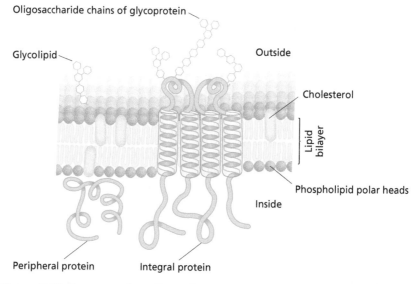

Figure 2.13 Structure of a cell membrane. Cell membranes are phospholipid bilayers that contain integral and peripheral proteins and cholesterol. The outer surface of a cell's plasma membrane can have carbohydrates bound to some of the proteins and lipids.

passage across it. As mentioned previously, proteins are also an important component of cell membranes (Figure 2.13). Some of the proteins are embedded in only one half of the membrane's bilayer and are known as peripheral proteins while integral proteins are embedded and extend through both halves of the membrane. Many integral proteins function as receptors for signaling molecules or as transporters that regulate the passage of molecules *across* the membrane that cannot simply pass *through* the membrane.

> **Discuss the importance of the relationship between structure and function in relation to an integral protein that serves as a receptor for a signaling molecule.**
>
> **Integral proteins function as transporters because they pass through the entire membrane, creating a hydrophilic path for the passage of charged or large polar molecules. What do you suppose are some functions of peripheral proteins?**

CELL STRUCTURE AND FUNCTION

Each type of cell in the body must be able to carry out its array of highly evolved and indispensable functions. Therefore, despite the commonalities that all cells share, each cell type has a unique composition and arsenal of specialized structures. Figure 2.14 is a composite illustration of a typical human cell with numerous phospholipid membrane-bound internal structures apparent. These membrane-bound structures allow for compartmentalization of the cell so that particular functions will occur within particular intracellular locations. These structures are analogous to tiny organs, and as such they are appropriately termed organelles.

Nucleus is the control center of the cell

The most prominent organelle is the nucleus (Figure 2.14). It is where the chromosomes are located and where the processes of DNA replication and transcription take place. After being synthesized in the nucleus, mRNA molecules are transported into the cytosol, the region of the cell between the nuclear and plasma membranes, through transport proteins located in the nuclear membrane. Found throughout the cytosol are ribosomes, the sites where mRNA molecules are translated into proteins (Figure 2.14).

Roles of endoplasmic reticulum and Golgi in modifying and trafficking proteins and lipids

The nuclear membrane is continuous with the membrane of the endoplasmic reticulum (ER) (Figure 2.14). The smooth endoplasmic reticulum, the first of two forms of the organelle, is involved in the

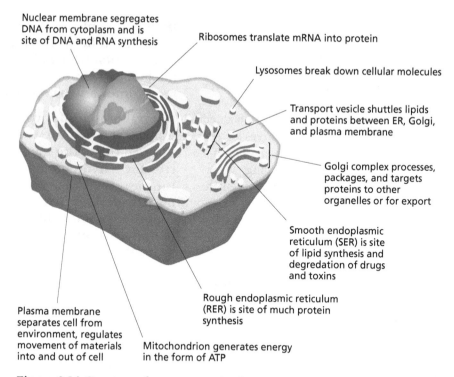

Nuclear membrane segregates DNA from cytoplasm and is site of DNA and RNA synthesis

Ribosomes translate mRNA into protein

Lysosomes break down cellular molecules

Transport vesicle shuttles lipids and proteins between ER, Golgi, and plasma membrane

Golgi complex processes, packages, and targets proteins to other organelles or for export

Smooth endoplasmic reticulum (SER) is site of lipid synthesis and degredation of drugs and toxins

Rough endoplasmic reticulum (RER) is site of much protein synthesis

Plasma membrane separates cell from environment, regulates movement of materials into and out of cell

Mitochondrion generates energy in the form of ATP

Figure 2.14 Structure of a representative human cell.

synthesis of various types of membrane lipids and steroid hormones as well as the degradation of drugs and poisonous molecules into nontoxic substances. The rough endoplasmic reticulum has a knobby appearance to the cytoplasmic side of the membrane when viewed under the microscope because ribosomes are bound to it. The proteins synthesized on these ribosomes either become embedded in the membrane or enter the lumen (i.e., internal space) of the rough ER. Proteins and lipids in the rough ER are modified by enzymes that add carbohydrates, a process called glycosylation, to form glycoproteins and glycolipids. Tiny vesicles that bud off of the rough ER membrane transport the modified lipids and proteins to the Golgi apparatus or complex (Figures 2.14, 2.15). The Golgi apparatus is a series of flattened membranous sacs positioned between the rough ER and the plasma membrane. Lipids and proteins are passed from sac to sac by way of a series of vesicles, small membrane-bound sacs carrying various types of molecules, that bud from and fuse to the golgi sacs. During the journey through the Golgi apparatus, the proteins and lipids are further modified by a variety of enzymes. Each form of modification brings the molecules closer to their proper functional states.

The vesicles that bud from the Golgi sac closest to the plasma membrane travel either to the plasma membrane, where the lipids and proteins are inserted or secreted out of the cell, or to lysosomes (Figure 2.15).

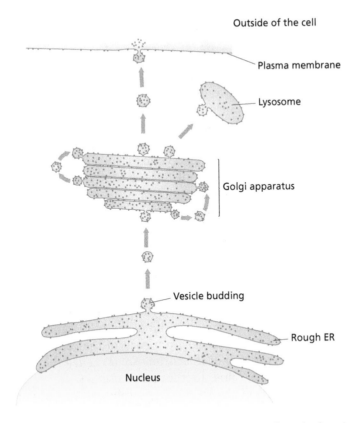

Outside of the cell

Plasma membrane

Lysosome

Golgi apparatus

Vesicle budding

Rough ER

Nucleus

Figure 2.15 Transport of proteins and lipids from the rough endoplasmic reticulum through the Golgi apparatus to the plasma membrane or lysosome. Proteins and lipids can be glycosylated in the endoplasmic reticulum, packaged into vesicles that bud from the ER and travel to the Golgi apparatus. As the proteins are passed from sac to sac of the Golgi appraratus they can be further glycosylated and then packaged into vesicles that travel either to the plasma membrane or to lysosomes.

Lysosomes are membrane-bound sacs that contain enzymes which break down large molecules that the cell has taken in or are no longer being used.

> **Discuss the importance of a lysosome having a semipermeable membrane.**

Mitochondria are sources of energy

Another important organelle is the mitochondrion, appropriately referred to as the powerhouse of the cell (Figure 2.14). The energy stored in carbohydrates and lipids is released as they are broken down within mitochondria and converted to the all important energy molecule ATP, a nucleotide that efficiently provides the energy needed for a wide variety of cellular reactions.

Cytoskeleton provides cell shape, stability, and a means of transport

Only those components of a typical human cell that are important to the function and treatment of cancer cells will be discussed in this section. Just as humans possess an internal skeleton, cells, too, possess a structure that serves a similar purpose. The cytoskeleton is an internal framework of proteins that provides shape and stability to a cell and also acts as a rail system that transports vesicles, such as those that bud from and fuse to the Golgi, throughout the cell. The three types of proteins that comprise the cytoskeleton are actin, intermediate filaments, and microtubules. The actin filaments are along the periphery of the cell just inside the plasma membrane (Figure 2.16a). Scattered throughout the interior of the cell but not extending into any of the organelles are the intermediate filaments (Figure 2.16b). The centrosome, a region adjacent to the nucleus, is the point of origin for the microtubules, which radiate outward toward the plasma membrane in a starburst fashion (Figure 2.16c). During cell division the centrosome is replicated and each one moves to an opposite end of the cell (Figure 2.16d). Some of the microtubules radiating out from each of the centrosomes bind to a particular region on each of the chromosomes called the centromere. The microtubules that bind to the chromosomes are referred to as spindle fibers. The spindle fibers act as a cable system to pull one copy of each of the replicated chromosomes toward each of the centrosomes in order to segregate them into each of the daughter cells during cell division.

RELATIONSHIP BETWEEN STRUCTURE AND FUNCTION IS IMPORTANT

Possessing a conceptual framework of how living systems are organized as well as the composition, structure, and function of cells is critical to an understanding of cancer or any other form of disease. An important and unifying concept that is pervasive throughout this chapter is the relationship between structure and function.
• The structure of a cell is related to its function.
• Cells are organized into tissues and tissues into organs in ways that allow them to communicate with one another and work together as a single cohesive unit in order to perform particular tasks.
• The structures of DNA and RNA enable genetic information to be efficiently stored and expressed.
• The function of a protein is highly dependent upon its structure.
• The structure and composition of cellular membranes allows for compartmentalization and segregation of functions.

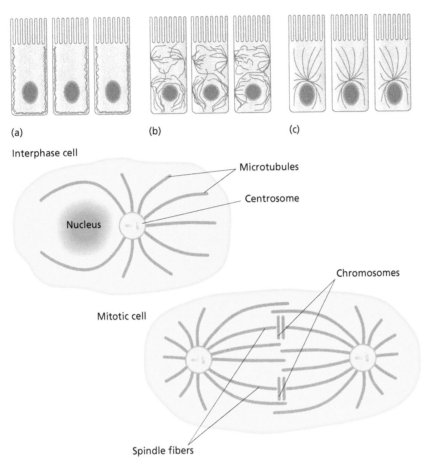

(a)   (b)   (c)

Interphase cell

Microtubules

Centrosome

Nucleus

Chromosomes

Mitotic cell

Spindle fibers

Figure 2.16 The cytoskeleton and mitotic/meiotic apparatus. (a) Actin filaments are located just below the plasma membrane. (b) Intermediate filaments are located throughout the cytosol. (c) Microtubules radiate outward from the centrosome. (d) During the processes of mitosis and meiosis the centrosome is replicated and each moves to an opposite pole of the cell. Certain microtubules, termed spindle fibers, bind to the proteins at the centromeres of the chromosomes enabling the chromosomes to be pulled to opposite poles of the cell.

The basics of cell structure and function learned in this chapter will be expanded upon in Chapter 3 in order to understand the events that transform a normal cell into one that is cancerous.

EXPAND YOUR KNOWLEDGE

1 Why do you think a cancer cell is considered a "rogue" cell?
2 If you were given the structure of a nucleic acid molecule, how would you determine if it was an RNA or DNA molecule?

3 The mRNA molecule produced from the transcription of a DNA molecule has the sequence ACGUAUGCCGAAAGUC. What is the double-stranded sequence of the DNA molecule that produced the mRNA? Which is the template strand?

4 Compare the bonds between the individual glucose molecules of starch and glycogen in Figure 2.9 with those between the glucose molecules in cellulose. Why do you think our digestive enzymes, which are proteins, can bind to and break the bonds present in starch and glycogen but not the bonds in cellulose?

5 What would be the consequence if a drug negatively affected the functionality of spindle fibers?

ADDITIONAL READINGS

Alberts, B., Bray, D., Hopkin, K., Johnson, A., Lewis, J., Raff, M., Roberts, K., and Walter, P. (2004) *Essential Cell Biology*, 2nd edn. New York, NY: Garland Science.

Braden, C. and Tooze, J. (1999) *Introduction to Protein Structure*, 2nd edn. New York, NY: Garland Science.

Bretscher, M. (1985) The molecules of the cell membrane. *Scientific American*, **253**(4): 100–108.

Campbell, N. A. and Reece, J. B. (2005) *Biology*, 7th edn. San Francisco, CA: Pearson Benjamin Cummings.

Cooper, G. M. and Hausman, R. E. (2007) *The Cell: A Molecular Approach*, 4th edn. Sunderland, MA: Sinauer Associates.

Marieb, E. N. and Hoehn, K. (2007) *Human Anatomy and Physiology*, 7th edn. Upper Saddle River, NJ: Prentice Hall.

Martini, R. and Bartholomew, E. (2007) *Essentials of Anatomy & Physiology*, 4th edn. Upper Saddle River, NJ: Prentice Hall.

Petsko, G. A. and Ringe, D. (2003) *Protein Structure and Function*. Sunderland, MA: Sinauer Associates.

Tanford, C. (1978) The hydrophobic effect and the organization of living matter. *Science* **200**(4345): 1012–1018.

3

The human genome and protein function

It is humbling for me and awe inspiring to realize that we have caught the first glimpse of our own instruction book, previously known only to God.

Francis Collins, MD, PhD, physician, geneticist, former Director of the National Human Genome Research Institute, and Director of the National Institutes of Health

CHAPTER CONTENTS

- The composition and function of the human genome
- Having a diploid genome has its advantages
- Proteins carry out diverse functions
- Expand your knowledge
- Additional readings

Cancer: Basic Science and Clinical Aspects, 1st edition. By C. A. Almeida and S. A. Barry. Published 2010 by Blackwell Publishing, ISBN 978-1-4051-5606-6.

The process of transcription produces RNA molecules by using the DNA nucleotide sequences of genes. Proteins, the work-horses of cells, are then synthesized from the RNA. It is the collective activities of these proteins that yield the true functionality of a cell. Since you are a collection of cells, you are who you are, in a sense, as a result of the expression of your genetic information. This chapter will delve more deeply into how DNA serves as a repository for genetic information and how changes in that information have widespread effects on cell function.

THE COMPOSITION AND FUNCTION OF THE HUMAN GENOME

Human cells have 46 chromosomes that can be arranged into 23 pairs. This is because there are two copies of each of the 23 chromosomes – a person inherits one chromosome in each pair from his or her mother and the other from his or her father. The pairing of the chromosomes

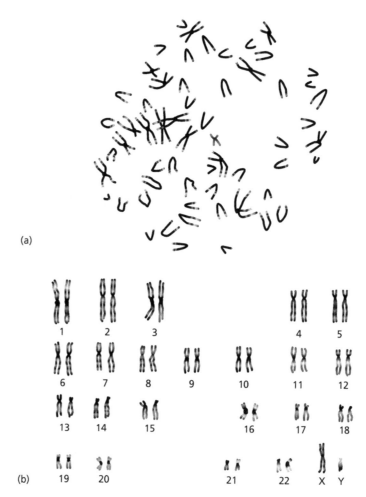

(a)

(b)

Figure 3.1 Human chromosomes. (a) The random distribution of chromosomes after having been isolated from a cell. (b) Arranging the 23 pairs of maternal and paternal chromosomes into pairs creates a karyotype. The 23rd pair is the sex pair; females possess two X chromosomes, whereas males possess one X and one Y chromosome.

can be easily observed following a lab analysis during which the chromosomes are isolated from a single cell, paired up using a set of criteria, and then arranged according to their sizes to yield a karyotype (Figure 3.1). The twenty-third chromosome pair is known as the sex pair and determines the gender of a person. Females have two X chromosomes while males have a single X and one Y chromosome. The smaller size of the Y chromosome allows it to be easily distinguished from the X chromosome.

When a sperm cell carrying 23 chromosomes fertilizes an egg containing 23

> **Which of the chromosome pairs is unique? Explain.**

chromosomes, the newly formed cell, a zygote, has the full complement of genetic information needed to form an entire human being. This set of genetic information, referred to as the human genome, consists of approximately 6.4 billion nucleotide base pairs. The sequence of nucleotides (As, Ts, Cs, and Gs) of one person's genome is *99.9% identical* to that of another person's.[3.1] This means, of course, that there is a *0.1% difference* in the entire sequence of each person's genome. Some of the differences are within the genes themselves while others are among the nucleotides between genes. This implies

> **If there is a 0.1% difference between the genomes of two people selected at random from the general population, then how many nucleotide base pairs does that represent?**

> [3.1] If you were to read the order of all 6.4 billion base pairs of the human genome at the rate of 1 nucleotide per second, it would take nearly 203 years!

that while we all have the same genes, not all of them will have the same sequence. Amazing as it may seem, it is that small 0.1% difference that is the source of the uniqueness of each person.

Significance of alternate versions of genes

We all have the same genes, yet for any given gene it is more than likely that not everyone will have the same sequence. This is what contributes to the 0.1% difference in each of our genomes. In an attempt to illustrate this point, assume that the sentence THE CAT SAW AND ATE THE RAT represents a gene. Just as a gene conveys the information a cell needs to make a protein to carry out a function, the sequence of letters in the sentence conveys information about the activities of the cat and the fate of the rat. Now suppose that the "C" in CAT is replaced with a "B". The same activities, seeing and eating, are being carried out but now by a different animal. The bat would certainly carry out the activities differently from the cat, but for the rat, the outcome is the same. One person may possess the CAT version of the gene while another person may possess the alternative BAT version. Such alternate versions of genes are called alleles. The familiarity most people have with alleles is related to eye color. For example, one allele of the eye color gene produces brown pigment while another produces blue pigment. The activity instructed by each allele is slightly different but, in the end, pigment is produced.

The sequence of one allele of a gene may differ from another by a single nucleotide substitution (e.g., G^C and G^B), but that may be sufficient, for instance, to specify the replacement of a negatively charged amino acid (present in the G^C protein) with a positively charged amino acid (present in the G^B protein). These two amino acids differ sufficiently enough in their chemical characteristics that they would not bond to other amino acids in the protein in the same way during the folding process and consequently, would result in a structural difference between the two proteins. Because of its altered structure and the inherently strong structure–function relationship that proteins have, it is likely that the G^B allele will produce a protein unable to function in the same way as the protein coded for by the G^C allele.

Consider a third allele of the previous example, THE MAT SAW AND ATE THE RAT. Intuitively, the same activity is happening and the rat experiences the same fate. A MAT, however, is not capable of seeing or eating and as a result, the sentence conveys an impossible series of events . . . it simply does not make sense. Certain alterations to a gene's sequence can result in meaningless information for the cell. If the information in the CAT allele helps regulate when a cell will divide, then the same is true for the RAT and MAT alleles. The extent to which the regulation is carried out may be different and that can have an effect on the rate or timing of when a cell divides. A cell that possesses the MAT allele may lack the regulation entirely and as a result, the cell will have a greater tendency to divide continuously.

> **Explain why one of the three gene sequences (sentences) above is truly unique from the others.**

Box 3.1

Are all cells within a person genetically identical?

In theory, the DNA sequence in every cell in our body should be identical since all 10 trillion of the cells originated from a single cell, the zygote, formed from the fertilization of an egg by a sperm. In reality, however, it is inevitable that there will be differences that arise over time. Prior to the zygote dividing for the first time, it must make identical copies of all 46 chromosomes so that a complete set can be partitioned into each of the new daughter cells. Replicating 6.4 billion nucleotides without making an error is certainly a daunting task, not to mention that this must occur every time cells divide. Cellular mechanisms have evolved to minimize changes to the sequence during DNA replication, but the process is not perfect and mistakes do occur.

Changes to the sequence of our DNA are not due solely to errors made during DNA replication. As mentioned in Chapter 1, DNA damage can result from exposure to chemicals that enter our bodies in the air we breathe and in the food we eat and drink, as well as from radiation that we are subjected to when we are in the sun, fly in an airplane, or undergo certain medical tests or treatments. Once the sequence of a chromosome is altered, also commonly referred to as mutated, then all of the descendants of the cell it is in will possess the alteration.

Alterations to DNA sequence that a person accumulates over a lifetime will not be passed along to his or her children unless the mutations occur within the cells that produce sperm or eggs. DNA mutations can play a critical role in the development and progression of cancer and the likelihood that the disease can be treated successfully.

Genes that control when a cell divides can be placed into two categories. Protooncogenes are those that promote cell division whereas tumor suppressor genes, as their name implies, prevent a cell from dividing. Over time, cells that accumulate mutations in these two classes of genes lose the ability to properly regulate when they divide and, as a result, undergo uncontrolled growth – in other words become cancerous. The role of these genes and the transformation process that cells undergo are the topics of Chapters 4 and 5.

> **Would an alteration in a proto-oncogene that inhibits the function of its protein product contribute to cancer? What about an alteration that enhances its function?**
>
> **Would an alteration in a tumor suppressor gene that inhibits the function of its protein product contribute to cancer? What about an alteration that enhances its function?**

Influence of genetic information and the environment

Is it only genetic information that makes a person unique? Absolutely not. If this were true, then identical twins should be the same in every single trait they exhibit . . . and that is certainly not the case.[3.2] Identical twins can differ in traits that they have hardly any or no control over (e.g., height, birthmarks, sense of humor, self-confidence, etc.) as well as traits that are certainly under their control (e.g., whether or not to dress alike or have the same hairstyles). This is because interactions with the environment, from

> 3.2 Identical twins arise when the two genetically identical cells formed by the division of the zygote separate from one another rather than remain together. Each cell on its own then continues through the developmental process producing two genetically identical persons. Fraternal twins arise when two individual eggs are fertilized by two separate sperm. This results in two nongenetically identical individuals.

the time spent in the womb through to old age and death, also influence who a person is as an individual. Consider a woman who is pregnant with identical twins and has the tendency to nearly always stand with the right side of her body closer to the stove when she cooks. The fetus being carried in the right half of her womb will regularly be exposed for periods of time to temperatures that are two or more degrees warmer than is the fetus on the left. Temperature is an environmental condition that affects speeds of reactions and stability of molecules. What is the consequence of these regular environmental temperature differences on the development of the two genetically identical individuals?

Also consider a set of identical twins who are separated at birth and grow up in two very different home environments: one is full of love, attention, and support whereas in the other home there is abuse, loneliness, and denigration. What is the effect of these two very different environments on the dispositions and personal outlooks on life for the two genetically identical individuals?

> Provide an example of a "nature" influence as well as a "nurture" influence.

The question is, to what degree is a trait or characteristic influenced by a person's genome and/or environment? This question is the crux of the age-old *nature vs. nurture* debate. To attempt to understand this ambiguity and to answer similarly intricate questions, we have to have a good understanding of the information stored in our genome. The first step in obtaining this understanding required the determination of the sequence of As, Ts, Cs, and Gs on each of the chromosomes. That task was the goal of the Human Genome Project, which was begun in 1990. Researchers in governmental, public, and private labs used robotics, new technologies, and supercomputers over the span of 13 years to determine the sequence of the entire human genome.

Identifying genes in the human genome

Computer programs are currently being used to assist in potentially locating where, among all of the sequence on each of the chromosomes, the estimated 20,000–25,000 genes are located. Once a potential gene location has been identified, researchers then do further analyses to verify whether or not it is in fact a gene. Following the identification of a gene is the determination of the function of its gene product, whether it be an RNA molecule or a protein that carries out a particular function in the cell. This task is certainly a challenge; greater than 50% of the genes discovered to date do not have known associated functions. Through this identification and discovery process, many genes will be discovered and associated with a wide variety of roles in health and disease. The more that is learned about the information contained in genes, the better the position we will be in to predict a person's predisposition to develop a disease, to confirm a diagnosis, to determine the most appropriate form of treatment method, and to predict the likelihood of surviving. The ultimate goal, of course, is to obtain the ability to prevent the onset of a disease.

> Go online and for two cancers of your choice determine if there are alleles of genes that have been identified that are associated with their development, aggressiveness (i.e., rapid rate of growth, likelihood of metastasis, and lack of response to treatment) and/or a recommended form of treatment.
>
> Discuss how you think a particular mutation would impart a decreased susceptibility to a certain form of treatment.

HAVING A DIPLOID GENOME HAS ITS ADVANTAGES

Both chromosomes in a pair contain the same genes in the same locations and are therefore considered to be homologous (*homo-* = same).

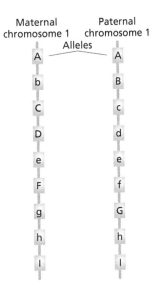

Figure 3.2 Homologous chromosomes. The maternal and paternal chromosomes within a single pair possess the same genes (A–I) but may not possess the same alleles (b vs. B; C vs. c, etc.) of those genes.

For example, if the maternal version of chromosome 1 contains nine genes labeled A through I, then so does the paternal copy of chromosome 1 (Figure 3.2). This means that there are at least two copies of every gene. If the nucleotide sequence of the two copies of a gene are identical then the alleles are homozygous, but if there is a difference in the nucleotide sequence then the alleles are heterozygous (Figure 3.2). A system that uses capital and/or lowercase letters as well as superscripted letters is typically used to differentiate alleles. For example, the allele "THE CAT SAW AND ATE THE RAT" can be designated by G^C; the "G" stands for "gene" and the "C" stands for "CAT". Alternative alleles of the gene, such as "THE BAT SAW AND ATE THE RAT" and "THE MAT SAW AND ATE THE RAT" can be represented by G^B and G^M, respectively. Remember, the two copies of gene G on chromosome 1 could be homozygous (G^C and G^C or G^B and G^B or G^M and G^M) or heterozygous (G^C and G^B, or G^C and G^M or G^B and G^M) (Figure 3.3). Allelic differences can produce gene products with varying levels of functionality and, as a result, it is the unique combination of alleles and interactions with the environment that is responsible for a person's unique physical and psychological characteristics.

An inherent advantage of having a diploid genome is that there can be a built-in backup in the event that one of the two copies of a gene produces a nonfunctional gene product after acquiring a mutation. This can be illustrated by comparing the effectiveness of various allelic combinations of a gene that codes for a protein. Proteins, which will be discussed in more detail in a later section of this chapter, carry out a great

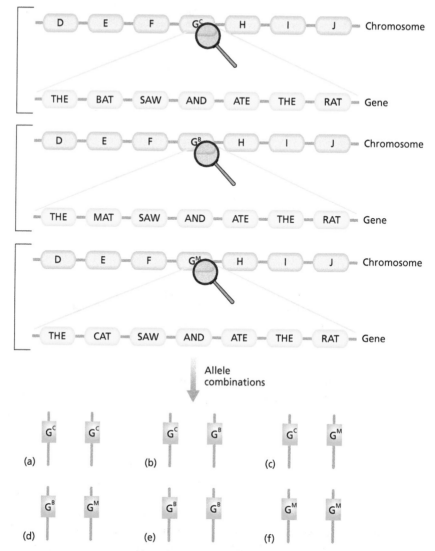

Figure 3.3 Numerous combinations of alleles are possible. Since there are three alleles of gene "G", then a person would possess one of six possible allelic combinations; three are homozygous (a, e, and f) and three are heterozygous (b, c, and d).

deal of the functions performed by a cell. In order for the minimal needs of a cell to be satisfied, the proteins must perform at a certain activity level, referred to as the **basal level**. Conditions are apt to change, requiring an increase in cellular activity above that of the basal level (e.g., when a person gets out of bed and walks).

Table 3.1 The efficiencies of allelic combinations at meeting cellular requirements

Allele combinations	Efficiency
$G^C G^C$	100%
$G^C G^B$	80%
$G^B G^B$	50%
$G^B G^M$	25%
$G^M G^M$	0%

A cell is negatively affected when the amount of activity provided does not meet the level required. For example, a cell homozygous for the G^C allele is capable of satisfying 100% of the cell's needs with regard to the activity of the protein coded for by the gene G (Table 3.1). A cell that is heterozygous and possesses alleles G^C and G^B may be able to satisfy the basal level of activity required by the cell, but when the requirements increase, the two versions of the protein coded for by gene G may only be 80% efficient at satisfying the cell's needs. The most deleterious case would be when a cell is homozygous for the G^M allele. In this scenario, it is unlikely that the cell's activity needs would be met, even at the basal level. How critical this protein's function is under any conditions a cell may experience, will determine whether or not it would survive.

What has been described to this point is the scenario in which certain alleles of a gene produce RNA or proteins that do not function as well as others. Alternatively, it is also possible that the product of an allele interferes with the functionality of the product of another allele. For example, if a cell is heterozygous and possesses the G^C and G^M alleles it is possible that there would be sufficient functionality from the G^C-encoded protein to satisfy the cell's needs under nonstressful conditions. It is possible, however, that the G^M protein interferes with the ability of the G^C protein to function. In this scenario it does not matter what the G^M allele is paired with, because its protein product will mask the expression of the other G allele.

Differential expression of the genome's information

Each cell does not require all of the information in the genome and therefore only expresses or uses those genes needed for its particular structure and functions. Still, there are certain components and functions that every cell must possess and perform. Those needs are satisfied by a select set of genes, known as housekeeping genes, that are expressed in every cell. The expression of additional genes at certain times during development, under specific conditions, or in particular tissues is referred to as

3.3 Differential gene expression to achieve specialized cell functions is similar to how computers in a company, loaded with the same operating system and software packages, work to perform particular tasks. Every computer will use the same operating system (i.e., housekeeping) software but each one will use only a certain subset of the available programs based on the task an employee needs to complete at a particular time.

differential gene expression.[3.3] Therefore, the alteration of a gene's sequence may only be significant in those cells that express that gene. If over the course of a person's life the sequence of a gene is mutated from the G^C to the G^M version, but that alteration only occurred in a cell that does not express that gene, then there will be no consequence. However, if that same gene mutation occurred in a cell within a tissue that expresses the gene, then there would likely be a negative effect.

PROTEINS CARRY OUT DIVERSE FUNCTIONS

No matter what the function of a protein is, it can be classified into one of several types (Figure 3.4). Some are enzymes that catalyze reactions that would otherwise occur too slowly to sustain life. There are *structural* proteins that provide the shape, stability, and flexibility to cells or body structures (e.g., ears and the tip of the nose). Others are *transport* proteins that move molecules from one location to another in the body or within a single cell. Some are *protection* proteins that provide us with a defense against infectious agents such as bacteria and viruses. The final group consists of *informational* proteins that allow communication to take place within or between cells and tissues.

More often than not, protein function depends upon its interaction with another protein or molecule. The two surfaces that make contact during the interaction must be complementary in their contours and chemical makeup. For example, if the surface of a protein has a positive charge, an appropriately sized molecule with a complementary shape and negative

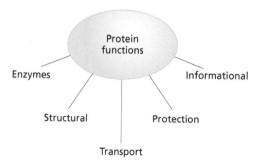

Figure 3.4 The types of functions that proteins can perform.

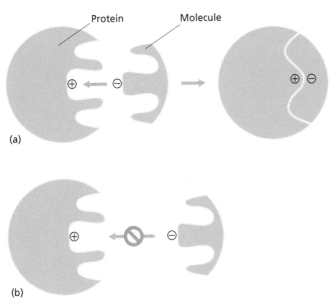

Protein   Molecule

(a)

(b)

Figure 3.5 Characteristics that influence binding to proteins. (a) A molecule will bind to a protein if it has a size, shape, and charge that is complementary to the protein's surface. (b) A molecule will not be able to bind to a protein if its size, shape, or charge is not complementary to the protein's surface.

charge would be attracted and bind to it. On the other hand, a molecule with a similar size and shape but possessing a positive charge would be repelled (Figure 3.5).[3.4] The shape of a protein is often altered when it interacts with other molecules. This is analogous to the shape changes that occur to the hands of two people prior to, during, and after a hand-shake. A protein, therefore, will have unbound and bound conformations or shapes depending on whether it is free or bound to another molecule. Notice in Figure 3.5 how the shape of the protein is different before and after it binds the small molecule.

> **Describe how a molecule binding to a protein is similar to what occurs when you catch a ball in your hand.**

[3.4] It is common for two people to use their right hands for a handshake because of their complementarity. A handshake between a right and a left hand lacks complementarity because the left hand is a mirror image of the right.

As was mentioned in Chapter 2, there is a strong relationship between the structure of a molecule and its function. Certain proteins may be non-functional when in the unbound conformation but become active upon the shape change that occurs when it interacts with another molecule. In other situations the exact opposite may be the case – a protein is active in the unbound state and inactive in the bound conformation.

The following are examples to illustrate how the structural, chemical, and binding characteristics of proteins can be used in the treatment of a disease. Consider the case when a disease is the result of a mutant protein that is overactive. Treatment may involve the use of a drug with a shape and charge that is complementary to the mutant protein. This enables the drug to bind to the mutant protein so as to alter its structure and function. In this case, the altered function would be a reduction in its activity. Alternatively, a drug could bind to a protein in such a way so that the molecule that would normally bind to the protein can no longer do so. The inability of the protein to interact with the molecule it normally does bind with will alter its functionality.

Cells receive information via receptor-dependent protein cascades

Mutations that negatively affect the cellular reaction pathways that tightly regulate the timing of cell division are strong contributors to the development of cancer. The process of cell division is typically initiated by specific signaling molecules binding to particular receptors that then set off a cascading series of reactions (Figure 3.6). The end result of those reactions is a change in the function of the cell that promotes those

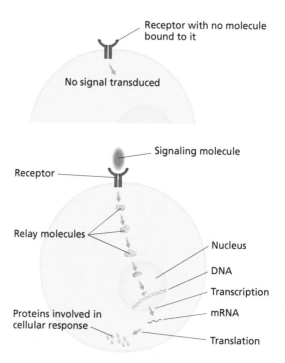

Figure 3.6 Signal transduction. The binding of a signaling molecule to its receptor on the surface of a cell triggers a series of cascading reactions that alters gene expression resulting in a changed cell function.

activities that must occur in order for division to proceed. The series of reactions that take place, one after the other, in response to the binding of a signaling molecule to a receptor is referred to as a signal transduction cascade pathway. The word "transduction" is included because the binding of a signaling molecule on the external surface of a cell must be transduced, or conveyed, across the membrane and into the cell. This is similar to people in a house being made aware that someone is outside by his pushing a doorbell (the receptor) to close a circuit to allow the flow of electricity (the signal transduction cascade) that sets off a chime (the response).

Receptor proteins are integral plasma membrane proteins that have a portion that extends away from the outer surface of the cell membrane as well as a portion that extends into the cytoplasm of the cell. A structural change occurs when a signaling molecule outside of the cell binds to the receptor. In its new bound conformation it can now interact with a component in the cell, which is most often an enzyme. The enzyme, now active as a result of undergoing a conformational change, can bind and modify or alter certain molecules (e.g., proteins) in a very specific way. In this case, the enzyme will begin producing molecules in the cell that were either not present previously or present only in very low concentrations.[3.5] The presence of these new molecules sets into motion a coordinated series of events as they interact with other molecules – one protein activates another protein which activates another and so on (Figure 3.6). A common consequence of signal transduction pathways is changes in gene expression – certain genes will be turned on to make certain proteins while the expression of other genes will be inhibited. Ultimately, the cell will be functioning differently than it was prior to the binding of a molecule to its receptor.

The membranes of cells throughout the body possess a specific protein termed epidermal growth factor receptor (EGFR) that binds the small protein called epidermal growth factor (EGF). The name of this signaling molecule indicates not only the type of cell that it affects but also the type of effect it has on those cells – epidermal cells that bind EGF grow or divide to produce daughter cells. As you may recall from Chapter 1, epidermal

[3.5] The original signaling molecule is referred to as the primary messenger while the molecule being produced by the enzyme inside the cell is referred to as the secondary messenger because it is produced in response to the binding of the primary signaling molecule to the receptor.

Two neighboring cells are nearly identical except that one possesses a mutation that causes the overexpression of a growth factor receptor gene. As a result one cell has twice the number of that type of growth factor receptor on its cell surface than the other cell. What is a potential consequence to this situation?

What would the consequence be to a cell that possesses a mutation that inhibits the production of growth factor receptors?

cells are the most common cell type that undergo cancerous transformation. The number of a particular type of protein receptor in the membrane of a single cell can vary from only a few to up to tens of thousands and determines the strength of the signal received by a cell; the greater the number of receptors there are for a particular signaling molecule, the greater the cell's capacity will be to bind that molecule and generate a signal that enters the cell. Certain cancer cells can be very aggressive or vigorous in their growth as a consequence of having abnormally large numbers of EGFRs in their outer membranes. Examples of various types of growth factors and their receptors as well as their associations with particular types of cancers will be discussed in later chapters.

Drug resistance may arise from defective transport proteins

The composition and structure of the cell membrane enable it to function as a semi-selective barrier. Some of the integral proteins in membranes function as transporters, allowing the cell to regulate the passage of molecules *across* the membrane that cannot *pass* through it on their own (Figure 3.7). The structure, size, and chemical makeup of each particular type of transport protein are unique, each forming a tunnel in its center that accommodates a specific molecule to be transported. This allows for the regulated movement of specific molecules across a cell's membrane.

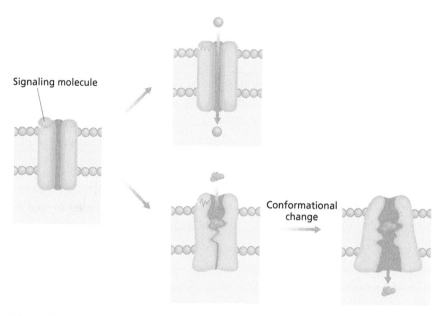

Figure 3.7 Integral membrane proteins can function as transport proteins. There are different types of integral proteins that can assist in the transport of molecules across a membrane.

An issue that patients may experience during the course of treatment is drug resistance – the loss of a drug's effectiveness after a period of time. One way that this can occur is when transporters in the outer membranes of the cells very quickly pump out the drugs shortly after they have entered the cell. This can occur as a consequence of a mutation in a transporter's gene that causes the protein product to be much more efficient in pumping the drug out of the cell. Alternatively, the cell may increase the rate of expression of the transporter gene so that many more of the transporters are made and put into the cell membrane, thus enabling more of the drug to be exported. Cancer cells that can accomplish either of these events are going to be less affected by the chemotherapeutic drugs . . . they will essentially be drug resistant.

> **Which types of macromolecules are found within cell membranes and what are their functions?**
>
> **Suggest a way in which such drug resistance due to defective transport proteins could be overcome.**

EXPAND YOUR KNOWLEDGE

1 It was learned in this chapter that the maternal and paternal chromosomes with each chromosome pair contain the same genes but not necessarily the same alleles. Which chromosome pair is an exception to this rule? Explain.

2 This chapter described how an overexpressed EGF receptor, or any protein in its signal cascade, can lead to uncontrolled division (i.e., cancer). Suggest another possible alteration in the receptor and or/signaling proteins that can lead to a similar result.

3 Suppose a mutation occurred in genes encoding antibodies (proteins that function in the body's immune response). In which cells would such a mutation have the greatest effect and why?

4 This chapter described two examples of how protein interactions are used to treat disease. Considering what you know of the specificity of protein interactions, which example do you think would be considered "competitive inhibition" of the protein, and which "noncompetitive inhibition"?

5 List any cellular components whose alteration would affect the ability of the cell to respond to an extracellular signal.

ADDITIONAL READINGS

Alberts, B., Bray, D., Hopkin, K., Johnson, A., Lewis, J., Raff, M., Roberts, K., and Walter, P. (2004) *Essential Cell Biology*, 2nd edn. New York, NY: Garland Science.

Braden, C. and Tooze, J. (1999) *Introduction to Protein Structure*, 2nd edn. New York, NY: Garland Science.

Bretscher, M. (1985) The molecules of the cell membrane. *Scientific American* **253**(4): 100–108.

Campbell, N. A. and Reece, J. B. (2005) *Biology*, 7th edn. San Francisco, CA: Pearson Benjamin Cummings.

Cooper, G. M. and Hausman, R. E. (2007) *The Cell: A Molecular Approach*, 4th edn. Sunderland, MA: Sinauer Associates.

Petsko, G. A. and Ringe, D. (2003) *Protein Structure and Function*. Sunderland, MA: Sinauer Associates.

4

Cell cycle, oncogenes, and tumor suppressor genes

Some cancer patients ask, 'Why me?' I ask, 'Why not me?'
Susan Parker, pancreatic cancer patient

Cancer: Basic Science and Clinical Aspects, 1st edition. By C. A. Almeida and S. A. Barry. Published 2010 by Blackwell Publishing, ISBN 978-1-4051-5606-6.

Cell division, a necessary and essential part of development, produces all of the 10 trillion cells required to form an adult human. In adults, the purpose of cell division has shifted from one of growth and differentiation to mostly replacement of cells that have either died because they reached the end of their lifespan, have been damaged beyond repair, or have been sloughed off the surface of the skin or digestive tract during daily activities. It is essential that cell division be a tightly regulated process to ensure the proper structure and function of tissues, organs, and organ systems. A loss in regulation can result in uncontrolled cell growth, a characteristic exhibited by cancer cells. There is no single event that is common to all cells that become cancerous. Rather, there are a small number of unique mutations in the genetic information that set into motion a series of events that causes the functions of cells to become increasingly abnormal. The mutations that occur may be related but do not have to be identical in all cancers. This diversity of genetic mutations that results in cancer can factor into the determination of the most appropriate treatment methods and into the prognosis for survival.

CELL DIVISION IN GERM-LINE AND SOMATIC TISSUES

All of the cells in the body are classified as belonging to either the germ-line or somatic tissue. Germ-line tissue produces gametes, the reproductive or germ cells commonly known as sperm and eggs. All other cells of the body belong to somatic tissues. Somatic cells possess 46 chromosomes and divide by the process of mitosis to produce two diploid daughter cells, each possessing 46 chromosomes. The daughter cells produced by mitosis are genetically identical to the parental cell (Figure 4.1a). Sperm and egg cells, on the other hand, are the products of meiosis, the form of cell division that segregates one chromosome from each pair into a separate daughter cell. Each germ cell is haploid because it contains only 23 chromosomes, one half of the genome of the germ-line parent cell (Figure 4.1b). During conception, the 23 maternal and 23 paternal chromosomes are brought together to yield a diploid cell with a full complement of 23 pairs of chromosomes.

A cell spends the majority of its time in interphase, the first of two major stages of its life. During this time, the cell may grow in size, carry out its activities, replicate its chromosomes, and prepare for division. When a cell is not preparing to divide, the chromosomes resemble linear pieces

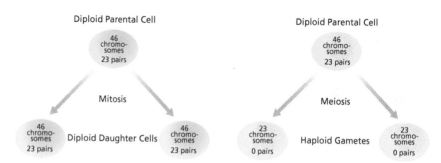

Figure 4.1 Products of mitosis and meiosis. (a) The parental human cell that undergoes mitosis contains 46 chromosomes in 23 pairs and produces two diploid daughter cells, each with the same chromosomes as the parent cell. (b) The parental human cell that undergoes meiosis contains 46 chromosomes in 23 pairs and produces haploid gametes, each with 23 chromosomes – one from each chromosome pair.

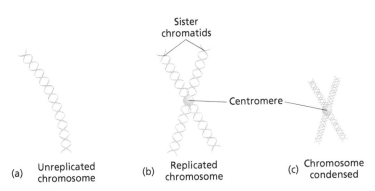

Sister
chromatids

Centromere

(a) Unreplicated
chromosome

(b) Replicated
chromosome

(c) Chromosome
condensed

Figure 4.2 Unreplicated and replicated chromosomes. (a) Prior to replication a chromosome is a single DNA molecule. (b) A replicated chromosome is composed of two sister chromatids that are identical in their nucleotide sequences. The sister chromatids are held together at a region on the chromosome known as the centromere which is also where the spindle fibers attach during cell division. (c) Just prior to cell division the sister chromatids are compacted.

of thread (Figure 4.2a). During interphase, genes along each of the chromosomes are accessible and can be expressed.

Events that occur during cell division

The binding of growth factor signals to receptors on the surface of a cell will set into motion a series of events that will culminate in cell division (Figure 4.3). Cells divide in two unique yet similar manners: germ-line cells divide by a process known as meiosis, and all other cells of the body (stromal cells) divide by mitosis. Most cellular responses to a growth signal are identical regardless of whether a cell will divide by mitosis or meiosis (Figure 4.4). When preparing to divide, a cell produces an identical copy of each chromosome and the two copies remain attached at a single point called the centromere (Figure 4.2b). Each chromosome bound to the centromere is referred to as a sister chromatid. The chromosomes then undergo a process of compaction that increases the efficiency and accuracy of their segregation into the daughter cells (Figure 4.2c).[4.1] The nuclear membrane breaks down, exposing the chromosomes to the cytosol (Chapter 2), and the chromosomes align in the center of the cell. The centrosome, a structure normally adjacent to the nucleus, replicates and

4.1 Segregation of the chromosomes in a noncompacted state would be difficult considering that the average body cell is only a few micrometers in diameter yet the total length of the 46 chromosomes, when aligned end to end, is approximately two meters in length (1 micrometer is equal to 1 millionth of a meter and 1 meter is equal to 3.28 feet).

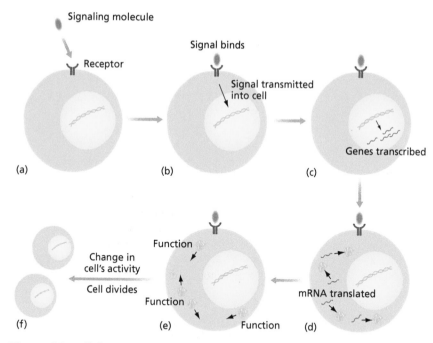

Figure 4.3 Cellular response to an external signal. (a) A cell with an appropriate receptor for a signaling molecule. (b) The signaling molecule binds to the receptor on the surface of the cell and the signal is transferred across the membrane into the cell. (c) In response to the signal the expression of certain genes is inhibited while the expression of other genes is promoted. (d) The mRNAs that were synthesized in response to the signal are translated to produce proteins. (e) The newly synthesized proteins carry out functions which change the cell's activity. (f) One possible result of a functional change may be the cell dividing.

4.2 What occurs during cytokinesis can be mimicked by wrapping a string around the center of a ball of dough so that the ends cross over one another. Pulling in opposite directions on the ends of the string will create a furrow in the dough. The more the string is pulled, the deeper the furrow will get until the string cuts completely through, yielding two separate balls of dough.

each one migrates to an opposite end of the cell. Radiating outward from each of the two centrosomes are numerous protein filaments called spindle fibers or microtubules that bind to the proteins on the centromeres of the chromosomes. The spindle fibers act as a cable system that pulls half of the chromosomes toward one centromere and half toward the other one. Once the chromosomes have been segregated to opposite ends of the cell, the actual process of cell division, also known as cytokinesis, occurs. A furrow (i.e., indentation) forms around the circumference of the cell and progressively deepens until the membrane pinches off forming two daughter cells (Figure 4.5).⁴·² Finally,

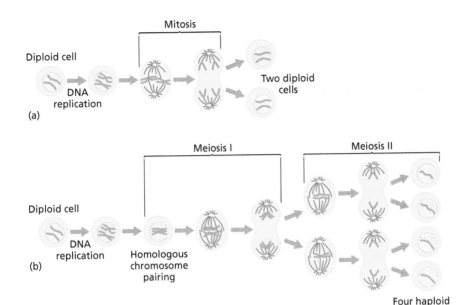

Figure 4.4 Processes of mitosis and meiosis. (a) Mitosis consists of a single cell division during which the chromosomes line up single file along the metaphase plate and the sister chromatids are pulled to opposite poles to yield two diploid daughter cells. (b) There are two cell divisions in meiosis. During the first division homologous chromosomes pair up along the metaphase plate and are pulled to opposite poles. The two daughter cells possess a haploid number of replicated chromosomes. The chromosomes line up single file along the metaphase plate during the second division of meiosis and sister chromatids are pulled to opposite poles yielding four haploid gametes.

Figure 4.5 Cell furrow formed during cytokinesis. A cell during mitosis with a deep furrow that will eventually separate the cell to yield two daughter cells.

nuclear membranes will reform in each of the daughter cells, separating the chromosomes from the rest of the cellular content, and the chromosomes will unwind into their interphase state.

Unique events of mitosis

Two significant differences that occur during mitosis and meiosis are the way in which the chromosomes line up in the center of the cell and the fact that mitosis has only one cell division, whereas meiosis has two. The goal of mitosis is the production of daughter cells with the same genetic information as the parental cell. Therefore, the 46 replicated chromosomes align *single file* in the center of the cell and the *sister chromatids* are pulled apart, segregated to opposite ends of the cell, and then enclosed within the nuclear membrane that develops in each daughter cell (Figure 4.4a). The two daughter cells have a full complement of chromosomes and are genetically identical – of course this assumes that each of the chromosomes was replicated without any errors and was not damaged during mitosis.

Unique events of meiosis

The objective of meiosis is the production of daughter cells that contain one member of each homologous pair (Chapter 3). This is accomplished by aligning the replicated chromosomes into *homologous pairs* in the center of the cell (Figure 4.4b). The *intact, replicated chromosomes* in each pair are pulled apart from one another to opposite ends of the cell. This differs from mitosis, which involves the separation of the *sister chromatids*. The consequence is that the two daughter cells will each have 23 replicated, not singular, chromosomes. The purpose of the second division of meiosis is to separate and then segregate the sister chromatids. In the second division, the 23 replicated chromosomes align single file in the center of the cell, similar to what occurs in mitosis. The spindle fibers then pull apart the sister chromatids to opposite sides of the cell and cytokinesis occurs.

> **Construct a bulleted list of events that occur during mitosis.**
>
> **Construct a bulleted list of events that occur during meiosis.**
>
> **Explain what would happen to gene expression during mitosis and meiosis.**
>
> **When in meiosis are the cells first haploid?**

Certain cancer therapies target the process of cell division

In 1963, an extract from the bark of the Pacific yew tree was found to not only inhibit the replication of cancer cells but also cause their death. It took the next four years to isolate the specific compound responsible

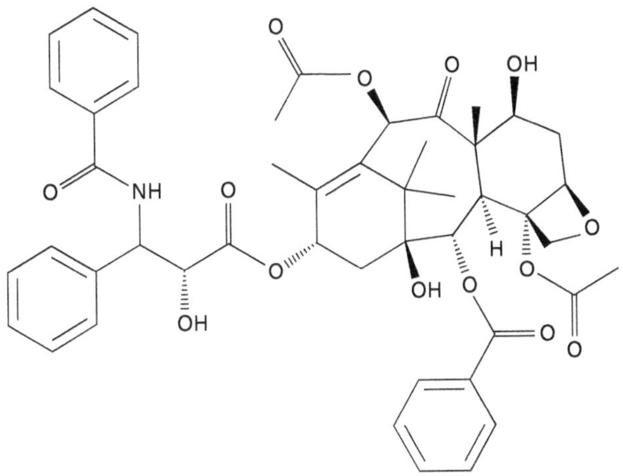

Figure 4.6 Structure of paclitaxel/Taxol.

for this effect from the many thousands present in the bark, and an additional four years to determine its structure (Figure 4.6). The compound was named paclitaxel and in 1979 a paper was published providing the mechanism for the drug's toxicity. Microtubules are polymers of the protein tubulin that are bound to one another in a chain-like fashion. Paclitaxel binds to tubulin molecules and stabilizes them, preventing the proteins from disassociating from one another and thus explaining the negative effect the drug has on cancer cells. The segregation of chromosomes during mitosis or meiosis is dependent on the disassembly of spindle fibers. The inability to segregate the chromosomes properly in the presence of paclitaxel would certainly interfere with cell division. Since cancer cells divide more frequently than most healthy body cells, they are more susceptible than other cells to paclitaxel. Since 1992, the drug, under the name Taxol, has been used for the treatment of lung, ovarian, and breast cancers, as well as late stage Kaposi's sarcoma, a type of skin cancer that occurs most often in people with suppressed immune systems.

The basic intent of all cancer treatments is twofold: to interfere with the ability of cancer cells to divide, and to take advantage of the poor ability of these cells (compared to healthy ones) to respond appropriately to and repair cell damage (Chapter 5). The objective of all forms of cancer therapy is to kill cancer cells at a higher rate than healthy cells; this may result in the patients becoming very sick during the course of treatment. Such side-effects of drugs or radiation are due to the negative effects they have on healthy cells. For example, the side-effects experienced by patients taking Taxol are due to the inhibition of healthy rapidly dividing cells in the body. There are areas of the body where cells are constantly being produced, such as the bone marrow, skin, digestive tract, and hair in order to replace the cells that are continuously lost due to death or because they are shed from the body. The microtubules in these rapidly dividing cells are being negatively affected by Taxol and other chemotherapeutic drugs just as they are in cancer cells.

CONSEQUENCES OF GERM-LINE AND SOMATIC TISSUE MUTATIONS

DNA mutations can arise in any cell of the body and, if not repaired or corrected, will be passed down to the cell's descendants. The consequence of inheriting these mutations is different for somatic versus germ cells. A mutation that occurs in a somatic cell possesses a high probability of affecting that cell and all of its descendants. Cells in the germ-line tissue that acquire mutations will also pass them along to their descendants. In this case, however, the daughter cells are sperm and egg cells. The significance of germ cell mutations is that they are heritable and can be passed down to the next generation. Only 5–10% of known cancers are the result of mutations inherited through germ cells, meaning that the vast majority of cancers are the consequence of mutations acquired during a person's lifetime.

CELL DIVISION, DIFFERENTIATION, AND MATURATION OCCUR TO FORM FUNCTIONAL TISSUES

The human body is composed of trillions of cells and they all arise from a single, original zygote. How is it that more than a couple hundred different cell types, varying in size, shape, and functions, develop from a single cell? The zygote is an undifferentiated cell, meaning that rather than having a specialized function, it is totipotent, possessing all of the potential necessary to produce each of the different types of cells found in an adult. Such undifferentiated cells are also called stem cells and have the ability to divide continuously to produce new stem cells as well as cells that will eventually become the various types of mature, functional body cells. The zygote and its descendants divide countless times during embryonic, fetal, childhood, and adolescent development. Furthermore, cells in different areas of the body make direct contact with only a few of the extraordinary number of cells present in the entire organism. As a consequence of interactions with their neighbors, cells receive a vast array of physical and chemical signals. These varying sensory experiences lead to differential gene expression (Chapter 3) and thus to the specific compositions, structures, and functions of cells.

An effect of differential gene expression during cellular development is the progression from an undifferentiated to a differentiated state. Terminally differentiated cells have a specific structure, carry out specific functions, and

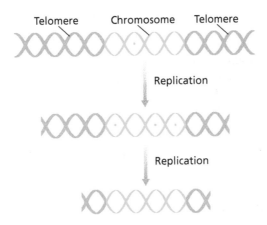

Figure 4.7 Telomere shortening. Each time a chromosome is replicated some of the nucleotides at the ends, telomeres, are lost.

divide infrequently if ever again.[4.3] Collections of differentiated cells in a particular environment will communicate with one another and work cooperatively to form tissues and organs.

> [4.3] Muscle and nerve cells are mature and terminally differentiated to the point that they no longer divide.

Cells possess a built-in system that keeps track of the number of divisions that have occurred within a single cell lineage. The two ends of each chromosome are known as telomeres. They do not contain any genetic information, but rather consist solely of many repetitions of the short nucleotide sequence TTAGGG. Every time chromosomes are replicated, a portion at the end of each telomere remains unreplicated (Figure 4.7). This loss of telomere sequence is a consequence of the mechanism cells use to replicate DNA. While there is a loss of *DNA* there is no loss of *genetic information* because telomeres do not contain any gene sequences. If there was no limit to the number of times a cell could divide, there would come a time when not only the entire telomere sequence would be lost, but genetic information would begin to disappear as well. Within a cell lineage, there is a maximum of approximately 50 rounds of division that a cell can undergo before reaching its limit. When a cell lineage has reached its limit of replications, the telomere sequence has been shortened to a particular length that effects a change in gene expression. The expression of genes that promote cell division is decreased while the expression of genes that inhibit the cell from dividing is increased. The result is an inability of a cell to divide again, thereby protecting against the potential loss of genetic information.

As mentioned previously, there are certain types of cells, such as skin, digestive tract, and blood, that must be produced continuously throughout

4.4 The stem cells present in mature tissues are referred to as adult stem cells, although their potency and ability to produce a variety of cell types is much less than that of embryonic and fetal stem cells. Adult stem cells are partially differentiated themselves in that they can only produce the types of differentiated cells found in the tissue in which they reside. For example, stem cells within the bone marrow produce only blood cells and platelets (cell fragments involved in clotting). To date, researchers have had limited success in the laboratory stimulating stem cells isolated from umbilical cords to differentiate into liver, heart muscle, and brain cells.

a person's life. The constant production of these cells over a person's lifespan certainly requires more than 50 divisions. Stem cells are not restricted to developing embryonic and fetal tissue, but are present in mature tissues of the body as well.[4.4] Stem cells in an adult divide on a continuous basis to produce new stem cells, as well as cells that will differentiate into the mature cell types required of the tissue in which they are located. A property unique to all stem cells is that they do not have a limit on the number of times they can divide. This is possible because stem cells produce an enzyme called telomerase. The name of this enzyme is appropriate because it binds to telomeres and prevents the shortening that typically occurs when the chromosomes are replicated. The lack of chromosome shortening enables division to continue without the loss of genetic information. Telomerase is not normally expressed in the vast majority of body cells but rather only in stem cells. One of the many abnormal events that occurs during the transformation of a healthy cell into a cancerous one is the inappropriate expression of the telomerase gene. The presence of telomerase in cancer cells provides them with the ability to divide continuously without incurring the inhibitory signal from the loss of telomere sequence.

CELL DIVISION IS UNDER THE REGULATION OF THE CELL CYCLE

When a person grows, it is primarily due to an increase in the number of cells. Growth at the cellular level, however, refers only to an increase in the size of a cell by increasing the number of proteins, carbohydrates, lipids, and size or number of organelles. A cell develops incrementally just as a person goes through different stages of development during growth. The process of cell growth and development is known as the cell cycle and it has five phases: G_0, G_1, S, G_2, and M. (Figure 4.8). Interphase, the term that represents the phases G_0, G_1, S, and G_2, is the time spent by cells carrying out their normal functions or those tasks necessary to prepare for division. Cytokinesis, or cell division, actually occurs during M phase. The "M" stands for mitosis if the dividing cell is a somatic cell, or meiosis if it is a germ cell producing eggs or sperm.

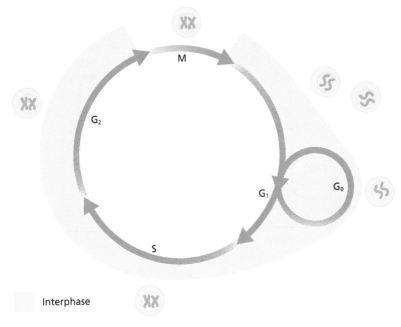

Figure 4.8 Cell cycle. The five phases of the cell cycle are G_0 (also known as the quiescent phase), G_1, S, G_2, and M.

The vast majority of somatic cells are capable of dividing but do so only when needed to replace damaged or dead cells. Most cells remain, at any given time, in G_1 or in G_0, the **quiescent** phase and an extension of G_1. Cells enter G_0 when there is no immediate likelihood of dividing and are essentially arrested in their growth. Those that are capable of dividing can exit G_0 and reenter G_1, the first growth phase. Depending on the type, the cell may remain in the G_1 phase for only a few minutes or for several hours before progressing through the remainder of the cell cycle. In preparation for division, a cell will progress from G_1 into the S phase, which is the DNA synthesis phase when the chromosomes are replicated. The cell then progresses into G_2, the second growth phase, and prepares for division by increasing its size and number of components. Following the progression from G_2 into the M phase, the cell divides to produce two daughter cells, each of which will immediately enter the G_1 phase.

While it takes the average dividing human cell 12–24 hours to make its way through the cell cycle, the time spent in M phase is approximately only 30 minutes to one hour. As previously mentioned, skin, digestive tract, and blood cells have a relatively high turnover rate and thus progress through the cycle at a fairly constant pace in order to replace those that have died and been shed from the body. Mature and terminally differentiated cells (e.g., muscle and nerve cells) reside nearly exclusively in G_0. Only rarely do these cells exhibit the ability to reenter the cell cycle and divide.

The timing and rate at which all cells progress through the cell cycle must be tightly controlled to preserve the structural and functional integrity of tissues, organs, organ systems, and ultimately, the person. Regulation of the cycle relies upon the proper interactions of cellular protein products of two types of genes, proto-oncogenes that promote and tumor-suppressor genes that inhibit a cell's progression through the cycle. Together, the competing activities of these proteins result in a system of checks and balances that ultimately will determine the action taken by a cell, either death or division.[4,5] The ability of the activity of one group of proteins to gain the upper hand over that of the other group is determined to a great extent by the signals a cell does or does not receive.

> [4,5] The proteins that promote and inhibit a cell's passage through the cell cycle act like a car's accelerator and brake, which control when it will and will not move forward.

Cell cycle regulation

Each phase of the cell cycle contains checkpoints where progression temporarily halts until a set of conditions is met (Figure 4.9). There are two checkpoints late in G_1. The first is more appropriately described as a restriction point and is a major regulatory point in the cell cycle that is largely under the control of external growth factors. In the absence of growth factors, a cell will not be able to pass through the restriction point and, therefore, will remain in G_1 or enter the G_0 phase. The majority of body cells possess growth factor receptors that enable them to proceed through the G_1 restriction point upon binding of the appropriate signaling molecule. Differentiated cells that remain in G_0 but are still capable of division will, upon receiving the proper signals, proceed out of G_0 into G_1 and then pass through the restriction point. Mature cells rarely, if ever, divide but remain in G_0 once they enter that phase, and are referred to as terminally differentiated.

> A terminally differentiated cell and a nonterminally differentiated cell are adjacent to one another. What is one possible difference between these cells that may be responsible for the fact that only the nonterminally differentiated cell will begin to divide in the presence of growth factors?

The checkpoints late in G_1, S, and late in G_2 are similar in that they all allow the cell to assess whether or not any of the DNA is damaged. It would be a waste of cellular resources and energy, and potentially dangerous for the organism, for a cell to enter S phase with damaged DNA, to replicate chromosomes that were damaged after the cell had passed through the G_1 checkpoint, or to undergo cell division and segregate damaged chromosomes into daughter cells. A cycle will be temporarily halted at these checkpoints until damage is repaired. If the

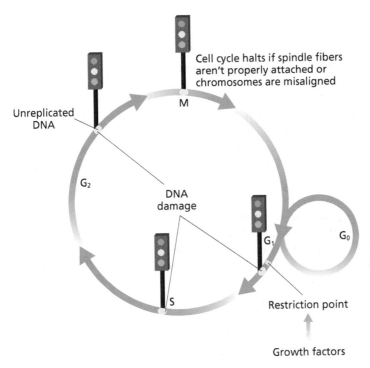

Figure 4.9 Cell cycle checkpoints. There are two checkpoints late in the G_1 phase. The first is a restriction point that halts the cell's progression through the cycle in the absence of growth factors. The second G_1 checkpoint and those in S and late in G_2, halt the cell's progression through the cycle in response to DNA damage. Cell progression through the cycle will also be halted late in G_2 if DNA synthesis is incomplete. The final checkpoint is in M and halts the cell cycle if chromosomes are not properly attached to the spindle fibers or are not properly aligned in the center of the cell.

damage is too extensive for repair, a cell may initiate the process of apoptosis, or programmed cell death.[4.6] Apoptosis can be viewed as an altruistic act that benefits the whole organism; cells that can no longer function normally because their genetic information has been corrupted, die rather than compromise the health of the organism. In addition to being triggered by extensive DNA damage, apoptosis also occurs to eliminate cells that function abnormally or are infected with bacteria or viruses.

[4.6] The peeling that occurs a few days after a sunburn is the loss of large numbers of skin cells that have undergone apoptosis as a result of extensive DNA damage from the sun's UV rays.

At the checkpoint late in G_2 the cell verifies that, in addition to being intact, all of the DNA in a cell has been replicated. It is essential that the cell does not exit G_2 and enter into M phase without having two complete sets of chromosomes, otherwise an incomplete set of genetic information would be segregated into one of the two daughter cells upon

Box 4.1

Apoptosis is a normal part of development

Apoptosis ensures the proper formation and functioning of the body. All vertebrates possess webbing between their fingers and toes during fetal development. In some animals, the webbing remains while in others, humans for example, it disappears later in development because the cells of this area undergo apoptosis. Approximately half the nerve cells in a newborn undergo apoptosis because they fail to properly interact with neighboring cells. Everyday our bodies produce tens of millions of white blood cells to protect us from getting sick. It is estimated that approximately 90% of our newly formed white blood cells undergo apoptosis during development because if they survived they would actually attack rather than protect our own body cells and result in an autoimmune disease such as lupus.

cell division. DNA synthesis will be completed if necessary and, in the absence of DNA damage, a cell will be allowed to pass through the G_2 checkpoint and enter M, the final phase of the cycle. Progression through M is halted if the chromosomes are not properly aligned in the center of the cell and/or not attached to the spindle fibers. If either of these events did not occur, it would likely result in inaccurate segregation of chromosomes. Daughter cells with either too many or too few chromosomes are likely to function aberrantly or die prematurely.[4.7]

4.7 Down syndrome is a condition that results from trisomy 21, the presence of three copies of chromosome 21.

What is the collective or overall function of the checkpoints in the cell cycle?

Explain how a loss in control of any of the checkpoints allows for the accumulation of mutations that are required for cancer formation.

Abnormalities in cancer cells often allow them to progress through cell cycle checkpoints regardless of whether or not the conditions required for passage are satisfied. A loss in the effectiveness of the cell cycle checkpoints is often what enables cells within a tumor to accumulate mutations and chromosomal abnormalities that are not permitted or tolerated during normal, healthy cell division. Tumors commonly contain cells with a variety of different genetic anomalies that result in different growth characteristics and/or responses to treatment.

Proto-oncogenes and oncogenes promote the cell cycle

Reproduction is the key to the survival of any species. Viruses are incapable of reproducing independently; that burden is placed on the host cells they infect. Viruses are infectious agents that possess only RNA or DNA for their genetic information and an outer protein shell called a capsid (Figure 4.10a). Some viruses are a bit more elaborate and contain a few enzymes within the protein coat and/or an outer phospholipid bilayer referred to as an envelope (Figure 4.10b). The function of a cell can be dramatically altered as a result of a viral infection. There is often a shift away from the performance of normal cellular functions and instead, the

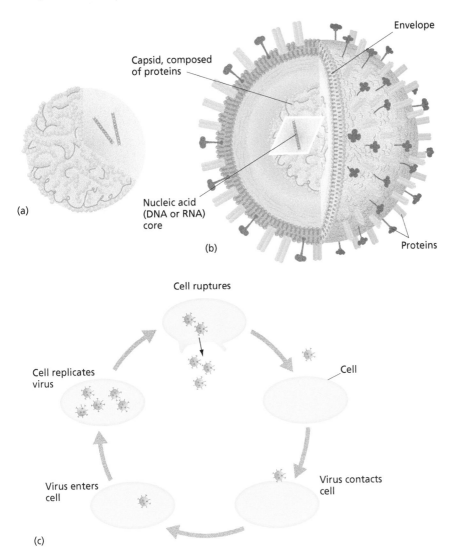

Figure 4.10 Structures and lifecycle of viruses. Cross-sectional views of (a) nonenveloped or naked and (b) enveloped viruses. (c) Viruses infect a cell, hijack control of the cellular machinery causing the activities of the cell to be focused on the synthesis of new viruses. Viruses released from the cell are available to infect other cells.

cell becomes a factory for the mass production of viruses (Figure 4.10c). The new viruses are released from the cell, often at the demise of the host cell, and infect additional cells that can serve as hosts for the making of even more viruses.

A series of experiments were conducted during the 1970s to determine *how* certain viruses cause cells to lose control over the cell cycle, divide continuously, and form a tumor. The Rous sarcoma virus (RSV) is one such oncogenic (*onco-* = cancer; *-genic* = producing) virus that induces connective tissue tumors in chickens. RSV possesses only four genes in its genome, three of which are essential for its ability to replicate inside cells. A strain of RSV that lacked the fourth gene was capable of infecting and reproducing within cells but did not produce tumors. This fourth gene is not essential to the survival of the virus yet necessary for disrupting the regulation of the cell cycle and transforming a cell into one that divides uncontrollably. This fourth gene was named *v-src* (*v* = viral; *src* = sarcoma) and was classified as an oncogene, a member of a group of genes that code for proteins that drive a cell through the cell cycle and promote the continuous cell division that may eventually lead to cancerous growth.[4.8] Possession of a nonessential oncogene enhances a virus's chances for survival because the product of this gene will induce the cell to not only produce new viruses but to also divide rapidly, increasing the number of cells that produce the virus. This may be a win-win situation for RSV but it certainly is not one for the chicken.

> [4.8] Remember that sarcomas are cancers of connective tissue.

Researchers have determined that the *v-src* gene was actually acquired by RSV from a host cell that it had infected long ago in the evolutionary history of the virus. The cellular version of *v-src* is termed *c-SRC* (*c* = cellular). The function of c-Src, the protein coded for by the *c-SRC* gene, is to promote passage through the cell cycle as a means to enhance survival. The protein produced from a mutated allele of the *c-SRC* gene is overactive and contributes to uncontrolled cell division and extension of the cell's lifespan. Since the *c-SRC* gene is a precursor of an oncogene it was termed a proto-oncogene (*proto-* = original). Proto-oncogenes can be converted into oncogenes by gain-of-function mutations.[4.9] These alterations enhance the function of a protein that promotes passage through the cell cycle in one of two ways. One way is that a gene's sequence is altered, changing the structure of the protein it codes for so that the protein's function is enhanced. Alternatively, a nucleotide change can increase the expression of a gene, thereby increasing the synthesis of

> [4.9] Proto-oncogenes are absolutely necessary for the normal functioning of cells. It is only when these types of genes are mutated in specific ways that they can contribute to the malignant transformation of cells.

Table 4.1 Human oncogenes and associated cancers

Oncogene	Type of cancer
ERBB2 (a.k.a. Her-2/neu)	Brain, breast, lung, ovarian, stomach
KRAS	Colorectal, lung, pancreatic
HRAS	Bladder, kidney, thyroid
NRAS	Leukemias, thyroid
BRAF	Colorectal, melanoma, ovarian, thyroid
SRC	Colorectal
BCR-ABL	Chronic myelogenous leukemia
C-MYC	Breast, Burkitt's lymphoma, lung
L-MYC	Small cell lung cancer
N-MYC	Lung, neuroblastoma
CYCD1	Breast, lymphomas
CDK4	Glioblastoma, melanomas, sarcomas
BCL2	Non-Hodgkin's lymphoma
MDM2	Sarcomas

Adapted from Kleinsmith, L. J. (2006) *Principles of Cancer Biology*. San Francisco, CA: Pearson Benjamin Cummings; and Cooper, G. M. & Hausman, R. E. (2007) *The Cell: A Molecular Approach*, 4th edn. Washington, DC: ASM Press; Sunderland, MA: Sinauer Associates, Inc.

the gene product. These types of mutations can produce a stronger than normal intracellular signal that enhances cell division and survival. Table 4.1 provides a list of some of the oncogenes that have been determined to play a role in the development of human cancers.

Dozens of proto-oncogenes are expressed throughout a cell's life and they perform a variety of functions, such as cell cycle regulation, cell differentiation, signal transduction, and apoptosis. One group of proteins encoded by proto-oncogenes are cyclins. There are several types of cyclins that promote progression through the cell cycle. The name of these proteins is derived from the cyclic rise and fall of their concentrations as a cell progresses through the phases of the cycle. (Figure 4.11). The concentration of cyclins is lowest when cells are newly formed and, in response to external growth factor signals, steadily increases as a cell proceeds through the cycle. An increase in intracellular cyclins will result in an increase in cell division. The concentration of intracellular cyclins peaks during M phase and decreases rapidly immediately prior to cell division, resulting in the low cyclin concentration found in the new daughter cells. Only if the daughter cells receive the proper growth signals will the cyclin concentration increase and the cells once again progress through the cell cycle.

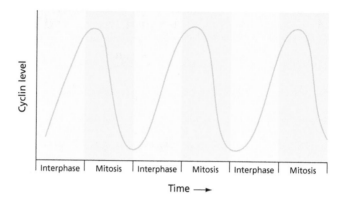

Figure 4.11 Cyclin concentration rises and falls through the cell cycle. The concentration of cyclins is lowest at the beginning of G_1 and increases as the cell progresses through the cell cycle. Cyclin concentration peaks in the M phase and then decreases. The two newly formed daughter cells possess the lowest concentration at the beginning of G_1.

The newly produced cyclins bind to a group of proteins called cyclin dependent kinases (cdks), another type of proto-oncogenic protein. As the name implies, the activity of the kinase proteins is dependent upon their interaction with cyclins. The cdk protein, which was inactive on its own, becomes active when bound to a cyclin due to changes in structure and function (Figure 4.12). When a cyclin–cdk complex binds to a target protein, the complex modifies it by adding a phosphate group (PO_4^{2-}) to one of the amino acids on the surface of the target protein in a process called phosphorylation. The bonding of the phosphate group, which has a strong negative charge, results in the target protein undergoing a structural change. The functionality of the many different target proteins that are affected in this way following the binding of growth factors and the increase in cyclin concentration is what drives the progression through the cell cycle.

> 4.10 The activity of transcription factors can be thought of as being a *factor* that determines whether or not a gene will be *transcribed*.

Cyclins regulate the activity of transcription factors, proteins that bind to genes and regulate whether or not they are expressed.[4.10] The E2F protein is one such transcription factor and it regulates the expression of S-phase genes (Figure 4.13). Those genes should be expressed only when a cell is replicating its chromosomes in preparation for division. During any other phase of the cell cycle, the expression of S-phase genes should be inhibited because they are not needed. During every phase of the cell cycle except S phase, the E2F protein is bound to another protein, Rb, forming the E2F–Rb transcription factor complex. The binding of the E2F–Rb protein complex to S-phase genes inhibits their expression (Figure 4.13a).

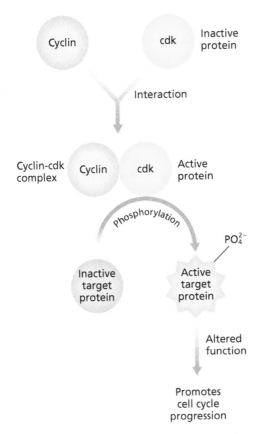

Figure 4.12 Function of cyclin and cdk. The interaction of a cyclin with a cdk results in the activation of cdk. Activated cdk in the cyclin–cdk complex is capable of phosphorylating a particular target protein. Once phosphorylated, the target protein is altered and its activity promotes the cells progression through the cell cycle.

The intracellular activity that occurs after the binding of growth factors to a cell in the G_1 phase will result in its progression into the S-phase. This means that the inhibition of the expression of S-phase genes by the binding of the E2F–Rb protein complex needs to be relieved. This is accomplished by the activity of a cdk that has become active as a result of its binding to a cyclin produced in response to growth factor signaling (Figure 4.13b). The activated cyclin–cdk complex phosphorylates the Rb protein causing an alteration in its shape. This conformational change no longer allows Rb to interact with the

> **Provide a function of proteins encoded by S-phase genes.**
>
> **Outline the steps necessary for cell division to occur, starting with the binding of a growth factor to the cell surface. Do the same for the specific Rb–E2F pathway.**

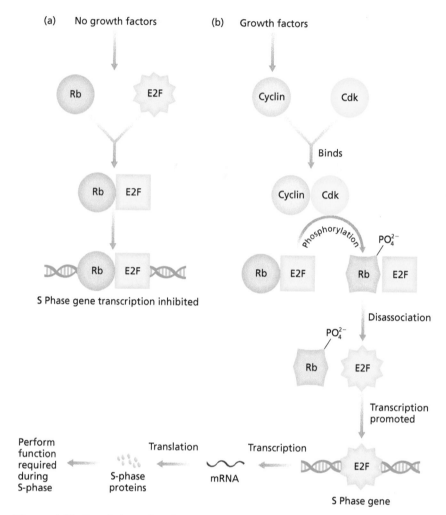

Figure 4.13 Regulation of S phase gene expression by Rb and E2F. (a) In the absence of growth factors Rb and E2F interact and the complex binds to S-phase genes and represses their transcription. (b) Cyclins synthesized in response to growth factor signals bind to cdks, which leads to their activation. A cyclin–cdk complex phosphorylates Rb resulting in a structural change that causes it to dissociate from E2F. The structure of unbound E2F enables it to bind to S-phase genes and activate their expression.

E2F protein. Although this free E2F protein binds to the same S-phase genes that it previously did when complexed with Rb, its new unbound conformation provides a new function – the activation, rather than repression, of the S-phase genes. Consequently, chromosomal replication and continuation of the cell cycle can proceed. The Rb protein is produced from the RB1 gene, which was discovered by researchers studying tumor cells obtained from children with retinoblastoma, a form of eye cancer (see Box 4.2).

Box 4.2

Retinoblastoma: a cancer of the eye most commonly diagnosed in children

Retinoblastoma is a cancer that develops in the retina, the collection of cells in the back of the eye that light strikes after entering through the pupil. This particular form of cancer affects approximately 250 children every year in the US and accounts for approximately 3% of cancers that affect children younger than 15 years old. The most prominent indication of a tumor present on the retina is an alteration in the appearance of the pupil. The pupil is normally black but when a bright light is shown on it, such as during flash photography, it appears red. When a bright light is shown on an eye with a tumor on the retina, the pupil shows a white discoloration, a condition called **leukocoria** (*leuko-* = white; *-kore* = pupil; see Figure 4.14).

Retinoblastoma is caused by mutations in the RB1 gene that inactivate the Rb protein. Both alleles of the RB1 gene must be mutated for the cancer to develop. When a child develops the disease in both eyes (bilateral retinoblastoma), it is likely that there is a family history of the disease. It is very rare, however, that a child would receive the mutated gene from each parent. Rather, a single mutated RB1 gene was inherited from one of the parents and the child then subsequently acquired a mutation in the second RB1 gene early in development, before the

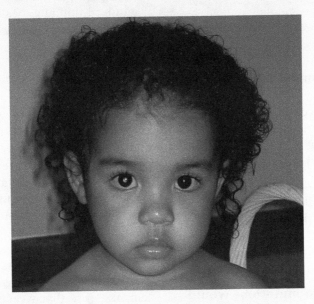

Figure 4.14 A case of leukocoria in the right eye of a child. Image courtesy of Liza Talusan.

eyes began to form, so that both copies of the gene in the retinal cells of the eyes produce inactive Rb protein. A child who develops the disease in a single eye (unilateral retinoblastoma) most likely acquired the mutations spontaneously later in development after cells have committed to becoming part of the retina, so that only cells of the retina in one eye possess the altered RB1 genes.

A tumor results as a consequence of the Rb protein being inactive and unable to bind to the E2F protein even when Rb is not phosphorylated. This results in free E2F protein that continuously and inappropriately binds to S-phase genes and activates their expression (Figure 4.13b). The constant expression of S-phase genes results in the cells of the retina dividing rapidly despite the absence of growth signals. This inappropriate growth of the retinal cells leads to the formation of a tumor.

The treatment for most cases of unilateral retinoblastoma is the removal of the eyeball (enucleation) and its replacement with a prosthesis. There has been success treating cases of bilateral retinoblastoma first with chemotherapy drugs to reduce the size of the tumor and then treating the remainder with either radiation, a laser, or extremely cold gas. Those who have been diagnosed and treated for bilateral retinoblastoma are at risk later in life for developing other forms of tumors, especially a variety of sarcomas. Many children, however, do not develop other cancers and lead fulfilling, healthy lives.

> **Provide an explanation for why those that develop cancer early in life usually have a family history of the disease.**

Tumor suppressor genes inhibit the cell cycle

Since proto-oncogenic proteins *promote* proliferation and survival it would make sense to balance their activity with a set of proteins that *inhibits* a cell's proliferation and promotes its death. Proteins that carry out these functions are coded for by tumor suppressor genes. Just as gain-of-function mutations to proto-oncogenes convert them to oncogenes that have negative consequences on cell cycle regulation, a similar consequence occurs when tumor suppressor genes are affected by loss-of-function mutations. This type of mutation alters the structure of a protein in a way that weakens its function or causes a decrease in the expression of a gene that yields less of a functional protein. Loss-of-function mutations in

> **Aside from Rb, what other components do you think would, if mutated, result in a similar consequence (i.e., retinoblastoma)?**
>
> **What kinds of mutations would they be, gain-of-function or loss-of-function mutation?**

Table 4.2 Human tumor suppressor genes and associated cancers

Genes	Type of cancer
Gatekeeper genes	
APC	Colorectal
RB1	Bladder, breast, lung, melanoma, retinoblastoma, osteosarcoma
p53	Brain, breast, colorectal, esophageal, liver, lung, sarcomas, leukemias, lymphomas
Caretaker genes	
ATM	Bladder, breast, ovarian, pancreatic, stomach
BRCA1	Breast
BRCA2	Breast, melanoma, ovarian, pancreatic, prostate

Adapted from Kleinsmith, L. J. (2006) *Principles of Cancer Biology.* San Francisco, CA: Pearson Benjamin Cummings; and Cooper, G. M. & Hausman, R. E. (2007) *The Cell: A Molecular Approach,* 4th edn. Washington, DC: ASM Press; Sunderland, MA: Sinauer Associates, Inc.

tumor suppressor genes result in a loss of control over cell growth and division and may lead to a diminished or complete loss of apoptosis initiation under appropriate conditions. There are a wide variety of cancers associated with mutant tumor suppressor genes (Table 4.2). One is retinoblastoma caused by loss of function mutations in the RB1 gene.

There are two types of tumor suppressor genes: caretakers and gatekeepers. The caretakers are involved in maintaining the integrity of the genome. The proteins coded for by the caretaker genes repair structural damage to chromosomes, correct mutations in the DNA sequence that arise during DNA replication, and sort the chromosomes into daughter cells during cell division. Members of the second type of tumor suppressor genes, the gatekeepers, produce proteins that function at the cell cycle checkpoints. The gatekeeper proteins prevent a cell from proceeding further along the cell cycle unless certain conditions have been met. In the presence or absence of specific signals or under certain conditions gatekeepers are capable of initiating apoptosis.

It is estimated that more than 50% of cancers possess a mutant allele of the *p53* gatekeeper gene, making it the most frequently mutated gene in human cancers. The amount of p53 in cells is normally very low, although its concentration dramatically increases in the presence of damaged DNA. The p53 protein functions at the checkpoints late in G_1, S, and late in G_2 (Figure 4.15) and acts as a transcription factor regulating the expression of many genes that promote cell cycle arrest, DNA repair,

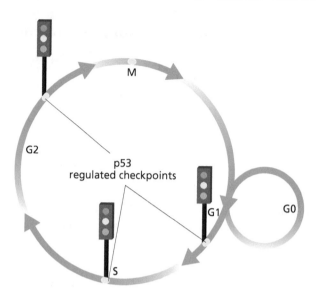

Figure 4.15 p53-regulated checkpoints. The p53 protein halts the cell cycle in response to DNA damage at the checkpoints late in G_1, S, and late in G_2.

and apoptosis. A mutation of this gene that causes the loss of this essential regulatory protein could potentially result in unchecked division of damaged or transformed cells, resulting in tumor formation.

Nearly all of us have experienced a sunburn at some point during our lives. The p53 protein is a major player in the cellular events that occur following the damage caused by exposure to ultraviolet (UV) radiation emitted by the sun and tanning lamps or beds. UV radiation is a relatively weak form of energy that can only penetrate a few of the outer cell layers of the skin. As UV radiation is absorbed by the skin, its energy is sufficient to damage DNA in its path. Cells increase their production of p53 protein in response to damaged DNA. As a transcription factor, p53 promotes the expression of particular genes to produce proteins that will halt the progress of a cell through the cell cycle and carry out the repairs necessary to fix the damage done to DNA.

The expression of the *p21* gene is increased by the activity of p53. The newly made p21 protein binds to and inhibits a variety of proteins. One of the target proteins of p21 is cdks. The binding of p21 to a cdk inhibits the cdk's ability to phosphorylate many other proteins, resulting in the arrest of the cell cycle. For example, the inhibition of cdks would prohibit Rb from being phosphorylated and result in the E2F–Rb transcription factor complex remaining intact, thus suppressing the expression of S-phase genes. This ensures that the cell remains in the G_1 phase. A loss-of-function mutation in the *p53* gene would result in the inability to produce the p21 protein in response to DNA damage. In the absence of p21, the cdks would remain active, Rb would be phosphorylated, and E2F would bind in its free state to S-phase genes and activate their expression. Inappropriate activation of S-phase genes is undesirable because the cell would produce the proteins that replicate the damaged chromosomes.

Another target of the p21 protein is DNA polymerase, a very large enzyme complex composed of multiple proteins that work together to synthesize DNA during S phase. The genes that code for the proteins that comprise the DNA polymerase complex are S-phase genes and as such are under the control of the Rb–E2F transcription factor complex. Once synthesized, DNA polymerase is regulated by the p21 protein, which binds to and inhibits one of the proteins of the DNA polymerase complex, effectively rendering the entire enzyme complex nonfunctional and so preventing the replication of damaged DNA.

In cells that have accumulated DNA damage that is too extensive for recovery, the activity of p53 inhibits a protein known as Bcl2. Because Bcl2 functions to block apoptosis, inhibition of Bcl2 activity allows for the series of events that leads to apoptosis. Mutation of the *p53* gene allows the Bcl2 protein to remain active in the presence of damaged DNA, preventing apoptosis and allowing the cell to progress through the cell cycle. This allows a cell with DNA damage that is functioning abnormally to not only live but reproduce. These types of cells are certainly on the road to becoming cancerous, if they are not already so, because of the mutations they possess.

LOSS OF CELL CYCLE CONTROL LEADS TO UNCONTROLLED CELL GROWTH

Following conception, countless cell divisions occur to produce all of the tissues of the body. The proper form and function of cells require that cell divisions occur in the appropriate locations at specific times. Also of great importance is the differential gene expression that provides the molecules required by cells to establish the necessary shapes, structures, and cell-to-cell and cell-to-extracellular matrix contacts necessary for proper function. Cell division is a complex and tightly regulated process carried out by a large number of proteins. The proteins produced from proto-oncogenes promote cell division while those from tumor suppressor genes inhibit cells from dividing. Multiple gain- and loss-of-function mutations in these genes cause cells to lose the ability to tightly regulate the progression through the cell cycle and potentially to divide in a rapid and uncontrolled manner.

There is the potential that once cells lose the ability to control their rate of division they will also lose their ability to perform other normal tasks. One example of this is the inability of the cell to maintain or repair the genome, which will lead to an increase in the amount of DNA damage and a number of mutated genes. The particular characteristics that are exhibited by cancer cells is the subject of Chapter 5. One trait they do possess, however, is the expression of telomerase that allows them to undergo an unlimited number of divisions and form a tumor. Tumors can be harmful and certainly even life-threatening because of their potential to negatively impact the ability of tissues, organs, organ systems, and the body as a whole to perform the functions essential for life. The continued loss of regulation and accumulation of mutations as a tumor grows often leads to a lack of genetic uniformity among the cells. This has a direct impact on the growth rate of the cancer, the ability to metastasize, the response to particular treatment methods, and the prognosis for the patient's survival, all of which will be discussed in subsequent chapters.

EXPAND YOUR KNOWLEDGE

1 What type of tissue present in men but not women is likely to express telomerase? Explain.

2 What type of tissue present in women but not men is likely to express telomerase? Explain.

3 The Rous Sarcoma Virus possesses an oncogene that, while not essential, assists in its survival. Would you be surprised to learn that certain viruses possess tumor suppressor genes that possess loss-of-function mutations? Explain.

4 A person lays in the sun, falls asleep for an hour, and wakes to find that he has a sunburn. Later that day he receives a paper cut on his thumb, which is sunburned. In response to both the sunburn and the paper cut, there is a release of growth factors to stimulate the division of skin cells to replace not only those that have undergone apoptosis from DNA damage caused by the excessive exposure to UV light, but also the cells that were sliced open by the paper cut. It is likely that some of the epidermal skin cells that receive the growth factor signals also possess some, but not extensive, DNA damage from the UV light. These cells are receiving conflicting signals.

    (a) What are the conflicting signals being received by the skin cells?

    (b) What action will be taken by the skin cells in response to the conflicting signals?

5 Why would those who are diagnosed with bilateral and not unilateral retinoblastoma be at a higher risk for developing other forms of cancer?

ADDITIONAL READINGS

Alberts, B., Bray, D., Hopkin, K., Johnson, A., Lewis, J., Raff, M., Roberts, K., and Walter, P. (2004) *Essential Cell Biology*, 2nd ed. New York, NY: Garland Science.

Cooper, G. M. (1995) *Oncogenes*, 2nd edn. Boston: Jones and Bartlett Publishers.

Cooper, G. M. and Hausman, R. E. (2007) *The Cell: A Molecular Approach*, 4th edn. Sunderland, MA: Sinauer Associates.

Evans, T. Rosenthal, E. T., Youngblom, J., Distel, D., and Hunt, T. (1983) Cyclin: a protein specified by maternal mRNA in sea urchin eggs that is destroyed at each cleavage division. *Cell* **33**: 389–396.

Kleinsmith, L. J. (2006) *Principles of Cancer Biology*. San Francisco, CA: Pearson Benjamin Cummings.

Potten, C. and Wilson, J. (2004) *Apoptosis: The Life and Death of Cells*. New York: Cambridge University Press.

Schiff, P. B., Fant, J., and Horwitz, S. B. (1979) Promotion of microtubule assembly *in vitro* by taxol. *Nature* **277**: 665–667.

Stehelin, D., Varmus, H. E., Bishop, J. M., and Vogt, P. K. (1976) DNA related to the transforming gene(s) of avian sarcoma viruses is present in normal avian DNA. *Nature* **260**: 170–173.

5

Tumor formation, growth, and metastasis

Growth for the sake of growth is the ideology of the cancer cell.
> Edward Abbey, American Writer

CHAPTER CONTENTS

- Tissue changes that occur in response to stimuli
- Feeding tumor growth by angiogenesis
- Characteristics of benign and malignant tumors
- Events that occur during the process of metastasis
- Expand your knowledge
- Additional readings

Cancer: Basic Science and Clinical Aspects, 1st edition. By C. A. Almeida and S. A. Barry. Published 2010 by Blackwell Publishing, ISBN 978-1-4051-5606-6.

Chapter 4 discussed the regulation of the cell cycle that is critical in maintaining the structural and functional integrity of all tissues. The inability to control passage of a cell through each of the cycle checkpoints can result in unwanted growth within a tissue. The growth not only can disrupt the function of that tissue but also of those nearby. The situation becomes much more serious if cells break free from the tumor and travel to other tissues where they may take up residence, multiply, and create additional problems. This chapter will examine the events that are involved in transforming normal cells into cancerous ones, allowing for the formation of tumors and the spreading of cancer cells from their original growth site to other locations in the body.

TISSUE CHANGES THAT OCCUR IN RESPONSE TO STIMULI

Our cells experience many different types of chemical and physical stimuli on an almost constant basis. For example, our cells are exposed to both beneficial and harmful chemicals in the air, food, and water we take into our bodies, hormones surging through our bloodstreams relaying messages to the cells to which they bind, and stresses and strains are applied when we move heavy objects. The type and strength of the stimuli cells receive or are subjected to affect the structural and functional changes they will undergo. The cellular changes that occur in response to stimuli are an indication of both the susceptibility to signals and the adaptability that cells exhibit in response to changes in their environment. It is logical to expect that if a certain stimulus causes a cell to change in a particular way, then the cell should revert back to its original condition upon removal of the stimulus. The transformation that cells in a tumor have undergone is often the result of changes brought about by certain stimuli. A unique aspect of tumor cells is that the cellular changes remain even after the stimulus that led to their transformation is no longer present.

Metaplasia is the presence of a normal cell type in the wrong location

Epithelial cells that line certain portions of the respiratory tract are known to undergo changes in appearance and function when exposed to noxious chemicals in the air. The pathway of air through the respiratory tract begins in the nose or back of the throat and travels down through the trachea or windpipe, the tube that leads from the back of the throat to the lungs (Figure 5.1). The base of the trachea branches to form two bronchi, one going to each of the lungs, which then progressively branch into many smaller tubes called bronchioles that spread to all areas of the lungs. At the ends of the bronchioles are tiny balloon-like sacs called alveoli where gas exchange between the area and bloodstream occurs. The surface layer of the trachea, bronchi and some of the bronchioles consists of pseudostratified columnar epithelial cells (Figure 5.2). These cells are somewhat rectangular in shape and aligned side by side in a single layer. The term pseudostratified comes from the fact that there appears to be more than one layer of cells present (*pseudo-* = false; *stratified* = layered) because the position of the nuclei in adjacent cells alternates from being centrally located to being at the bottom of the cell.

The respiratory epithelium serves a protective function. Among the columnar cells are specialized goblet cells that secrete thick, sticky mucus that coats the epithelium. The sticky mucus traps particulate matter in the air, such as dust and microorganisms, preventing it from getting deeper into the lungs. The exposed surface of a columnar epithelial cell possesses cilia, short hair-like structures that beat back and forth to

> **Provide two possible negative consequences that could occur if particulate matter reaches the lower lungs.**

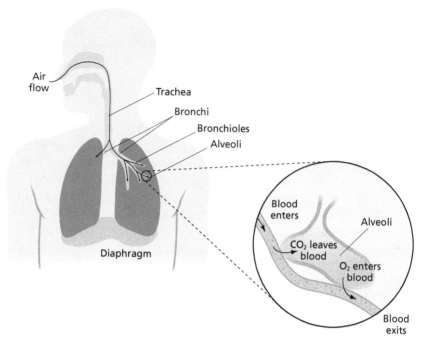

Figure 5.1 Respiratory system. Air that is inhaled passes through the trachea to the bronchi then through the bronchioles to the alveoli. Oxygen in the air diffuses across the membrane of the alveoli to enter the red blood cells in the circulatory system as carbon dioxide leaves the blood and enters the alveoli to be exhaled.

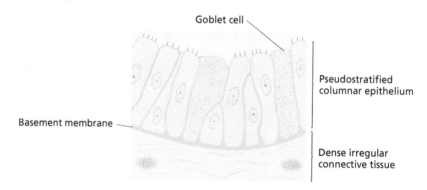

Figure 5.2 Pseudostratified columnar epithelium. Respiratory epithelial cells possess cilia on the exposed surface and the locations of the nuclei in adjacent cells alternate giving the false appearance that there are multiple layers of cells.

sweep the mucus and anything trapped in it upward to the back of the throat. The material brought up by the so-called ciliary escalator can be either swallowed and destroyed in the highly acidic environment of the stomach or expectorated (spat out).

Smoke is a mixture of many different types of chemicals, liquids, and solids and is an irritant of the respiratory epithelium. Over time, smoke paralyzes the cilia of the respiratory epithelium, allowing mucus to build up in the airways and material to travel deeper into the lungs. Also observed in tobacco smokers is the replacement of pseudostratified columnar epithelium with stratified squamous epithelium (Figure 5.3). The multiple layers of flattened cells in this form of epithelium protect underlying tissues against abrasion. The cells in the outer layers are regularly sloughed off and replaced by the replication of the cells in the lower layers. Stratified squamous epithelium is normally present in the outer layer of skin and the inner lining of the digestive tract, but not in the respiratory tract.

The previously described condition is often exhibited in the airways of smokers. It is an example of metaplasia, which is the change of mature and differentiated cells from one *normal* cell type to another *normal* cell type. It is important to note that what is abnormal about metaplastic

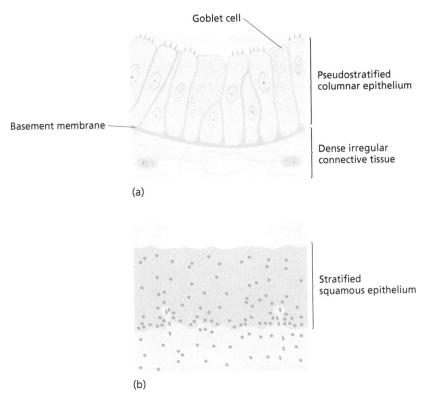

(a)

Goblet cell

Pseudostratified columnar epithelium

Basement membrane

Dense irregular connective tissue

(b)

Stratified squamous epithelium

Figure 5.3 Two types of squamous epithelia. (a) Respiratory epithelium normally contains pseudostratified columnar epithelium with cilia at the surface, but under certain conditions the cell type can change to (b) stratified squamous epithelium.

tissue when observed under the microscope is not the presence of abnormal cells, but rather the presence of a type of cell that is normally found in other types of tissue. The stratified squamous epithelium that may be present along the airway of a smoker can have a normal appearance. The problem is that it should not be present in that location.

Metaplasia is not a normal process that cells undergo; there must be an inciting stimulus that triggers the structural and functional changes that occur. A typical characteristic of metaplasia is that it is reversible – when the signal that initiated the changes is no longer present, healthy cells should revert back to their original form. A concern arises when the inciting stimulus that resulted in the metaplastic changes is no longer present, yet the cells do not revert back to their normal structure and function. In some cases, the permanent alterations are the result of genetic mutations that have negatively affected oncogenes and tumor suppressor genes. These types of cellular mutations certainly enhance the likelihood that cells will become cancerous.

Hypertrophy and hyperplasia are forms of tissue growth

Metaplasia is *not* a form of growth, which means either an increase in individual cell size or an increase in the number of cells. The purpose of the majority of new growth that occurs between conception and adulthood is to form the variety of differentiated body tissues. In adults, tissues are mature in their size, structure, and function, and the primary role of cell division is the replacement of those cells that have either died of old age, are lost due to abrasion (occurs to outer layer of skin and inner lining of the digestive tract), or are damaged beyond repair.

> **Why is metaplasia not a form of growth?**

There are, however, times when growth does occur in adults. For example, the goal of resistance or weight training is the growth of muscle tissue. A muscle is composed of individual muscle cells known as muscle fibers, and each fiber contains bundles of particular proteins that are responsible for contraction and relaxation. The lifting of heavy weights damages the protein bundles. During the recovery or repair period, the cells destroy and replace the damaged proteins. The cells, in an attempt to be stronger and prevent similar damage from occurring again, increase the number of protein bundles. This form of growth, which is due to an increase in *size* but not number, is known as hypertrophy (Figure 5.4a). Similar to metaplasia, hypertrophic growth occurs in response to an inciting stimulus and is reversed when that stimulus is no longer present. When resistance training is stopped, there will be a loss in muscle size and tone since there is no longer a need to maintain the greater number of protein bundles.

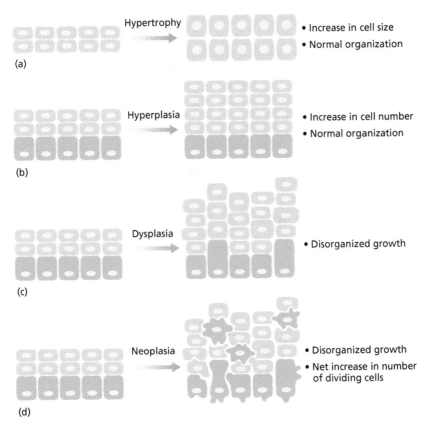

Figure 5.4 Types of tissue growth. (a) Hypertrophy occurs when the size, not the number of cells change. (b) Hyperplasia is an increase in the number of cells. (c) Dysplasia is hyperplasia with the cells being disorganized in their arrangement. (d) Neoplasia is dysplasia with an increase in the number of cells that are able to divide.

Another example of growth that occurs in adulthood is the increase in breast size that occurs during pregnancy. This form of growth is known as hyperplasia and is the result of an increase in the *number* of cells in a tissue (Figure 5.4b). The combinations and levels of hormones produced during pregnancy act as an inciting signal for the cells of the breast's mammary tissue to progress through the cell cycle and divide. This growth results in the development of mammary glands and ducts that produce milk to nourish the newborn child. A mother will continue to lactate (produce milk) for as long as the child breast feeds. When a woman stops breast feeding, there is an absence of the inciting stimuli (i.e., pregnancy hormones and infant suckling) that led to the development,

> **What characteristics do metaplastic, hypertrophic, and hyperplastic growth have in common?**

activity, and maintenance of the mammary tissues. The result is that the cells formed in response to the stimuli now undergo apoptosis, effectively reverting the tissue to its original state.

Hyperplastic growth can also occur in the absence of proper stimuli. This occurs when there is a loss of regulation at the checkpoints of the cell cycle. The most common cause for a loss of cell cycle regulation is an accumulation of gain-of-function mutations within proto-oncogenes, converting them into oncogenes, and/or loss-of-function mutations in tumor suppressor genes (Chapter 4). Mutations in caretaker tumor suppressor genes may cause structural and functional changes that do not allow cells to interact with one another in an organized fashion.

Dysplasia can lead to neoplasia

A Pap test is often part of a woman's routine gynecological exam. The test entails obtaining cells from the inner surface of the cervix and the lower portion of the uterus, followed by their examination under a microscope. Less than 5% of Pap tests display dysplasia, a disorganized arrangement of cells, which is typically reported as mild, moderate, or severe (Figure 5.4c). Mild cases often clear up on their own and are typically followed up with repeat Pap tests every 3–6 months. Moderate and severe cases require the use of treatment methods to remove the abnormal cells. The stimuli that result in cervical dysplasia, which is most common in women between 25 and 35 years of age, are unknown, although women who are infected with the human papilloma virus have an increased risk of exhibiting the condition (Chapter 10).

Dysplasia is an indication that the cells are not functioning properly, and is considered a pre-cancerous condition. The risk associated with dysplasia is the potential for the cells to progress to a state of neoplasia. A neoplastic growth has, in addition to a disorganized arrangement of cells, a larger than normal number of cells capable of dividing (Figure 5.4d). A neoplasm is also known as a cancer or tumor because its growth is the result of disruptions to the normal regulation of the cell cycle resulting in uncontrolled progression through the cell cycle and cell division.

> Construct a table that would be useful in comparing the characteristics of metaplasia, hypertrophy, hyperplasia, dysplasia, and neoplasia.

FEEDING TUMOR GROWTH BY ANGIOGENESIS

The formation of new blood vessels, a process known as angiogenesis (*angio* = blood and lymph vessel; *genesis* = production), first occurs during embryonic development and continues until early adulthood. Expansion of a blood supply is the result of the division and proliferation of the cells of the blood vessels currently in a tissue. Therefore, it is a form of growth. As mentioned previously, growth of new tissue is not a regular occurrence in adults but typically occurs only during the repair of injured tissue.[5.1]

Similar to the way that the cell cycle is regulated by balancing a set of opposing signals from the activities of proto-oncogenic and tumor suppressor proteins, angiogenesis is under the control of competing signals from many activator and inhibitor molecules. To date, there are more than two dozen proteins and small molecules that have been identified as angiogenic activators and inhibitors. In adults, the concentration of angiogenic inhibitors is higher than that of activators, thus restricting angiogenesis. A shift in the balance so that the concentration of the activators is higher than that of the inhibitors will have an opposite effect and result in the formation of new blood vessels.

> [5.1] The lining of the uterus is nourished and developed for a few days each month when a woman ovulates. This tissue growth, supported by angiogenesis, occurs in anticipation of the possibility of having to support the growth of an embryo.

Capillaries, the smallest blood vessels, are abundant in tissues to ensure that the nearby cells are provided with a continuous supply of essential nutrients and a way to remove the metabolic waste products. New tissue growth without a concomitant expansion of the blood supply is limited to $1-2$ mm^3 in size, which is approximately the size of the head of a pin. Tumors are able to exceed that growth limit by stimulating angiogenesis.

In an attempt to support growth, tumor cells may secrete the potent angiogenic activator vascular endothelial growth factor (VEGF; *vascular* = pertaining to vessels) (Figure 5.5a). This protein diffuses to the endothelial cells of a nearby blood vessel. The binding of VEGF to the appropriate receptors in the outer membranes of endothelial cells initiates a signal transduction cascade (Chapter 3) within the cells that results in changes in gene expression and cell function. For example, the expression of proto-oncogenes is enhanced while that of tumor suppressor genes is inhibited so that the cells will progress through the cell cycle and divide. As the endothelial cells divide, they will form a bud that protrudes from the blood vessel wall into the surrounding tissue (Figure 5.5b). As the number of endothelial cells increases, the bud elongates and the endothelial cells produce matrix metalloproteinases (MMPs). MMPs are enzymes that breakdown the extracellular matrix proteins (Chapter 2) to enable the growing blood vessel to migrate between the tissue cells toward the cancer cells. Once established, a tumor's blood supply will grow along with the tumor, nourishing it and removing its wastes (Figure 5.5c).

> **What would occur in order for a cell that normally does not produce VEGF to begin producing it?**

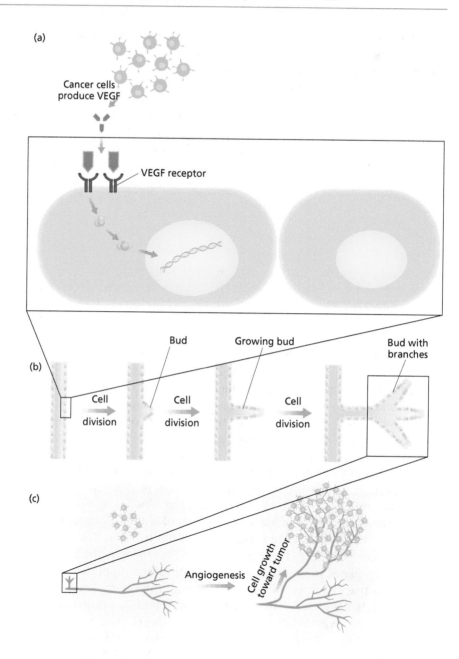

(a)

Cancer cells
produce VEGF

VEGF receptor

(b)

Bud   Growing bud   Bud with
branches

Cell   Cell   Cell
division   division   division

(c)

Angiogenesis   Cell growth
toward tumor

Figure 5.5 Tumor angiogenesis. (a) Cancer cells secrete vascular endothelial growth factor (VEGF), an angiogenic activator, which binds to VEGF receptors on endothelial cells of a capillary causing a change in gene expression within the endothelial cells of the capillary. In response to VEGF signaling an endothelial cell will divide and secrete matrix metalloproteinases (MMPs). (b) Many rounds of endothelial cell division produce a bud off of a capillary that grows and forms additional branches. (c) Endothelial cell growth toward a tumor supports its growth.

Angiostatin and endostatin are two angiogenic inhibitors that have been used in extensive animal studies and human trials. Tests conducted with mice have indicated that the treatment of tumors with angiogenic inhibitors are effective at inhibiting their growth and can limit the number of secondary tumors that may form. The use of angiogenic inhibitors as a potential method of cancer treatment will be discussed more thoroughly in Chapter 7.

CHARACTERISTICS OF BENIGN AND MALIGNANT TUMORS

Neoplasms are classified into two broad categories, benign and malignant. The classification of a tumor is most often done by a pathologist, a physician who specializes in interpreting and diagnosing changes in bodily fluids and tissues that occur in response to disease. The assessment of a neoplastic growth is based on a biopsy (*bio-* = life; *-opsy* = look or appearance), a macro- and microscopic examination of either a portion of or an entire tumor that has been surgically removed. Microscopic analysis provides a number of distinguishing features that are key to differentiating between benign and malignant tissue.

A benign tumor is noncancerous and classified as *in situ*, or contained solely within the tissue in which it originated; the abnormal cells have not spread to surrounding tissues or other areas of the body. In fact, there is typically a well-defined border between a benign neoplastic growth and normal tissue. Benign neoplastic growths are usually slow growing and although they are generally not life-threatening, they can become dangerous based on their location and whether or not their growth disrupts or interferes with normal healthy tissue functions. Their self-contained nature is an added benefit that often allows the entire tumor to be surgically removed, unless it is in an inoperable position, such as within an organ rather than on the surface or adjacent to major blood vessels or the spinal cord.

A concern with benign growths is that they can progress into the far more serious malignant or cancerous neoplasms. Principal among the distinguishing features of malignant tumors is that they are not contained solely within the tissue in which they originally developed. This means that a portion of the tumor has grown into one or more of the surrounding tissues or has spread to a distant location in the body. Metastasis, the process by which malignant cells travel from the original (primary) tumor to other (secondary) sites in the body, is often accomplished through the use of either the circulatory or lymphatic systems (Figure 5.6).

Normal tissues consist of differentiated cells performing specific functions. Malignant tissue typically exhibits anaplasia, the presence of undifferentiated cells that bear no resemblance to the cells normally found in that location. The presence of undifferentiated cells is a reflection of what is

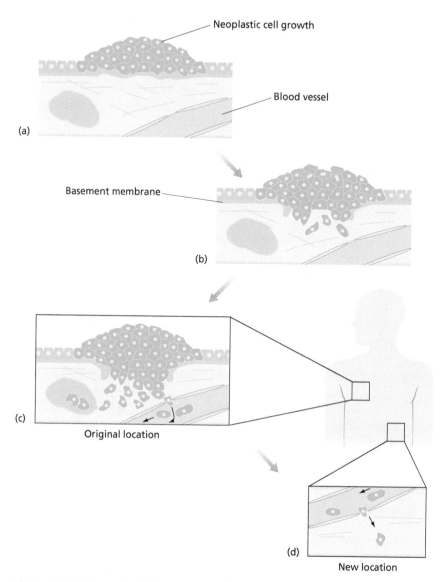

Figure 5.6 Metastasis. (a) Neoplastic cells grow, (b) produce proteases that breakdown the basement membrane, and then invade the surrounding tissue. (c) Malignant cells can gain access to the circulatory or lymphatic system, and then (d) exit and take up residence elsewhere in the body.

normally present during embryonic and fetal development when tissues are going through their formative stages. Since undifferentiated cells are involved in tissue formation they divide frequently. As a result, malignant tissues often exhibit a high **mitotic index**, the ratio between the number of cells undergoing mitosis and the total number of cells within the field of view. This accounts for the faster growth rate of malignant tumors.

Malignant tumors are considered life-threatening because of the rapid growth and production of undifferentiated cells that are invasive and disruptive to the structure and function of surrounding tissues. In addition, the anaplastic nature of malignant cells is an important factor in the likelihood that they will metastasize and wreak similar havoc on other locations in the body. Surgery alone is not a sufficient form of treatment for malignant tumors because of the possibility that some of the cells have spread to locations throughout the body. Chemotherapy, the use of toxic drugs (Chapter 7), is a more systemic form of treatment used to target the destruction of undetectable metastatic cancer cells in an attempt to prevent the growth and formation of new tumors.

EVENTS THAT OCCUR DURING THE PROCESS OF METASTASIS

The structural and functional changes that occur to the cells within a tumor can be a consequence of external growth factor signals and/or mutations to DNA. It is common for tumors, particularly those that are malignant, to exhibit tumor progression – the cells mutate independently of one another as they grow, thereby generating a collection of genetically different subpopulations. The genetic differences between cells result in their unique growth and metastatic potentials (Figure 5.7). It is only the more

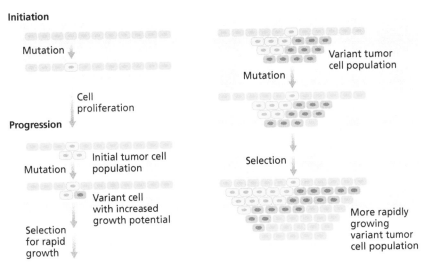

Figure 5.7 Tumor progression. An initial genetically different cell functions abnormally and grows to produce a collection of cancer cells. Over time mutations occur within cells independent of one another producing groups of cells within a tumor that are genetically distinct from other groups of cells within the same tumor.

potent cells that are likely to have the ability to invade the surrounding tissues, gain access to the circulatory or lymphatic systems, travel to and invade a tissue in a new location in the body, proliferate, stimulate angiogenesis, and form a secondary tumor.

Characteristics of metastatic cells

5.2 An additional function of the basement membrane is to help the epithelium withstand potentially damaging stretching and tearing forces.

Matrix metalloproteases (MMPs) are a group of proteases secreted by malignant cells. MMPs also play a role in what other process discussed in this chapter?

How is that process similar to the invasive growth performed by certain malignant cells?

Approximately 90% of all human cancers are carcinomas, which means that epithelial cells have undergone a neoplastic transformation. All epithelial tissue is attached to a basement membrane, a supporting layer of extracellular material composed of a variety of glycoproteins (proteins with sugars bound to them) and carbohydrates (Figure 5.8). The membrane provides a defining boundary between the epithelium and an underlying layer of connective tissue.[5.2] In order for cells within a carcinoma to invade surrounding tissue, they must be able to maneuver between other cells of that tissue and among the extracellular matrix, as well as degrade the basement membrane (Figure 5.6). These functions are the result of specific cellular changes that occur during the malignant transformation and tumor progression processes. Certain cells develop the ability to secrete proteases, the enzymes that destroy the proteins involved in cell-to-cell and cell-to-extracellular matrix connections as well as those of the basement membrane.

Associated with the fixed location most cells in the body have within tissues is an absolute requirement for anchorage dependence –

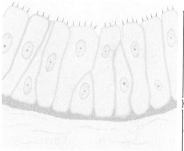

Pseudostratified columnar epithelium

} Basement membrane

Dense irregular connective tissue

Figure 5.8 Basement membrane of epithelium. The basement membrane is a collection of extracellular proteins that create a boundary between epithelial tissue and underlying connective tissue.

they adhere themselves to neighboring cells and the extracellular matrix. Red and white blood cells are the exception to the rule since they circulate freely in the blood stream. The ability to migrate through tissues depends on having a reduced need to be anchored.[5.3] Malignant cells that encounter and gain entrance to capillaries and lymphatic vessels by migrating between their outer layer of endothelial cells can be carried to secondary tissue sites (Figure 5.6d).

[5.3] Migrating malignant cells are similar to white blood cells that have the ability to leave the circulatory system and wander through tissues in order to scavenge cellular debris resulting from an injury or microorganisms in an attempt to prevent infection.

Outline the series of steps needed for a tumor to progress from a benign to malignant state.

Lung is a common location for metastatic tumors

It would seem that if malignant cells enter the circulatory or lymphatic systems, secondary tumors arising from metastasis should appear in all areas of the body, but this is not the case. In 1889 the English surgeon Stephen Paget published in the prestigious British medical journal *Lancet* his observation that metastatic tumors had a tendency to arise in certain organs of the body, such as the lungs, liver, and bones. In the article, he proposed the "seed and soil" theory of metastasis. The theory states that in order for a cancer cell (the seed) to grow, it must settle in a recipient organ (the soil) that is capable of supplying the necessary factors to support its growth.

For most cancers, the majority of secondary tumors develop in the lungs. This occurs primarily because of the pattern of blood flow in the circulatory system (Figure 5.9). The width of capillaries is approximately that of a single blood cell and so they pass through the vessels in a single file fashion. This enhances the exchange of gas (oxygen is delivered to the tissues and carbon dioxide is removed from the tissues), delivery of nutrients, and removal of wastes between the blood and tissue cells. Deoxygenated blood leaves a tissue and travels through capillaries that continually merge with one another to form larger blood vessels called venules. Venules from multiple capillary beds merge to form veins, which ultimately deliver blood to the heart. From the heart, the deoxygenated blood is pumped to the lungs through a large artery.[5.4] In the lungs, the blood vessels begin to branch and get progressively narrower, first forming arterioles and

[5.4] The direction blood flows in vessels of the circulatory system with respect to the heart can be remembered by the first letter of key words. "Arteries" and "arterioles" carry blood "away" from the heart; each word begins with the letter "A". Blood is carried "toward" the heart in "veins" and "venules"; the letters "T" and "V" are close to one another in the alphabet.

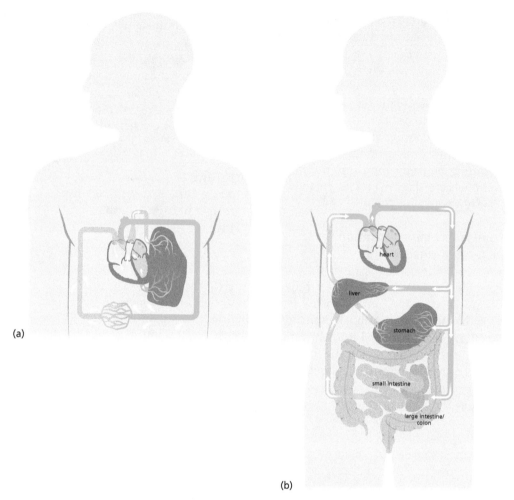

(a)

(b)

Figure 5.9 Circulation patterns of blood. (a) Blood coming from most tissues in the body is delivered to the heart and then to the capillary beds in the lungs. (b) Blood leaving the stomach and colon typically travels to the liver before going to the heart.

then a bed of capillaries, similar to any other tissue of the body. In the capillaries of the lungs, carbon dioxide from the tissues is exchanged for oxygen from air that was inhaled. The oxygenated blood then travels through a series of venules and veins back to the heart, which pumps it back out to the tissues to complete the cycle.

Cancer cells that enter a capillary would be carried along with the flow of blood. Since malignant cells are generally larger than blood cells, they may easily become lodged in a capillary, and the first capillary bed that blood typically flows through after leaving a tissue is in the lungs. The degree of anchorage independence a cancer cell possesses will determine whether it will have survived the journey thus far. A cell may then exit the circulatory system by squeezing between the endothelial cells of the capillary and maneuver its way among the surrounding lung cells. If the cancer cell is capable of interacting with the cells in the new site and grows, it can form a secondary tumor.[5.5]

Reasons for metastatic sites other than the lungs

There are certain types of cancers that have a tendency to metastasize to sites other than the lungs and this occurs for two reasons. The first is that the first capillary bed after the tissue it originates from is not in the lungs. For example, the stomach and colon are involved in the digestion of food and absorption of nutrients and water. The nutrient-rich blood supply leaving those organs travels next to the liver (Figure 5.9). As the blood passes through the capillary beds of the liver, the nutrients exit the blood to be metabolized by the organ before being used by the rest of the body. Stomach and colon cancers, therefore, are more likely to metastasize to the liver rather than to the lungs.

A second explanation for metastatic tendencies is that a particular type of tissue provides a more favorable growth environment for certain types of cancer cells. The liver is a favorable site for metastases because it is a rich source of growth factors. Prostate cancer, however, has a propensity to metastasize to bone tissue because this tissue produces a particular growth factor that is beneficial to the growth of malignant prostate cells.

[5.5] Studies have shown that a primary tumor weighing only a few grams (3 g = 0.1 ounce) is capable of releasing several million cells each day. The chances of a cell surviving the journey through the circulatory system to take up residence at a secondary site is estimated to be anywhere between 1 in 10,000 to 1 in 1 million.

What would you predict to be likely sites for secondary tumors that had metastasized from a primary lung tumor? Explain.

An experiment was done in which cells from a tumor that had not yet metastasized were placed into the tissue of a rabbit. After it metastasized, they found that the majority of the secondary tumors were found in the lung. They then took cells from the secondary tumors in the lung and placed them in the tissue (in the same location as the first time) in another healthy rabbit. A significantly greater number of secondary tumors were found in the lungs of this rabbit. In fact, with each subsequent reimplantation of secondary tumor cells into new rabbits, a greater number of secondary tumors were found. Account for this observation.

Appropriate designations for metastatic tumors

Once a primary tumor has spread to one or more secondary locations, it is still considered a cancer of the original site. The treatment of a metastatic tumor, like a primary tumor, is dependent upon the type of cancer it is

> **Is a bone tumor that is the result of a breast cancer metastasis a sarcoma or a carcinoma? Explain.**

and not simply its location because each type of cancer has unique characteristics (Chapter 7). For example, it is important to identify metastatic cancer in the bones as secondary prostate cancer and not primary bone cancer, and treat it appropriately.

Lymph node analysis for determination of possible metastasis

A second means of metastasis for a cancer cell is to travel through the lymphatic system. The circulatory and lymphatic systems are associated with each other (Figure 5.10). The role of the lymphatic system is best understood by first being aware of what happens to the fluid in the circulatory system as it passes through a capillary bed. Every time the heart contracts, pressure is created that pushes the blood through the vessels of the circulatory system. The pressure increases as the diameter of the blood vessels narrows and is greatest on the arteriole side of a capillary bed. Consequently, most of the liquid portion of the blood (plasma) is squeezed out of the blood vessels between the endothelial cells and into the surrounding tissue. This liquid, which bathes the tissue, is called interstitial fluid. The plasma lost to a tissue is drawn into the vessels on the venule side of a capillary bed, which is under low pressure. This results in the tissues being constantly bathed with an ever changing fluid – as the fluid enters on the arteriole side of a capillary bed it is being withdrawn on the venule side. The amount of fluid lost on the arteriole side, however, is greater than the amount of fluid drawn back into the circulatory system on the venule side. When this is not remedied it results in edema, or tissue swelling, because of a net increase in the amount of interstitial fluid.

Accompanying the capillaries on the venule side of a capillary bed are lymphatic vessels that take up the remainder of the fluid not reabsorbed by the venule capillaries (Figure 5.10). The lymphatic fluid, known as lymph, is transported from the capillary bed through a series of merging lymphatic vessels. The fluid is eventually returned to the circulatory system by draining into the superior vena cava, the largest vein that returns deoxygenated blood to the heart.

The lymphatic vessels from multiple capillary beds converge at certain areas of the body (e.g., arm pits, neck, groin) to form regional lymph nodes (Figure 5.11). The function of these tiny bean-sized structures is to filter and trap unwanted material that had been taken up in the fluid removed from the tissues. White blood cells are present in lymph nodes to engulf and, if necessary, stimulate an immune response against microbes present in the lymphatic fluid.[5.6]

> [5.6] Lymph nodes often become swollen when a person is sick because of all of the activity occurring within them by the white blood cells as they attempt to destroy the cause of the infection.

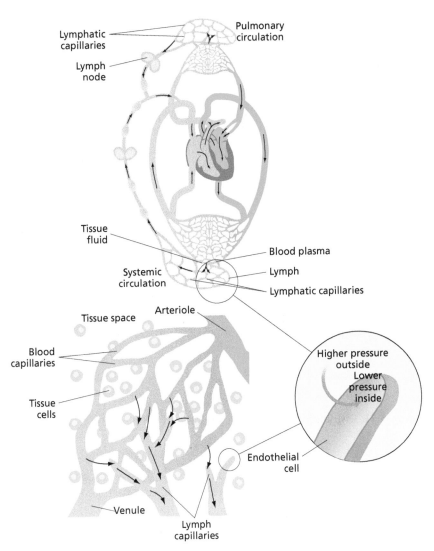

Figure 5.10 Circulatory and lymphatic systems. Blood travels from the heart to capillary beds in the lungs, back to the heart, to capillary beds in the tissues, and returns to the heart. Associated with the venous side of capillary beds are lymphatic capillaries that remove excess interstitial fluid. The lymph is filtered as it passes through lymph nodes on its way to being returned to the circulatory system by entering the superior vena cava.

It is likely that cancer cells that have broken free from a tumor would get trapped in the regional lymph nodes that filter the interstitial fluid of the tissue in which the tumor resides. Consequently, during surgical excision of a tumor, one or more regional lymph nodes are usually removed for biopsy in order to assess the likelihood of metastasis. The detection

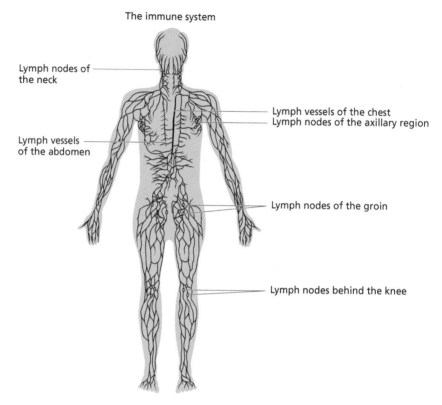

The immune system

Lymph nodes of the neck

Lymph vessels of the chest
Lymph nodes of the axillary region

Lymph vessels of the abdomen

Lymph nodes of the groin

Lymph nodes behind the knee

Figure 5.11 Distribution of lymphatic vessels and lymph nodes in the body.

> **Why does detection of cancer cells in lymph nodes indicate only a possibility of metastasis? In other words, why isn't it a guarantee that metastasis has occurred?**

of one or more cancer cells in lymph nodes is an indication of *possible* metastasis and can influence the method of treatment beyond the surgical removal of the primary tumor. The failure to detect cancer cells in lymph nodes is a good indication, but certainly no guarantee, that metastasis has not occurred: malignant cells may be present in lymph nodes other than those removed for biopsy, white blood cells may have already destroyed cancer cells that were trapped in the nodes, or metastasis may have occurred by way of the circulatory system.

The removal and analysis of regional lymph nodes during a biopsy procedure provide valuable information but it may come at a cost. The excision of lymph nodes creates a break in the lymphatic system that decreases the efficient drainage of interstitial fluid from the area. Over time, interstitial fluid builds up causing the area to swell, a condition known as lymphedema. In an attempt to minimize the risk of lymphedema while not compromising the value of the information that can be gained from lymph node analysis, a revised lymph node biopsy procedure has been devised. The first lymph node or two that take up a blue dye and weak radioactive molecule injected together into the area of a tumor are referred to as the sentinel node(s). Cancer cells from the tumor that enter the lymphatic system would most likely be found in the sentinel node(s). Rather than remove a collection of regional nodes that drain the tumor area, only a sentinel node(s) biopsy is performed (Chapter 8). The removal of only one or just a few lymph nodes minimizes the damage done to the portion of the lymphatic system draining the affected area. The sentinel node biopsy is a form of tumor assessment performed on breast cancers and melanoma, a particularly aggressive form of skin cancer. These forms of cancer are amenable to a sentinel node biopsy because the tumors are in locations that can be injected with the dye and radioactive tracer molecules, unlike tumors that are in the abdominal or chest cavities or within the brain.

EXPAND YOUR KNOWLEDGE

1 Compare and contrast the terms metaplasia, hypertrophy, hyperplasia, dysplasia, neoplasia, and anaplasia.

2 Provide a flow chart that delineates the typical series of events that occur to cells during the normal to malignant transformation process.

3 What characteristics are used by physicians to determine if a growth is benign or malignant?

4 Explain why metastatic tumors may be more aggressive (grow at a more rapid rate and less responsive to treatment) than the primary tumor from which they originated.

5 White blood cells circulate in the bloodstream and are also capable of migrating into and out of tissues in performance of their normal function of defending the body against infection and cleaning up cellular debris. Why is it logical that leukemias – cancers arising through mutations that cause excessive production of white blood cells – have an earlier than average age of onset than do other cancers?

ADDITIONAL READINGS

Abdollahi, A., Hahnfeldt, P., Maercker, C., Grone, H. J., Debus, J., Ansorge, W., Folkman, J., Hlatky, L., and Huber, P. E. (2004) Endostatin's antiangiogenic signaling network. *Molecular Cell* **13**: 649–663.

Alberts, B., Bray, D., Hopkin, K., Johnson, A., Lewis, J., Raff, M., Roberts, K., & Walter, P. (2004) *Essential Cell Biology*, 2nd ed. New York, NY: Garland Science.

Browder, T., Butterfield, C. E., Kraling, B. M., Shi, B., Marshall, B., O'Reilly, M. S., and Folkman, J. (2000) Antiangiogenic scheduling of chemotherapy improves efficacy against experimental drug-resistant cancer. *Cancer Research* **60**: 1878–1886.

Campbell, N. A. & Reece, J. B. (2005) *Biology*, 7th edn. San Francisco, CA: Pearson Benjamin Cummings.

Cooper, G. M. and Hausman, R. E. (2007) *The Cell: A Molecular Approach*, 4th edn. Sunderland, MA: Sinauer Associates.

Folkman, J. (1971) Tumor angiogenesis: therapeutic implications. *New England Journal of Medicine* **285**: 1182–1186.

Folkman, J. and Kalluri, R. (2004) Cancer without disease. *Nature* **427**: 787.

Marieb, E. N. and Hoehn, K. (2007) *Human Anatomy and Physiology*, 7th edn. Upper Saddle River, NJ: Prentice Hall.

Martini, R. and Bartholomew, E. (2007) *Essentials of Anatomy & Physiology*, 4th edn. Upper Saddle River, NJ: Prentice Hall.

Satchi-Fainaro, R., Mamluk, R., Wang, L., Short, S. M., Nagy, J. A., Feng, D., Dvorak, A. M., Dvorak, H. F., Puder, M., Mukhopadhyay, D., and Folkman, J. (2005) Inhibition of vessel permeability by TNP-470 and its polymer conjugate, caplostatin. *Cancer Cell*, **7**: 251–261.

Satchi-Fainaro, R., Puder, M., Davies, J. W., Tran, H. T., Sampson, D. A., Greene, A. K., Corfas, G., and Folkman, J. (2004) Targeting angiogenesis with a conjugate of HPMA copolymer and TNP-470. *Nature Medicine* **10**: 255–261.

6

Cancer screening, detection, and diagnostic procedures and tests

I have cancer. Don't be sad. Talk with me. Laugh with me. Celebrate life with me.
Barbara Kennedy, ovarian cancer patient

CHAPTER CONTENTS

The possibility exists that anyone can develop one or more types of cancer. This risk may be a result of lifestyle choices (e.g., sunbathing, smoking, diet), environmental exposures (e.g., radon, asbestos, second-hand smoke), or genetics. It is important to have a wide variety of reliable screening tests available to detect benign and cancerous growths at the earliest possible stage of development. If the result of a screening indicates a suspicious growth, one or more additional tests are required to diagnose whether it is benign or malignant. The earlier that detection is sought and a diagnosis is made, the better the chance that a cancer has not metastasized and can be successfully treated or controlled. This chapter covers various types of screening and diagnostic tests and procedures currently approved for use in the general population or being tested in clinical trials.

Cancer: Basic Science and Clinical Aspects, 1st edition.
By C. A. Almeida and S. A. Barry. Published 2010
by Blackwell Publishing, ISBN 978-1-4051-5606-6.

FACTORS THAT DETERMINE THE ACCURACY OF A DIAGNOSTIC TEST OR PROCEDURE

It cannot be emphasized enough that the key to successful treatment of cancer is early detection. The identification of a cancer at an early stage paves the way for a favorable prognosis, or recovery outlook, because the cancer is more likely to be small and well contained, and less likely to have metastasized. A small and confined tumor is likely to be surgically removed in its entirety with minimal damage to surrounding tissue, or treated with smaller, more localized doses of radiation.

A number of screening tests can be performed by an individual on a regular basis, and by a healthcare professional as part of a routine medical checkup in order to detect cancer at early or even pre-cancerous stages. Some screening tests are examinations of external areas of the body while others require imaging equipment to observe internal structures or are laboratory tests performed on specimen samples (Table 6.1). In most cases, people being screened are asymptomatic (they do not exhibit any symptoms of disease) but are at risk for one reason or another. Some screening tests are also used to confirm a diagnosis in response to characteristic symptoms as well as to track the progress of treatment. The quality or reliability of a test is determined by a number of factors, mainly its sensitivity, specificity, and rates of false positives and negatives. Cost is another factor that is also given serious consideration.

Sensitivity is the probability that a test result will be positive for a person who has the disease. Another way of looking at the significance of this term is that it is a measure of a test's ability to detect a molecule

Table 6.1 Some cancer screening tests and procedures

Gender	Test or procedure
Men	Testicular exam Digital rectal exam PSA test
Women	Breast exam Mammogram Pelvic exam Pap test
Men and women	Skin exam Fecal-occult blood test Colonoscopy Sigmoidoscopy

at a particular concentration, or a growth of tissue that has reached a certain size. If the mass size or the concentration of the molecule is below the detection limit, then the result will be negative, even though what is being analyzed for is present. Two tests may be used for the same purpose, but one of them may be more sensitive than the other and will detect a much lower concentration or tumor size.

Specificity is the probability that a test result will be negative for a person without the disease and positive for one with it. Essentially, the specificity is an indication of how reliable the test is at distinguishing the desired target molecule or tissue type from others. It is undesirable to have a test detect something that has a similar structure, chemical characteristic, or anatomy. As the selectivity of a procedure increases, so does its specificity.

Both the sensitivity and specificity of a test factor into its overall accuracy by providing a measure of its rates of false positives and false negatives. A false positive occurs when a result is reported as positive when in fact it should have been negative, and a false negative is when a negative result should have been positive.[6.1]

> **What are some consequences of false negatives and false positives?**

> [6.1] A false negative analogy is when an anti-spam filter on an e-mail account does not classify an incoming spam message as such and delivers it to the in-box. A false positive would be when a message is classified as spam when it really is not and it is delivered to the junk-box rather than the in-box.

There are a number of factors that affect a procedure's sensitivity, specificity, and rates of false positives and negatives: the time between when a specimen is collected and subsequently analyzed, the temperature at which a sample is stored or tests are performed, the freshness of the solutions used, and/or the experience the technologist has in performing the appropriate calculations, conducting the test or scan, and interpreting the results or radiographic image.

The necessary levels of sensitivity and specificity of a procedure are dictated by the population being tested. Diagnostic tests are performed on patients who exhibit certain symptoms and are highly likely to have a particular disease. The high incidence of disease in this population decreases the likelihood of false positive or negative results of a diagnostic procedure. By contrast, screening tests are used with asymptomatic yet susceptible patients and must be highly sensitive and specific to minimize error rate.

COMMON SCREENING TESTS

Cancer screening tests are performed based on risk factors such as gender, age, lifestyle choice, occupation, etc. Screening tests specifically for women are those used to detect breast and cervical cancers. Beginning at the age of 20, women should become familiar with the feel of their breasts by performing monthly breast self exams and have a clinical breast exam performed once every three years by a healthcare provider (Chapter 8). Beginning at the age of 40, the clinical breast exam should be performed annually along with mammography, an X-ray examination of the breast tissue. Three years after a woman becomes sexually active or at the age of 21, whichever comes first, she should have a regular Pap smear annually or the liquid-based Pap test every two or three years (Chapter 10) to screen for abnormalities in the lining of the cervix.

Screening tests specifically for men are those used to detect testicular and prostate cancers. Testicular cancer is the most common cancer in men between the ages of 15 and 35. Men should become familiar with the normal feeling of their testicles by performing a testicular self exam on a regular basis (see Chapter 12). Annual screening for prostate cancer should begin at the age of 50. The prostate is a gland that is positioned just below the bladder and in front of the rectum and produces a component of semen. Prostate cancer screening involves a physician or other healthcare provider performing a digital rectal exam and prostate specific antigen test (Chapter 11).

It is recommended that both men and women perform a skin self exam of their bodies once a month in order to detect skin cancer lesions. A monthly examination enables a person to become familiar with any

---

Box 6.1

A clinical trial to determine if pre-symptomatic screening reduces the mortality rates of prostate, lung, colorectal, and ovarian cancers

The National Cancer Institute is currently running a large long-term study titled the *Prostate, Lung, Colorectal, and Ovarian* (PLCO) Cancer Screening Trial to analyze the influence of certain screening tests on mortality rates. Collectively, these four types of cancers represent almost half the number of cancer diagnoses and deaths each year. The objective of this trial is to determine whether or not particular screening tests that can detect the presence of cancer prior to a person experiencing symptoms will reduce the number of fatalities associated with these cancers when treatment occurs in the pre-symptomatic stage. Nearly 155,000 healthy men and women between the ages of 55 and 74 have been randomly assigned into one of two groups: those that receive routine checkups from their health providers, and those that receive a series of examinations that specifically screen for prostate, lung, colorectal and ovarian cancers. Study participants will be monitored through 2017 to determine if these screening tests ultimately increase the survival rates of such patients.

moles or freckles so that one can detect changes in the size, shape or color of existing pigmented spots or any new ones that have developed (Chapter 13). Any changes to areas of the skin should be brought to the attention of a doctor. Beginning at the age of 50, men and women should be screened for cancer of the colon and rectum and a yearly fecal occult blood test is performed to detect blood in the stool (Chapter 15). A sigmoidoscopy, an examination of the inner lining of only the last third of the colon and the rectum performed with a small video camera on the end of a flexible tube, is recommended every 5 years. A colonoscopy, an examination of the inner lining of the entire colon and rectum using a flexible camera similar to that used for a sigmoidoscopy but longer, is recommended every 10 years.

DIAGNOSTIC PROCEDURES FOR THE CONFIRMATION OF A DISEASE

Diagnostic tests differ from screening tests in that their purpose is to *confirm* that a person has a particular condition or disease.[6.2] The latest laboratory tests and radiological equipment, as well as knowledge of patient history, have made it possible to make a diagnosis in a shorter period of time. Less time spent making a diagnosis provides more time to perform additional tests to obtain a better understanding of the exact condition, or to seek a second opinion, if necessary, regarding the diagnosis and treatment options.

> [6.2] As always, diagnostic testing varies depending on the site of the body that is affected and the required form of analysis that yields the most valuable form of information.

Blood and urine tests can indicate changes in health

Laboratory tests performed on blood and urine are among the most critical of all clinical tools used by healthcare professionals in order to determine or rule out the presence of illness in the body. Virtually non-invasive in comparison to other procedures, they yield an enormous amount of highly specific information allowing healthcare providers to gain an understanding of how the cancer is affecting the rest of the body. Although some cancers release protein markers into the blood to confirm a diagnosis, the majority of cancers do not or the markers have not yet been identified. Patients who have symptoms such as a lump, bleeding, pain, or discomfort, or who have cancer itself, will usually have some routine clinical laboratory tests performed to rule out certain conditions, to confirm a diagnosis, or to follow the effectiveness of treatment.

A commonly ordered blood test is a complete blood count (CBC), a procedure that examines the numbers and kinds of white blood cells

Table 6.2 Reference values for a complete blood cell count (adults)

Type of cell	Normal count (per mm^3; cubic millimeter)
White blood cells	4,300–10,800
Red blood cells	4.2–5.9 million
Platelets	150,000–400,000

(leukocytes), as well as the numbers of red blood cells (erythrocytes) and platelets (thrombocytes). Variations in the values of any of these types of cells (Table 6.2) from those of a healthy person can be indicative of particular disease states and can also provide important information about the status of a patient during the course of a treatment regimen. Abnormalities may indicate anemia (decreased red blood cells), infection (increased white blood cells), or abnormal bleeding (decreased platelet count). While CBCs are not always needed to make a diagnosis, they can be instrumental in ruling out other diseases or in assessing how the cancer or treatment is affecting normal cells in the body.

Another commonly ordered blood test is a chemistry panel, a series of between 5 and 35 tests that evaluate the electrolyte (salt), blood glucose, and cholesterol levels, as well as kidney, liver, and cardiac functions. Not only do these tests signify the status of certain organs, but the panel can also help determine the effect a cancer is having on the overall health of a patient.

A urinalysis is considered to be a painless biopsy of the kidney. While it is technically not an actual biopsy, the information obtained is useful in informing a physician about the status of the kidneys, bladder and urinary tract as well as other organs that produce molecules excreted from the body into the urine. A variety of chemical tests as well as a microscopic examination are performed on 5–10 mL (1–2 ounces) of urine. Urinalysis results can assist in the diagnosis of, and the monitoring of treatment effectiveness for a variety of conditions, including cancer.

Imaging techniques that utilize X-rays

Prior to the development of more sophisticated imaging systems, physicians relied on exploratory surgery and X-ray imaging to see inside the body of a patient. Surgery is an invasive procedure that can be painful and has its associated risks. From the clinician's and patient's points of view, the taking of an X-ray is a noninvasive procedure, although the radiation does penetrate bodily tissues. Dense tissue, such as bone, shows up well on an X-ray image, while organs, which are made of soft tissues, are unable to be viewed with much clarity. Currently, there are a variety of imaging techniques that provide clear and detailed internal views of the body. This is done in one of two ways. The first method detects how energy is altered as it passes through body tissues with different densities. The second method involves introducing into the body a radioactive molecule, referred to as a radiopharmaceutical or radiotracer, that has a tendency to accumulate in abnormal tissue, and then detecting where it is in the body. The energy detected in either method is converted by a computer into an image that can be analyzed.

During traditional X-ray imaging, a cone-shaped X-ray beam is passed through a broad area of the body, such as the chest. The traditional method for detecting the radiation that has passed through the body tissues entails the placement of a piece of radiographic film on the side of the body opposite the X-ray source. When an X-ray strikes an area on the radiographic film, a miniscule burst of light is produced that exposes it. The amount of light generated in any given area of film is therefore proportional to the amount of radiation it receives. Dense tissues, such as bone, appear white to a light shade of gray on an X-ray image because they prevent a large amount of the X-rays from passing through to the film (Figure 6.1). Softer tissues and organs appear in

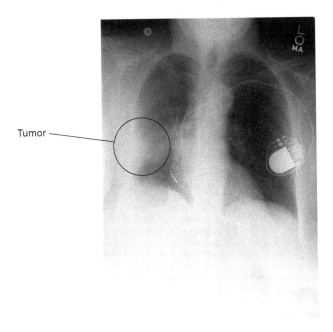

Figure 6.1 Chest X-ray. Note that the lungs, which are soft tissues filled with air, are dark gray to black. Image shows a tumor in the right lung. A pacemaker is visible on the left side of the chest. Image courtesy of David Gilmore at Beth Israel Deaconess Medical Center.

varying shades of gray whereas areas that contain large amounts of air, such as the lungs or colon, appear black because a large amount of radiation is able to pass through them.

Digital X-ray imaging is an alternative method that utilizes a sensor connected to a computer, rather than film, to detect the radiation. The amount of radiation that strikes a particular area of the sensor will be interpreted by the computer to generate a black and white image with shades of gray, similar to that of a traditional X-ray image. Digital radiography has several advantages over the traditional method; image development occurs in a matter of seconds rather than minutes because there is no film to chemically process, less radiation is used to obtain a digital image that is comparable in quality to a traditional image, computer manipulation enhances digital images (e.g., zoom in, enlarge, and adjust contrast levels) and thereby improves the accuracy of analysis, and image files can be electronically stored and transferred.

Computerized tomography (CT), also known as computerized axial tomography (CAT), is a procedure that takes digital X-ray imaging to the next level. This technique uses X-rays to produce multiple, detailed, cross-sectional images of inner body structures.[6.3] A CT scanner is shaped like a large doughnut with a hole in the center that a patient slides through while laying on an examination table (Figure 6.2). The X-ray source tube of a CT scanner rotates 360 degrees around the patient while emitting a thin, triangular-shaped X-ray beam that passes through a single plane of the body. A radiation detector is positioned so that it is always opposite the X-ray beam as it rotates around the patient. The result is a series of images of the same thin cross-section of the body with each one having been taken at a slightly different angle. Each individual image provides a slightly different view of the internal structures within that specific body plane. A computer assembles the series of overlapping images to create a single detailed cross-sectional image of the body plane. The table the patient is laying on then advances a very small interval and the scanner takes another set of 360 degree images to produce a new cross-sectional image that is adjacent to the previous image. Repeating this process results in a series of cross-sectional images that can be assembled one after another to create a single 3-D image (Figure 6.3).[6.4] A complete CT scan may take 5–30 minutes to perform depending on the size of the body area to be examined.

[6.3] Tomography comes from the Greek words *tomo*, meaning "to cut" or "to section", and *graphien*, meaning "to write". The American physicist Alan Cormack and British electrical engineer Sir Godfrey Hounsfield were awarded the Nobel Prize in Physiology and Medicine in 1979 for co-inventing the CT scanner.

[6.4] The CT computer assembles the scans of each body plane to obtain a single 3-D image in much the same way that individual slices of bread can be assembled to create an entire loaf of bread.

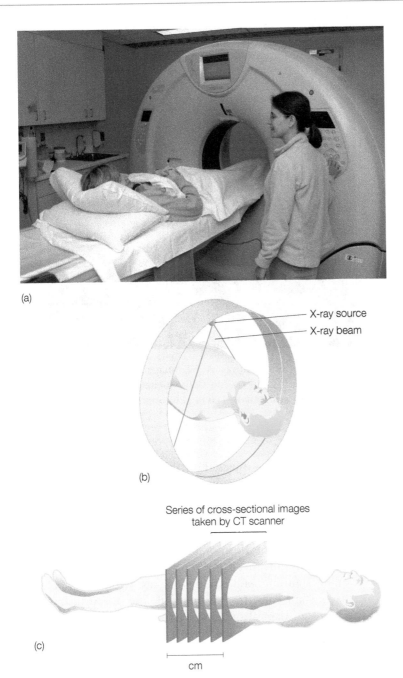

(a)

X-ray source
X-ray beam

(b)

Series of cross-sectional images
taken by CT scanner

(c)

cm

Figure 6.2 A CT scan. (a) The patient lies on an examination table that will slide through the doughnut-shaped scanner. Photo courtesy of David Gilmore at Beth Israel Deaconess Medical Center. (b) The X-ray beam emitter and detector travels in a circle around the patient. (c) This will be repeated multiple times as the table moves forward incrementally producing multiple cross-sectional images of that area of the body.

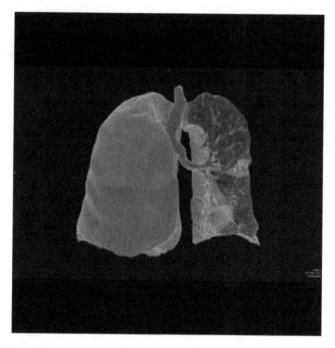

Figure 6.3 A CT image of the lungs. A 3-D image of both lungs with a portion of the left lung (shown on the right) digitally altered in order to view its interior. Image courtesy of David Gilmore at Beth Israel Deaconess Medical Center.

Contrast material is often injected into the patient to alter the passage of X-rays through the body. This enables better discrimination of the less dense tissues of the major body organs and blood vessels, as well as of abnormal tumor growths. A physician is also able to rotate the 3-D images in all directions, which assists in obtaining critical information about the size and specific location of a tumor.

A more technically advanced form of CT, called a spiral CT scan, has the scanner spinning and taking images while the table moves forward at a constant rate. In this method, the X-ray beam proceeds through the body in a spiral or helical pathway rather than in a series of individual planar cross-sections (Figure 6.4). This method enables a large area to be scanned in 20–60 seconds, much less time than required by a traditional CT scan. Shorter scan times can result in more accurate images due to less movement by the patient. Digitized CT images can be manipulated in a variety of ways. Particular areas of the black and white images can be highlighted by the addition of color (Figure 6.5).

What effect would the movement of a patient during a CT scan have on the results?

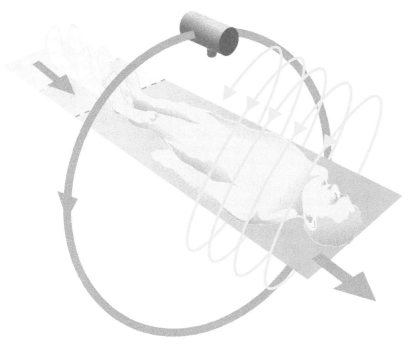

Figure 6.4 Spiral or helical CT scan. The patient on the examination table moves through the scanner as the X-ray beam travels in a circle creating a spiral or helical path.

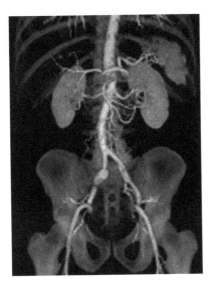

Figure 6.5 Spiral CT scan image. This is a computer-enhanced, 3-D image of the kidneys, blood vessels, and spine taken with a spiral CT. Image courtesy of David Gilmore at Beth Israel Deaconess Medical Center.

Magnetic resonance imaging utilizes radio waves

Another imaging technique is magnetic resonance imaging (MRI), which uses a magnetic field and radio waves (instead of X-ray radiation) to produce pictures of organs and structures inside the body. During the procedure a patient lies on his back on an examination table that slides into an enclosed tubular space surrounded by a strong magnet (Figure 6.6). The MRI unit emits pulses of radio waves, a form of energy, that are absorbed by the nuclei of certain atoms in the body. Those atoms then release energy in the form of different radio waves that are recorded by a detector. A computer collects and compiles the returned radio waves and constructs an image of the body tissues (Figure 6.7). MRI is superior to X-rays and CT scans in that it produces clearer, more detailed images of soft tissue and it enables the detection of small abnormalities or changes in body structures. The technique is often used to obtain images of soft tissues, such as muscles, tendons, brain, spine, chest, abdomen, pelvis, uterus, ovaries, prostate, blood vessels, and even the inside of bones. While an MRI test provides valuable information, there are downsides to it. Patients may experience claustrophobia due to the confining interior of the tube and it is a more expensive procedure than the other diagnostic tests due to the complexity of the equipment.

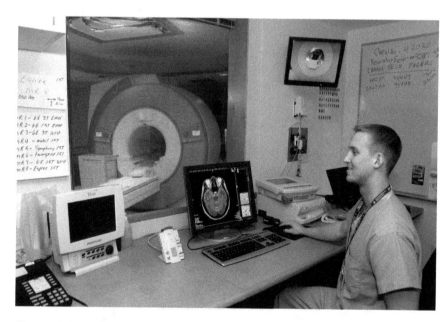

Figure 6.6 An MRI scanner. A patient lies on an examination table that slides into the inner chamber of an MRI scanner. Photo courtesy of David Gilmore at Beth Israel Deaconess Medical Center.

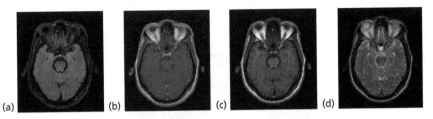

(a)   (b)   (c)   (d)

Figure 6.7 (a–d) A series of MRI images. The images are a series of cross-sectional images through a patient's head with the first plane (a) passing through the upper jaw with each successive plane (b–d) being closer to the top of the head. Images courtesy of David Gilmore at Beth Israel Deaconess Medical Center.

There are no known risks associated with being subjected to a magnetic field as strong as that produced by an MRI. The magnetic field, however, can create a dangerous situation due to the effect it can have on metal in and on the body. For example, the magnetic field can cause a pacemaker to malfunction or can move the position of metal shrapnel or staples if they are in soft tissues. MRI technologists interview the patient and administer a detailed questionnaire prior to the procedure to ensure that no adverse events occur.

Sound waves can be used to produce images

An ultrasound procedure, also known as ultrasonography or sonography, uses high-frequency sound waves to view tissues and organs inside the body. An ultrasound machine uses a probe that produces and receives sound waves (Figure 6.8a).[6.5] When the probe is placed against the body, the sound waves travel until they strike tissue or bone that cause them to be reflected back, or echo, to the probe (Figure 6.8b,c). The computer converts the reflected sound waves into a black and white image. The frequency (the wavelength) of the sound waves will determine how deep they penetrate into the tissues, and they can be adjusted to enhance the level of detail that is obtained in the images. This safe and painless procedure is preferred for soft organs or those that are filled with fluid. The technique can be used to visualize a variety of tissues (such as the colon and rectum, prostate, and breast) in an attempt to detect cancer at an early stage or investigate precancerous or malignant growths. Clinicians oftentimes order this test to determine if a suspicious lump is a cyst, a fluid-filled sac, rather than a mass

[6.5] Some probes are used for external analysis, while others can be inserted into the esophagus, vagina, or rectum in order to obtain clearer, more detailed images of the esophagus or internal organs such as the uterus and prostate gland. Changing the position and angle of the probe allows a technician to obtain different views of the area under study.

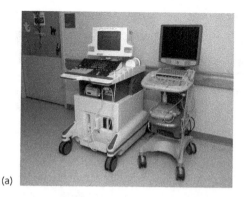

(a)

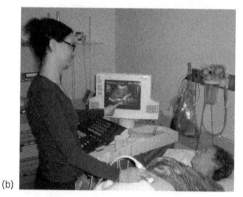

(b)

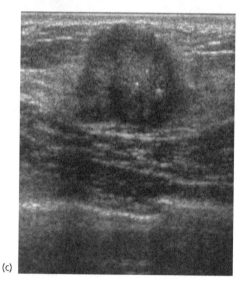

(c)

Figure 6.8 Ultrasonography. (a) an ultrasound machine consists of a monitor, computer, and sound probe. (b) Sound waves produced by the probe pass through the tissues and echo back. (c) The dark areas within the sonogram of a breast are tumors. Photos and image courtesy of David Gilmore at Beth Israel Deaconess Medical Center.

of cells. Sonograms can also be used as a guide, such as when a needle is being inserted into the body to remove a sample of tissue for examination. In recent years, 3-D ultrasound imaging has been developed that takes traditional ultrasonography, which produces only a 2-D image, to the next level.

Positron emission tomography detects differences in metabolic activity levels

Positron emission tomography, or PET scan, is a diagnostic examination that assesses the metabolic activity levels of tissues in an attempt to pinpoint the location of cancerous tissue. A PET scan is sometimes classified as a nuclear scan or a form of nuclear imaging because it analyzes the physiology or biochemical activity rather than anatomy of the body. Nuclear scans use radioactive molecules known as tracers that accumulate in areas of the body, creating "hot" or "cold" spots on the computer image depending on whether more or less of the tracer is taken up by the tissue.

One characteristic that distinguishes malignant cells from normal ones is that they typically have a higher metabolic rate in order to satisfy the energy required for their rapid growth. Prior to a PET scan, a patient receives an injection of a radioactive form of the sugar glucose, the preferred energy source of cells. Cancer cells require greater amounts of glucose than do normal cells to satisfy their higher energy needs and therefore take up a larger amount of the chemically modified sugar. As the PET scanner passes over the body, it detects the radioactivity released from the sugar. A computer collates the information to create an image that indicates the locations where the radioactive sugar has accumulated (Figure 6.9). PET is most commonly used to detect malignant lesions, evaluate the effectiveness of therapy, or assess changes in the activity levels of cancerous lesions.

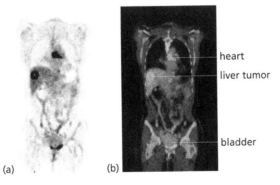

(a)   (b)

Figure 6.9 PET scan. (a) A traditional PET scan image. (b) A computer-enhanced PET scan image indicates the presence of a doughnut-shaped tumor in the liver. (It is normal for the heart and bladder to have high metabolic rates.) Images courtesy of David Gilmore at Beth Israel Deaconess Medical Center.

Radioactive tracer molecules can be targeted specifically to particular types of cancer cells by linking them to antibodies, proteins that have the ability to bind to specific proteins known to be present on the cell membranes of certain types of cancer cells. For example, if a patient has liver cancer, a specific radioactively labeled antibody can be used in a nuclear scan to determine where the cancer is present within the organ (Figure 6.10). There is also the potential of locating liver cancer cells that have metastasized to local lymph nodes or to other locations in the body.

Biopsies allow for detailed analysis of tissues

In the vast majority of cases, a definitive diagnosis of a malignant tumor requires a biopsy, the removal of tissue for subsequent examination by a pathologist. Once a specimen is removed, it is placed in a jar of preservative

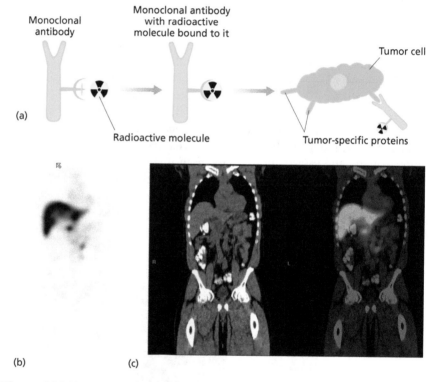

Figure 6.10 Targeting radioactivity with monoclonal antibodies. (a) A radioactive isotope is attached to a monoclonal antibody which is specific for a protein on the surface of a tumor cell. (b) A PET scan shows hot spots in the body where the radioactive monoclonal antibodies have bound to cancer cells. (c) A CT scan image is on the left and a PET/CT scan image is on the right clearly indicating that the radioactive antibodies are bound to cancer cells in the liver. Images (b,c) courtesy of David Gilmore at Beth Israel Deaconess Medical Center.

and delivered to the pathology lab. Specimens are processed in a variety of ways and then observed under the microscope to assess particular cellular characteristics. In some cases, a specimen may be large enough so that its overall appearance may be obvious enough to show signs of malignancy.

There is a wide variety of different types of biopsies and, depending on the extent of the biopsy, local or general anesthesia is administered so the patient does not feel any discomfort or pain. An excisional biopsy is the removal of an entire tumor, whereas an incisional or core biopsy is the removal of a portion of the tumor. A blade is used during a shaved biopsy to remove the outer layer of skin from an area suspected of containing a cancerous lesion. The use of a needle to extract a tissue sample is referred to as a fine needle aspiration biopsy or needle core biopsy, which is sometimes used to obtain specimens from lumps within a breast or prostate gland. Usually an ultrasound, X-ray, or CT scan is used during the procedure to guide the needle to the proper location within the tissue. Endoscopic biopsies are utilized to examine the respiratory, gastrointestinal, or urogenital tracts using an endoscope, a long, flexible tube whose end possesses a video camera as well as an instrument that removes tissue samples for examination.

> Why would an incisional biopsy be done to remove only a portion of the tumor and not the entire thing? Provide an example of such a situation.

During a surgical biopsy, one or more lymph nodes that drain the area in which the tumor is located are also likely to be removed. A lymph node biopsy or dissection is performed to determine whether or not cancer cells have detached from the tumor, migrated into the lymph system, and have become trapped in a nearby lymph node. If this has occurred, then the potential exists that cells have also entered the circulatory system and metastasized to other areas of the body. When a pathologist does not detect cancer cells within biopsied lymph nodes, it is very likely that metastasis has not occurred. This result is not a guarantee, and periodic follow-up examinations are required.

A wide variety of body fluids can also be sampled and analyzed for the presence of cancer cells. Some examples of fluids used are urine, sputum (phlegm), cerebrospinal, pleural (space around the lungs), pericardial (space around the heart), and peritoneal (abdominal cavity). Testing of fluids for the presence of cancer cells is referred to as a cytology test.

TUMOR GRADE AND STAGE FACTOR INTO THE TYPE OF TREATMENT REGIMEN AND PROGNOSIS

The degree to which cells appear abnormal when observed during the microscopic pathological examination is referred to as the tumor grade.

> **What types of tissue organization may be observed in a malignant specimen?**

The size and shape of the cells as well as the degree to which the cells are differentiated (Chapter 4) and organized (histological grade; Chapter 5) are assessed in addition to the size and shape of nuclei and the tissue's mitotic index (nuclear grade; Chapter 5). This information is valuable in determining whether the specimen is benign or malignant as well as determining the growth and metastatic potentials of the tumor.

Tumors are typically graded on a scale of 1–4 with the degree of abnormality and the potential for growth rate and metastasis increasing as the numbers rise. The American Joint Commission on Cancer has established the following characteristics to determine the grade of a tumor:

Grade	Description
GX	Grade cannot be assessed; undetermined grade
G1	Well-differentiated; low grade
G2	Moderately differentiated; intermediate grade
G3	Poorly differentiated; high grade
G4	Undifferentiated; high grade

This grading system is not universal. For example, prostate cancer is graded using the Gleason scores, breast cancer is graded using the Bloom–Richardson system, and kidney cancer is graded by the Fuhrman system.

Tumor grade should not be confused with the **staging** of a tumor, which categorizes the degree to which a cancer has developed and evaluates its potential aggressiveness and likelihood of metastasis to local or distant tissues. This type of information is critical to the treatment process and can help determine one's prognosis. Staging is also important for determining the eligibility of a patient to enter a clinical trial or research study. In addition to a physical examination of tissue, a pathologist may also consult CT, MRI, and/or PET images of the area where the specimen was taken from and also other parts of the body where particular cancers metastasize.

Although each cancer has its own classification, the one used most frequently by medical facilities is the **TNM system** of the American Joint Committee on Cancer (2002). This refers to:

T – the size of the *tumor*, and whether or not it has spread to nearby tissues and organs.
N – the involvement of lymph *nodes*.
M – an indication of whether the cancer has *metastasized* to other organs.

Further classification of staging uses numbers, most commonly 0–4, or letters after the T, N and M designations. While most cancers are unique and are staged as such, some examples may be:

The primary tumor (T)
TX – Tumor cannot be evaluated
T0 – There is no evidence of a primary tumor
Tis – Carcinoma *in situ*
T1–T4 – Determines the size and/or extent of the primary tumor (4 representing the largest)

Regional lymph nodes (N)
NX – Nodes cannot be evaluated
N0 – No cancer in the lymph nodes
N1–N4 – Determination of involvement of the lymph nodes (4 representing greatest involvement)

Metastasis (M)
MX – Metastasis cannot be evaluated
M0 – No distant metastasis
M1–M4 – Determination of distant metastasis (4 representing the greatest level of metastasis)

A breast tumor, for example, staged as T3 N2 M0 is large and has spread outside the breast to nearby lymph nodes, but not to other parts of the body. Prostate cancer staged as T2 N0 M0 is confined to the prostate and has not spread to the lymph nodes or elsewhere.

> **Describe a colon tumor with a staging of T2 N3 M1.**

Another form of staging uses the Roman numerals 0–IV. Generally, the lower the number, the more localized the cancer. The number IV indicates a more serious and widespread cancer. The following list explains the stages:

Stage	Description
Stage 0	Carcinoma *in situ* – early cancer that manifests itself only in a single group of cells that has not penetrated the basement membrane
Stages I, II, III	Larger concentration of affected cells; potentially more virulent; possible metastasis to lymph nodes and/or organs nearest to the primary tumor
Stage IV	Metastasis beyond the original site and compromising another organ

Some cancers are not classified by the TNM system, such as those of the brain and spinal cord, but rather are identified by their cell type and grade. In addition, cancers such as leukemia and lymphomas use a unique staging system (Chapter 16). The grade and stage of a tumor are critical pieces of information that are taken into account when deciding which treatment regimen would be best for a particular tumor as well as when determining the prognosis for a patient. Chapter 7 will cover various treatment options available once a diagnosis is made.

EXPAND YOUR KNOWLEDGE

1 Compare and contrast the following medical procedures. Give examples of when one is preferred over the other.
- (a) X-ray
- (b) CT scan
- (c) Spiral CT scan
- (d) MRI
- (e) Ultrasound
- (f) PET scan

2 What particular question would a patient who is about to undergo an MRI have to be asked that would not have to be posed to a patient about to have a CT scan?

3 Which imaging test would be performed to determine if prostate cancer has spread to the bones or the lungs?

4 How would PET scans be useful to monitor the effectiveness of a cancer therapy regimen?

5 What types of tests would a nuclear medicine technologist perform?

ADDITIONAL READINGS

Braden, C. and Tooze, J. (1999) *Introduction to Protein Structure*, 2nd edn. New York: Garland Science.

Christian, P. and Waterstram-Rich, K. (2007) *Nuclear Medicine & PET/CT: Technology & Techniques*, 6th edn. St. Louis, MO: Mosby.

Lab Tests Online. A public resource on clinical lab testing from the laboratory professionals who do the testing: http://labtestsonline.org

Mettler, F. A. and Guiberteau, M. J. (2006) *Essentials of Nuclear Medicine Imaging*, 5th edn. Philadelphia, PA: Saunders.

Radiological Society of North America: http://www.rsna.org

Turgeon, M. L. (2007) *Linné & Ringsrud's Clinical Laboratory Science: The Basics and Routine Techniques*, 5th edn. St. Louis, MO: Mosby.

US Drug and Food Administration. Home and lab tests: http://www.fda.gov/cdrh/oivd/consumer-lab.html

Ziessman, H. A., O'Malley, J. P., and Thrall, J. H. (2006) *The Requisites: Nuclear Medicine*, 3rd edn. St. Louis, MO: Mosby.

7

Cancer treatment modalities

There's something incredible when you see the cancer melt away and the person recovers and they can go on and have a normal life.
  Dr Patricia Ganz, professor at the UCLA Schools of Medicine and Public Health

CHAPTER CONTENTS

- Surgery: the oldest and most commonly used treatment method
- Radiation kills by causing extensive DNA damage
- Cytotoxic effects of chemotherapeutic drugs
- Side effects and risks from the use of cytotoxic drugs
- Hormonal deprivation treatment: used for estrogen- and androgen-dependent cancers
- Can cancer growth be controlled by inhibiting angiogenesis?
- Additional enzymes targeted for inhibition
- Biological therapy stimulates the body's ability to fight cancer
- Expand your knowledge
- Additional readings

Cancer: Basic Science and Clinical Aspects, 1st edition. By C. A. Almeida and S. A. Barry. Published 2010 by Blackwell Publishing, ISBN 978-1-4051-5606-6.

One of, if not the most, critical decisions to be made following the diagnosis of cancer is the determination of the most appropriate treatment method. There are a number of factors considered when deciding the form of treatment to use, such as the type of cancer, its stage, whether or not there is evidence of metastasis, the age and health of the patient, and in some cases, the presence of certain genetic mutations. Treatment is most commonly administered with the intent of achieving a cure, but if that is not a likely outcome, then the goal becomes managing the disease as a chronic condition. The latter option attempts to minimize the level of discomfort experienced while also maximizing the time a person lives with the disease. This chapter addresses the three traditional forms of treatment – surgery, radiation, and chemotherapy – as well as other more novel methods that are being utilized more frequently. Surgery and radiation are most often used in the treatment of tumors that are confined to specific locations, while chemotherapy is a systemic method. A combination of one or more treatment methods is often used when combating the disease to eliminate as much, if not all, of the cancer as possible and to reduce the likelihood of a future recurrence.

SURGERY: THE OLDEST AND MOST COMMONLY USED TREATMENT METHOD

Surgery, a procedure used to repair or remove a damaged or diseased portion of the body, is most often used in the treatment of tumors that are confined to specific locations. A cancer patient undergoes surgery with a number of objectives in mind. It offers the greatest hope for a complete cure of most cancers, especially those that have not yet metastasized. The amount of tissue that a surgeon removes is dependent upon the size of the tumor and evidence of metastasis. A person who has been determined through genetic testing to possess one or more specific alleles of certain genes that predispose her to a specific form of cancer may choose to have prophylactic (preventative) surgery to reduce the risk of developing cancer.[7.1] Tissue samples are removed for biopsy during diagnostic or staging surgery in order to determine if the tumor is benign or malignant and whether or not there are signs that indicate the growth has penetrated the basement membrane.

> [7.1] For example, a woman who has certain alleles of the BRCA1 and/or BRCA2 genes is predisposed to developing breast and/or ovarian cancer, and may choose to undergo a preventative double radical mastectomy, the removal of both breasts, and/or an oophorectomy, the removal of the ovaries.

> **Define palliative surgery and explain why it is not a curative procedure.**
>
> **Provide an example of reconstructive surgery that may be performed following the removal of a tumor.**

Most often, the primary goal of the surgeon is to remove an entire tumor. Sometimes the position of a tumor does not allow it to be removed in its entirety because doing so would severely compromise the function of the affected or adjoining organ or tissue. In such a case, debulking surgery would remove a portion of the tumor with the intent that reducing its size will increase the effectiveness of radiation or chemotherapy. Palliative surgery, the removal or debulking of cancerous masses that may be compressing nerves causing pain or interfering with the function of nearby organs, may be performed in order to improve the quality of a person's life. An example of this may be the excision of a portion of a lung tumor that is restricting the ability of the heart to beat properly. Reconstructive surgery may be offered following the removal of cancerous tissue in order to attempt to restore a natural appearance to a person's body.

In recent years, the development of a variety of surgical techniques, including those that are less invasive, have enhanced the ability to not only remove a tumor but to do so with fewer side effects or complications. Precisely focused, high-energy beams of light are used in laser surgery to target, superheat, and essentially vaporize cancerous cells while sparing nearby healthy ones. On the opposite end of the spectrum is cryosurgery, which kills malignant cells by exposing them either to liquid nitrogen that freezes them to a temperature of −195.5°C (−320°F), or to an ultra-cold probe. This is a surgical method used occasionally in the treatment of skin, cervical, and prostate cancers. Cancers of the mouth or skin are often treated by electrosurgery that uses high frequency electrical currents to kill cells.

There are various forms of laparoscopic surgeries that are minimally invasive because they require only small incisions for the laparoscope, a fiber optic camera, and other surgical instruments to be inserted (Figure 7.1). During a laparoscopy, a surgeon is able to remove tissue for biopsy, treatment, or symptom relief. The procedure is designed to be minimally invasive, resulting in a limited area available to be explored. Robotic surgery is performed similarly to laparoscopy except the surgeon controls all of the instruments indirectly through a set of hand controls.

In an attempt to kill any metastatic cancer cells that are present prior to and/or following surgery of a primary tumor, a patient may undergo neoadjuvant or adjuvant therapy. Neoadjuvant therapy is given prior to surgery in an attempt to increase its success, whereas adjuvant therapy is given after surgery. For example, a physician may decide that a woman

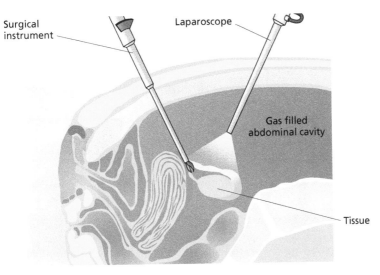

Surgical
instrument

Laparoscope

Gas filled
abdominal cavity

Tissue

Figure 7.1 Laparoscopic procedure. A physician performs surgery using instruments inserted through small incisions.

with a particular form of breast cancer would be best served by receiving neoadjuvant therapy in the form of radiation in an attempt to shrink the size of the tumor prior to its surgical removal. Following surgery, the adjuvant therapy would be chemotherapy to ensure the death of any metastatic cells.

Can surgery trigger metastasis?

There is indirect evidence that surgically removing a primary tumor may trigger the growth of metastatic tumors. The occurrence of this phenomenon has led to a fairly common misconception that when air contacts tumor cells during surgery it triggers their spread to other areas of the body. It would be a serious mistake with life-altering consequences for a person to refuse surgery based on a belief that it would trigger metastasis.

The appearance of multiple secondary tumors only after the surgical removal of a primary tumor is a phenomenon known as **tumor dormancy** (Figure 7.2). In this instance, the secondary tumors grow from cells that had broken free from a primary tumor and traveled to distant sites in the body *prior* to the removal of the primary tumor. Why do the secondary tumors form only after the removal of the primary tumor? The theory is that the primary tumor simultaneously produces both angiogenic activators *and* inhibitors. *Angiogenic activators* are only capable of stimulating blood vessel growth in the region of the primary tumor because they are quickly destroyed once they enter the bloodstream. *Angiogenic inhibitors*, however, are stable and can travel long distances through the bloodstream. Tumors secrete higher concentrations of the activators than of the inhibitors. As a consequence, even though the endothelial cells of the

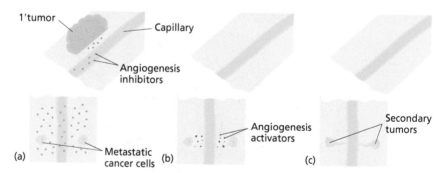

Figure 7.2 Tumor dormancy. (a) A primary tumor will secrete angiogenic inhibitors that travel throughout the body and prevent angiogenesis from occurring where there are metastatic tumor cells. (b) The inhibitory signal is lost when the primary tumor is removed allowing the angiogenic activator signals produced by the metastatic cells to stimulate angiogenesis. (c) Angiogenesis at metastatic sites enables the malignant cells to form tumors.

capillaries in close proximity to the primary tumor receive both activating and inhibitory signals, the growth signal is stronger and angiogenesis that supports the growth of the tumor will occur.

Distant metastatic cancer cells can grow to the size of the head of a pin. The amount of angiogenic activators these cells produce is not high enough to effectively compete against the stable, long-range angiogenic inhibitor signals released by the primary tumor. As long as the primary tumor is present in the body, angiogenic inhibitory signals produced will suppress the growth potential of the metastatic cells. Once the primary tumor is removed, however, there is no inhibitory signal to compete with the activator signals produced by the metastatic cancer cells (Figure 7.2b,c). The result is that the metastatic tumors are capable of stimulating the angiogenesis necessary to fuel their growth.

RADIATION KILLS BY CAUSING EXTENSIVE DNA DAMAGE

Approximately 50–60% of people diagnosed with cancer undergo some form of radiation therapy, which uses X-rays to destroy tumors. This method, like surgery, targets a distinct mass of tissue. The high-energy X-rays damage the DNA molecules as they pass through the cells of a tumor. The objective is to have the high-energy X-rays cause such an extensive amount of damage to the DNA in the cancer cells they pass through, that they are no longer able to function and will die. This is done in a way that minimizes the amount of damage done to the healthy cells adjacent to a tumor so that they are able to recover, creating fewer side effects.

Radiation treatment is based on the premise that actively dividing malignant cells typically have a reduced ability to provide protection against or repair DNA damage compared to healthy cells. Normally, progression through the cell cycle is halted when damage occurs to DNA. This occurs due to the damage triggering a series of reactions that ultimately increases the amount of p53 protein made from its gene (Chapter 4). During this time, the damaged chromosomes can be repaired prior to their replication and before the cell divides. In the event that the damage is so extensive and beyond repair, p53 is also capable of triggering the process of apoptosis, programmed cell death (Chapter 4). More than half of all human cancers, however, possess mutations in the *p53* gene that do not allow the p53 protein to function properly. These loss-of-function mutations certainly contribute to a tumor's aggressiveness and rate of growth and to its resistance to radiation treatment.

> **Explain how a loss-of-function mutation in the *p53* gene contributes to cancer cells having an aggressive growth rate and resistance to radiation treatment.**
>
> **Why are healthy cells better able to recover from radiation therapy?**

Box 7.1

Cell damage by free radicals and cell protection by antioxidants

It is common knowledge that radiation exposure can be harmful, especially in high doses. X-rays possess enough energy to not only break bonds within a molecule they strike but to also cause the molecule to lose an electron, which has a negative charge. This results in the molecule having one too many protons, which have a positive charge. Molecules with a net positive charge are ions, specifically cations. Molecules with one or more electrons than protons are also ions, specifically anions, because they have a net negative charge. X-rays are a form of ionizing radiation because they are capable of generating cations.

A molecule that has lost an electron is called a free radical (Figure 7.3). Free radicals are a source of cellular damage because they are unstable. In an attempt to regain the lost electron, a free radical will scavenge one from another molecule, damaging it in the process and turning it into a free radical. The consequence of the generation of free radicals is that it sets off a chain reaction very similar to one caused by the knocking over of the first in a line of dominos – a series of molecules is altered as a consequence of an initial molecule having been damaged.

The term for the chemical process by which an electron is lost from a molecule is oxidation. Antioxidants, such as vitamins C and E, are molecules that are capable of donating an electron to a free radical. This activity minimizes the number of cellular molecules that will be damaged by terminating the free radical-generating chain reaction.

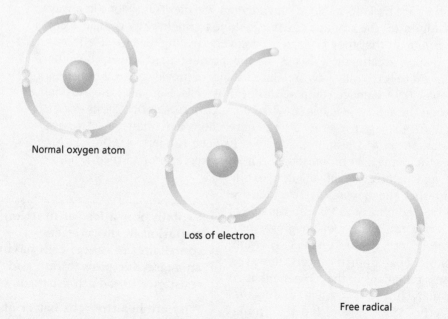

Normal oxygen atom

Loss of electron

Free radical

Figure 7.3 Free radical formation. The loss of an electron from an atom produces a free radical.

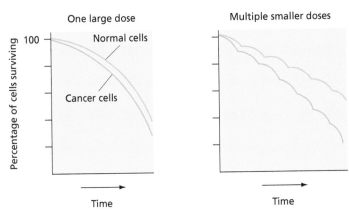

Figure 7.4 Survival differences with different radiation treatment schedules. Radiation treatment with multiple low dose exposures result in more cancer cells dying over time than healthy cells as compared to when a single high dose exposure is used.

Radiation therapy can be delivered either in a single high dose or multiple smaller doses with the total amount of radiation received being equal to that of the single high dose. The latter method has been proven to be more effective (Figure 7.4). By spacing out the radiation treatments, the slight advantage that healthy cells have over cancerous cells in repairing and recovering from damage can be exploited over time. The form, dose, and frequency of radiation therapy are based on the type of cancer, a person's health, and whether the treatment is being administered with a palliative or curative objective.

Radiation therapy can originate from an external or internal source

External-beam radiation therapy (EBRT) delivers an X-ray beam directly at a tumor. The first step in the EBRT procedure is to use CT, MRI, and/or PET scans to create a precise 3-D map of a tumor within a patient's body. The computer generated images are used to determine three types of tumor volumes: (i) the gross tumor volume – all areas within a region that are known to contain cancer cells; (ii) the clinical target volume – all of the area within the gross tumor volume, as well as adjacent areas where there is a high probability that cancer cells reside but are presently undetectable; and (iii) the planning tumor volume – includes a margin just beyond the clinical target volume to compensate for the possibility of a slight shift in the position of the diseased tissue that may occur as the result of a patient's movements between, or even during, treatment. It is absolutely critical that the beam is positioned accurately in order to target and kill the cancer cells while minimizing the damage done to surrounding healthy cells. Small tattoo-like ink marks

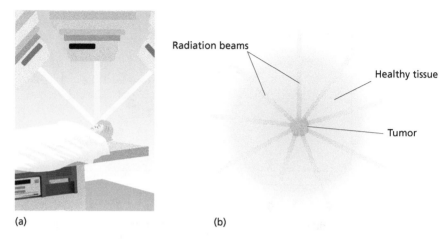

(a)   (b)

Figure 7.5 External beam radiation. (a) The radiation source rotates around a patient. (b) The cancerous area receives the highest radiation dosage because it is at the point where the multiple radiation beams converge. The surrounding healthy tissue receives less radiation.

> [7.2] In a linear accelerator, which is also known as a linac, the speed at which electrons are traveling is increased prior to their colliding into metal, which releases high-energy X-rays.

> [7.3] The outer edges of tumors are typically uneven because the growth rates of the cells are different.

are placed on the patient's skin to be used as references at each treatment. Just prior to receiving radiation, the patient lies on the horizontal table of a linear accelerator, a machine that produces the X-rays (Figure 7.5a).[7.2] The ink marks and a set of alignment lasers are used to properly position the patient.

Manipulating the size and shape of the X-ray beam as its source rotates in an arc over the patient allows the beam to conform to the edges of the tumor.[7.3] The cancerous tissue is located where the individual beams in the arced pathway converge (Figure 7.5b). This maximizes the amount of radiation the tumor receives while it minimizes the exposure level by the surrounding healthy tissues. The less damage the adjacent tissue receives in a treatment session, the faster it will be able to repair and recover. This has a significant effect in reducing the undesirable side effects and complications a patient may experience. Patients typically receive EBRT 5 days a week for up to 8 weeks, with each appointment taking about 10–30 minutes. The actual administration of the radiation, however, takes only a few minutes. Thus, the majority of the time during a radiation appointment is not spent administering the radiation, but is instead spent ensuring that the patient is properly aligned.

An alternative to delivering radiation from an external source is to implant radioactive pellets or "seeds" the size of rice grains, directly into

Figure 7.6 Internal radiation therapy. Brachytherapy seeds shown in comparison to the size of a penny.

or adjacent to cancerous tissue (Figure 7.6). This method is termed internal radiation therapy, or brachytherapy, and allows the treatment of a smaller area with a higher dose of radiation. Brachytherapy is used to treat certain forms of prostate, ovarian, uterine, cervical, vaginal, breast, gallbladder, and head and neck cancers.

The placement of the pellets is done using a needle guided by images from MRI or CT scans or ultrasound (Figure 7.7). The pellets may be placed either permanently or temporarily in the targeted area and the number used depends on the size of the cancerous region. The permanently placed pellets contain the elements palladium-103 or iodine-125, which decay to a nearly undetectable level of radioactivity in about 3–6 months time.[7.4] This is referred to as low-dose rate (LDR) brachytherapy. The seeds do not have to be removed since the patient does not experience discomfort due to their small size. The pellets implanted on a temporary basis contain iridium-192, which emits a larger amount of radioactivity and thereby provides a higher dose, thus the term high-dose rate (HDR) brachytherapy. The pellets are removed after a period of 5–15 minutes and the procedure is repeated up to three times over a series of consecutive days.

[7.4] Radioactive decay refers to the reduced amount of radiation given off by radioactive material over time.

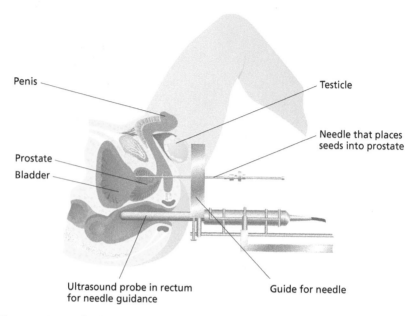

Penis

Testicle

Needle that places
seeds into prostate

Prostate

Bladder

Ultrasound probe in rectum
for needle guidance

Guide for needle

Figure 7.7 Brachytherapy procedure. The implantation of radioactive seeds into a prostate is done using a needle that is guided by the use of an ultrasound image. Source: www.stuarturology.com

Brachytherapy is typically performed in combination with EBRT that uses a lower external beam dose than if it were used alone. The combination approach enables the cancer cells to receive a very high dose of radiation compared with that received by the healthy tissue nearby. The Nuclear Regulatory Commission (NRC), the governing body that provides guidelines for the use of radioactive materials, has not established any precautions for people with permanently implanted seeds. This is largely because the surrounding tissues block the vast majority of the radiation and keep it contained. Patients are likely to be told, however, to limit their close daily contact with small children and pregnant women until the radioactivity has fully decayed. On rare occasions, permanent seeds have migrated from their original position and either were excreted into the urine or traveled to distant parts of the body, but generally this does not cause any harm.

> **What is the main advantage in using brachytherapy?**

Potential side effects and risks of radiation therapy

A patient who undergoes radiation therapy is likely to experience side effects, which are usually considered tolerable. They depend on the area of the body exposed to radiation as well as the dose received. The early or acute side effects during the course of treatment include skin irritation marked by a redness that resembles that of a sunburn, hair loss in the treatment area, mouth sores, and infertility if the testes or ovaries are irradiated. Radiation damage to the ovaries can cause a woman to enter menopause early, which is known as artificial radiation menopause, due to a dramatic decrease in the ability to produce the hormone estrogen. Late side effects, those after the course of treatment, include thickening of the skin and/or damage to nerves, organs of the GI and urogenital tracts, lymphatic system, and the heart and lungs.

There is a small possibility that damage to the genetic information in healthy tissues exposed to the ionizing radiation may be repaired incorrectly or not at all. If this occurs, it may contribute to the development of a new cancer. The risk that the radiation used to treat one tumor will lead to the development of another one is very low and is far outweighed by the benefits of the treatment.

CYTOTOXIC EFFECTS OF CHEMOTHERAPEUTIC DRUGS

Surgery and radiation are *localized* forms of treatment because they involve the removal or treatment of cancerous tissue in one or more specific locations of the body. The use of cytotoxic drugs, those that inhibit the growth of or kill cells, to treat cancer is referred to as antineoplastic drug therapy or, more commonly, as chemotherapy.[7.5] It is a *systemic* form of treatment because the drugs are able to travel throughout the entire body by way of the circulatory system. The distinct advantage of this is that chemotherapy can halt the growth of or kill cancer cells that are not in the form of tumors (e.g., leukemias and lymphomas) or that are metastatic and have not yet become detectable masses.

Advancements are made on a nearly continuous basis that improve our understanding of the essential molecules involved in the malignant transformation process that drives the growth and survival of cancer cells. This is critically important knowledge to have in order to determine the best ways to not only reduce the risks of developing the disease but also to effectively treat it. By targeting the activities of such molecules, chemotherapy

[7.5] Technically, chemotherapy is defined as the treatment of any illness or condition with chemicals or drugs. For example, taking an antibiotic to treat a bacterial infection is a form of chemotherapy. The term, however, is more commonly associated with the use of cytotoxic drugs to specifically treat cancer.

Provide five cellular processes or events important to the development, growth, and/or metastasis of cancer cells that could be targets for chemotherapeutic drugs.

can negatively affect the molecular processes required by cancers and potentially significantly impair their ability to divide or even live.

The objective of chemotherapy depends on the type of cancer being treated as well as its stage. Optimally, it may be used to kill all of the cancer cells present and result in a complete cure. When used following surgery, it can reduce the chance of a re-growth of the primary tumor or inhibit the development of metastatic tumors in the future. If a cure is impossible then it may be employed to extend a person's life by reducing the size of a tumor and preventing it from growing and metastasizing. In the event that the cancer is at an advanced stage, chemotherapy may not be successful in extending the life of a patient, but may have value in a palliative form to improve the quality of that person's life.

Most chemotherapeutic drugs are nonspecific and cause side effects due to their effect on healthy dividing cells in addition to cancer cells. The abnormal function of cancer cells, however, has a tendency to make them more susceptible to the effects of cytotoxic drugs. The treatment of cancer with chemotherapy often entails walking a fine line between providing a high enough drug dosage to be effective against the malignant cells, and minimizing or eliminating side effects that are intolerable or cannot be managed with the use of other drugs or forms of therapy.

Actively proliferating cells are most susceptible to chemotherapeutic drugs

There are over 100 chemotherapeutic drugs approved for use in North America and Europe and many new drugs are constantly undergoing development or being studied in clinical trials. The anticancer drugs are classified according to how they damage or interfere with the function of DNA, by the type of cellular process that is disrupted by their activity, or by the sources from which they are derived. The decision to use one or more specific drugs in order to achieve the greatest likelihood of a favorable outcome depends on factors such as a person's age and health status, the particular type of cancer, the cancer stage and aggressiveness, and, more recently, the knowledge of specific mutant alleles present in a patient's cancer.

Many chemotherapeutic drugs are toxic in the form in which they are administered. Some of the drugs, however, are nontoxic initially but become poisonous once they are metabolized either in the liver, a major site for the chemical modification of a wide variety of molecules, or in another organ or tissue. The majority of chemotherapeutic drugs are most effective against actively growing cells by binding to specific molecules and interfering with their ability to carry out certain functions in particular stages of the cell cycle. Oftentimes, it is the structure or function of the chromosomes that is affected by the drugs, preventing further progression through the cell cycle or inhibiting the utilization of the genetic information for survival.

Chemotherapeutic drugs affect the synthesis, structure, functions, and segregation of chromosomes

During the time when the benefits of using mustard gas to treat lymphoma were being realized (Chapter 1), the determination of metabolic pathways was also progressing. A metabolic pathway is a series of chemical reactions that converts an initial (starting) molecule, through a series of intermediate molecules, to the final (product) molecule, and each reaction is catalyzed by a different enzyme. Once researchers determined the structures of the molecules in a metabolic pathway, they began to synthesize analogs, compounds that are structurally similar to the naturally occurring metabolic molecules. The theory was that the enzymes that catalyze the metabolic reactions would use the synthesized analog rather than the natural molecule because the enzymes would not be able to distinguish between them. Consequently, the enzyme would not be able to function properly, thereby disrupting the cell's ability to function. These types of molecules are known as antimetabolites, and their development and synthesis continues today.

Most chemotherapeutic antimetabolites interfere with the synthesis and repair of DNA. Cells that are unable to replicate their chromosomes will not be able to divide and their rate of growth will decrease. The ability to repair damaged chromosomes is essential to maintaining the structural and functional integrity of the genetic information from one generation to the next. The drugs methotrexate and aminopterin are folic acid analogs and inhibit the enzyme dihydrofolate reductase (Figure 7.8). Folic acid, a type of B vitamin, is a molecule involved in the metabolic pathway that synthesizes the DNA bases A, G, C, and T, which are essential building blocks for chromosomes (Chapter 2). There are other common antimetabolite chemotherapeutics that are analogs of each of the four DNA bases. These "imposter" molecules may either bind to and interfere with the function of enzymes involved in chromosome replication and repair or alter the sequence of a DNA or RNA molecule in which they become incorporated.

Alkylating drugs modify the DNA bases (see Chapter 2) by attaching alkyls, molecules composed of one or more carbon and hydrogen atoms. When chromosomes become alkylated, cell functions are usually disrupted in one of three ways. The first method is the indirect destruction of the DNA. In an effort to repair the chemically modified DNA, the cell produces DNA repair enzymes that attempt to remove and replace the alkylated bases by initially cutting them out of the DNA molecule. When so many of the DNA bases have been alkylated, the chromosomes become fragmented during the repair process, essentially destroying their ability to function (Figure 7.9a).

In a second mechanism of disrupting cellular function, alkyl groups present on a DNA strand can result in this strand's inability to appropriately

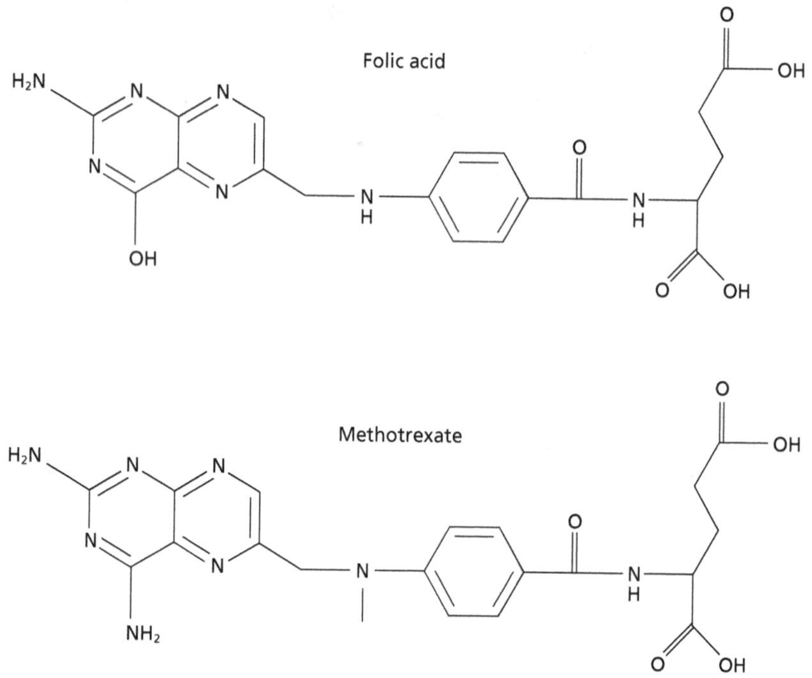

Figure 7.8 The chemical structure of methotrexate is similar to that of folic acid.

base pair with its complementary strand (Figure 7.9b). Normally, the base A pairs with T and C base pairs with G, but the alkylation of a G may result in its pairing with a T, which can lead to alterations or mutations in the DNA sequence. Nontraditional base pairings within genes expressed in a particular cell can influence the function of that cell and can be passed along to future generations of cells, affecting their ability to function as well.

The final mechanism of cellular disruption is the formation of permanent crosslinks between the two strands of a DNA molecule (Figure 7.9c). The presence of multiple crosslinks would prevent the two strands from separating, a necessary process during gene expression and chromosomal replication. It is even possible for alkylating agents to form crosslinks between the strands of two separate DNA molecules.[7.6]

> 7.6 Platinating agents are members of a class of drugs that contains the atomic element platinum. They form crosslinks between DNA molecules.

In Chapter 2, the structure of a DNA molecule was compared to that of a rope ladder that is twisted to form a helix. In order for a gene to be expressed from a chromosome or for it to be replicated prior to cell division, the DNA helix must be unwound

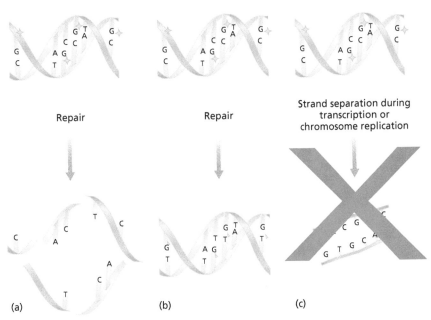

Figure 7.9 Mechanisms of alkylating agents. (a) Alkylating compounds add alkyl groups to DNA bases. Repair enzymes result in the fragmentation of the DNA molecule in an attempt to remove the damaged bases. (b) The addition of alkyl groups to DNA bases results in mismatch pairing of the bases. (c) Alkyl groups can form crosslinks between two DNA strands preventing them from separating during the processes of transcription and DNA replication.

and the two strands must be separated from one another. This allows them to serve as template molecules to synthesize RNA or DNA molecules with complementary sequences. Topoisomerase, the enzyme that unwinds the DNA helix, is another target of chemotherapy drugs. The use of topo-isomerase inhibitors assures that the two strands of a DNA molecule will remain twisted in the form of a helix. This effectively prevents a cell from replicating its DNA and transcribing its genes to make the proteins it needs to function.

If the chromosomes have been replicated, then during cell division they are pulled to opposite poles of the cell and delivered into the daughter cells by the activity of the spindle fibers, protein filaments made of tubu-lin (Figure 7.10). There are chemotherapeutic drugs that are classified as mitotic inhibitors because they interfere with the assembly of tubulin into spindle fibers. Cells exposed to these toxic compounds are capable of dividing, although the chromosomes are not segregated properly into the daughter cells. It is likely that the newly formed cells will die since

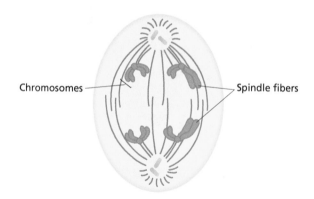

Figure 7.10 Spindle fibers pull chromosomes to opposite poles of a dividing cell.

they will possess either too few or too many chromosomes. One member of this class of drugs is Taxol, which initially was extracted from the bark of the endangered pacific yew tree. Now made semi-synthetically using cultivated yew trees, Taxol is used in the treatment of breast and ovarian cancers (Chapters 8 and 9).

The final class of chemotherapeutics is a group of molecules that are more commonly referred to as antibiotics. An antibiotic is a compound produced by a microorganism that inhibits the growth of bacteria. It may also be a chemically modified form of a naturally occurring molecule that is synthesized in the lab. It has been demonstrated that some of these molecules also have the ability to control the growth and survival of cancer cells. They function by alkylating DNA, inhibiting topoisomerase, positioning themselves between adjacent base pairs within a DNA molecule so as to interfere with the process of transcription, and generating breaks along one or both strands of DNA molecules.

SIDE EFFECTS AND RISKS FROM THE USE OF CYTOTOXIC DRUGS

Both radiation and chemotherapy are used to attack cancer cells because, compared to healthy cells, they are deficient to some degree in their ability to repair the DNA damage caused by drugs. This difference gives healthy cells an intrinsic advantage over cancer cells in their ability to recover from the intended effects of the treatment methods. Chemotherapy, a systemic form of treatment, can negatively affect a much larger and more diverse population of healthy cells than does radiation, which is a localized form of treatment. The side effects that result from the use of chemotherapeutic drugs are mostly a consequence of their effect on the healthy, rapidly dividing cells in the bone marrow, that line the gastrointestinal (GI) tract, that form hair follicles, and that comprise the testes and ovaries.

There is a constant turnover of red and white blood cells in the body. Certain white blood cells have a lifespan of only a few days while others can remain in the body for years. Red blood cells turn over every 120 days and platelets, which are involved in clotting, have a lifespan of 10 days. Chemotherapy destroys the cells in the bone marrow that produce the red and white blood cells and platelets. Those cells that are mature are not dividing and are less susceptible to the drugs. A depletion of white blood cells places a person at a higher risk of infection. Some white blood cells are also termed "granulocytes" because when they are observed under the microscope they appear to have many small granules. Following chemotherapy, the administration of the growth factor granulocyte

---

Box 7.2

Helping cancer patients by donating blood and platelets

Cancer treatment such as chemotherapy often leads to critical side effects associated with anemia (shortage of red blood cells), leukocytopenia (decreased white blood cells) and/or thrombocytopenia (not enough platelets). Since there is no artificial substitute for any of these bodily components, people are encouraged to become blood donors to help replace what has been damaged by treatment or to ensure that there will be a supply available when it is needed. Red blood cells are only viable for up to 42 days and platelets for 5 days, creating an ongoing need. Donor eligibility guidelines are established by the US Food and Drug Administration (FDA) and state regulatory agencies. Minimal requirements include: at least 17 years of age (or 16 with written parental consent), a minimum weight of 110 pounds (50 kg), and good health. Potential donors are asked verbal and written questions relating to their medical and travel history before they begin the process.

Blood donation is a simple process involving the insertion of a sterile needle into the arm that removes about 1 pint (<500 ml) of blood and directs it into a sterile bag. The process takes about 10–15 minutes. In apheresis, the procedure for donating platelets, a machine collects blood from one arm and separates the platelets from the other blood components, which are then returned to the donor. The entire procedure takes about 90 minutes. All blood and platelet donations undergo extensive testing to ensure the safety of the blood supply. A healthy individual's body replaces the fluids within 24 hours and the blood cells within 6 weeks.

A willing participant can donate blood every 8 weeks. Platelets may be donated every 2 weeks, up to 24 times a year. Contact the American Red Cross or local hospitals or blood drives for additional information.

colony-stimulating factor (GCSF) has been demonstrated to increase the number of these white blood cells, thereby decreasing the risk of infection. A decrease in the number of platelets is associated with a reduced ability to form blood clots, leading to increased bruising and a higher risk of bleeding. A decrease in the number of platelets can be remedied by either a platelet transfusion or the stimulation of the body to produce additional platelets by taking GCSF. A reduction in the number of red blood cells leads to anemia, a deficiency in the ability to transport oxygen to the tissues that results in a person feeling tired and sluggish. Erythropoietin is a medication that mimics a natural hormone that stimulates the production of red blood cells and counteracts the anemia.

The lining of the digestive tract is called the GI mucosa. It produces mucus that provides a protective barrier against the low and high pHs and degradative enzymes that are involved in the digestion of food. The cells of the GI mucosa are continually being sloughed off as digested material and wastes move across it, and so they must also be continually replaced. Since these cells are rapidly dividing, they are susceptible to the inhibitory growth effects of chemotherapy. The reduced ability to replace the lost cells is likely to result in irritation and inflammation that leads to nausea, vomiting, intestinal distress, and diarrhea. The nausea and vomiting that most patients experience during the first few hours after chemotherapy are the result of the effects that the cytotoxic drugs have on the brainstem, the area of the brain involved in stimulating the vomiting reflex. There are antiemetic (antinausea) medications available that lessen these side effects and make treatment easier to endure.

After approximately 2 weeks of chemotherapy, a patient may experience partial or complete loss of hair over the entire body. This is due to the cytotoxic effects on the dividing cells of hair follicles. The hair usually grows back after the completion of treatment, and for some it may even grow back prior to the end of treatment, indicating the presence of drug resistant cells within the hair follicles. Sometimes, the hair that grows back is a different color and/or has a different texture than that of the original hair.[7.7] Men and women who choose to wear a toupee or wig may not have to pay for it since some health insurance policies cover the cost when it is "prescribed" by a patient's oncologist.

7.7 For some, especially women, the loss of hair is a traumatic event. Some patients may choose to take a position of control and empowerment and shave off their remaining hair when it first begins to fall out rather than wait for the chemotherapy to do so.

A final group of cells that are affected and lead to side effects are the cells that produce sperm and eggs. In men, the production of sperm, or spermatogenesis, is a process carried out by particular cells in the testes that continually divide and are therefore susceptible to anticancer drugs. It is typical for a man to become infertile as a result of a reduction in his sperm count when he is undergoing chemotherapy. In most cases, the infertility is permanent even after the course of treatment is complete. Men who are young and may desire to have children in the future are counseled to have their sperm maintained in a sperm bank prior to their starting chemotherapy.

In women, the follicular cells in the ovary that produce a fertile egg each month are affected by the cytotoxic drugs. As a result, premeno-pausal women do not have their menstrual periods while undergoing chemotherapy and are likely to experience symptoms common to meno-pausal women. Whether or not a woman remains infertile after a chemo-therapy regimen depends on her age, the drugs used, how long they were used, and at what concentrations. Young women are counseled similarly to young men to have some of their eggs frozen in the event that they want to have children in the future. It should be noted that some chemotherapy patients may retain some degree of fertility and so they are advised to take appropriate measures to reduce the likelihood of pregnancy during the treatment period.

> **Why is it recommended that chemotherapy patients take precautions to avoid a pregnancy while receiving treatment?**

Not all cell mutations are repaired or the repair process is not performed correctly. It is possible that chemotherapy may result in mutations that may be the first of a set of alterations in DNA sequence within a normal cell that will lead to it becoming cancerous in the future. One or more tumors that develop after a period of remission following the completion of a treatment regimen may be metastatic, or may be associated with the treatment received for the primary cancer. The risk of mutations is greatest for those patients who are long-term cancer survivors and who have undergone treatment of certain forms of cancer that are typically especially sensitive to drugs (lymphomas, myeloma, and carcinoma of the ovary).

HORMONAL DEPRIVATION TREATMENT: USED FOR ESTROGEN- AND ANDROGEN-DEPENDENT CANCERS

Dr Charles Huggins (1901–1997) received the Nobel Prize for Physiology or Medicine in 1966 for discerning that the manipulation of hormones could be used as a means to treat certain cancers. The growth of primarily prostate and breast cancers, but others as well, is stimulated by androgens and estrogens, the male and female sex steroidal hormones. Thus, these cancers are referred to as hormone dependent. Following positive results in animals, Dr Huggins' treatment of men with advanced prostate cancer consisted of the surgical removal of both testicles, known as a double orchiectomy, and the administration of the female hormone estrogen. More than half of these men exhibited a reduction in the size of the tumor and an improvement in their health. A comparable surgery in women is an oophorectomy, the surgical removal of the ovaries.

Dr Huggins' newfound association between hormones and cancer triggered the development of drugs that could achieve similar or better results without the use of a drastic and irreversible procedure such as castration. An alternative to the use of chemotherapy for the treatment of hormone responsive cancers is hormone deprivation, also known as hormone ablation or depletion. This entails either eliminating the production of the androgens or estrogens that stimulate the tumor's growth, or preventing them from binding to their receptors in the prostate and breast tissue cells. The principal advantage of hormone deprivation therapy over chemotherapy is that it is less toxic and more specific in the type of cell it targets.

> **Explain the principle behind the use of hormone deprivation drugs as a form of chemopreventative therapy in women who are at high risk of developing breast cancer.**
>
> **Explain how a drug could function to prevent the binding of hormones to their receptors on cancer cells.**
>
> **What do you think are some possible side effects of hormone deprivation?**

The objective of hormone deprivation is to slow the growth or reduce the size of tumors and enable surgery or radiation therapy to be performed with greater ease and efficacy. Hormone deprivation can also be used as an adjuvant therapy when the cancer has metastasized, or as a follow-up treatment when surgery or radiation proves ineffective.

GnRH antagonists block the receptors on pituitary cells that stimulate the secretion of LH and FSH

The hypothalamus, a glandular structure the size of an acorn, is found at the base of the brain and regulates the endocrine (hormone) system. One of the signaling proteins it produces is gonadotropin-releasing hormone (GnRH) which influences the pituitary gland, a pea-sized tissue also located at the base of the brain (Figure 7.11). This gland produces and secretes a variety of hormones into the circulatory system, two of which are gonadotropin hormones: luteinizing hormone (LH) and follicle stimulating hormone (FSH). Both LH and FSH stimulate cells within the gonads (testes and ovaries) to produce androgens and estrogens. The receptors that LH and FSH bind to are targets for therapy.

The levels of testosterone and estrogens can be reduced if the pituitary cells can be prevented from producing and/or releasing LH and FSH. GnRH antagonists are synthetic chemicals that bind to the GnRH receptors on pituitary cells that normally bind the signaling molecules from the hypothalamus. The binding of an antagonist blocks the stimulation of the

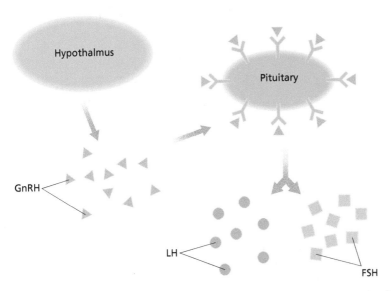

Figure 7.11 Hormones from the hypothalamus influence hormone production by the pituitary gland. The hypothalamus releases gonadotropin releasing hormone (GnRH) that binds to GnRH receptors on pituitary cells stimulating them to produce and release leutenizing hormone (LH) and follicular-stimulating hormone (FSH). LH and FSH stimulate the growth of breast and prostate tissues.

cellular response by GnRH (Figure 7.12). GnRH antagonists, therefore, prevent the release of LH and FSH from the pituitary, which in turn eliminates the stimulatory signal for the testes and ovaries to produce androgens and estrogens.

> **Why is hormone therapy not appropriate for all types of cancer?**

GnRH agonists suppress secretion of LH and FSH resulting in a decrease in androgen and estrogen production

An alternate method of lowering testosterone levels is the manipulation of LH and FSH levels with the use of GnRH agonists. An agonist is a chemical that is structurally and functionally similar to another molecule. Thus, an agonist is able to bind to and stimulate the mimicked molecule's receptor, triggering the same type of response.

Drugs have been synthesized that are agonists of GnRH. The concentration of an agonist used during treatment is higher than that of the GnRH normally produced by the body. This helps to ensure that each of the hundreds to thousands of GnRH receptors present on each pituitary cell is bound by either an agonist or the natural hormone (Figure 7.13). The large number of bound receptors sends a strong signal into the cells that

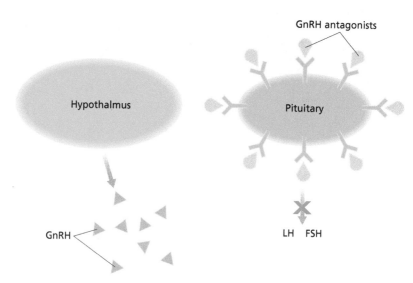

Figure 7.12 GnRH antagonists bind to GnRH receptors on pituitary cells preventing GnRH from binding. The result is that the pituitary cells are not stimulated and therefore do not produce LH and FSH.

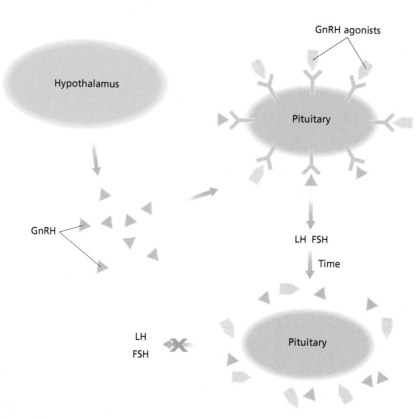

Figure 7.13 GnRH agonists and GnRH bind to GnRH receptors on pituitary cells. After an initial release of large amounts of LH and FSH the pituitary stops producing the hormones.

causes them to release large amounts of LH and FSH. This results in an increase in androgen and estrogen production by the gonads. Over the next few days the pituitary cells undergo an adaptation to the abnormally strong GnRH agonist signal and no longer produce the GnRH receptor.[7.8] The lack of GnRH receptor on the pituitary cells prevents them from being capable of binding either the natural GnRH hormone or the agonist. The cells, without the proper means to respond to the stimulatory signal, will not produce or release LH and FSH, which means that the gonads will not receive the inciting signal that will lead to their production and release of androgens and estrogens. The drop in androgen and estrogen concentration that is typically achieved with GnRH agonists is comparable to that attained by surgical castration.

> [7.8] This type of adaptation to a strong or constant stimulus is commonly used by the body. Another example is the fact that you are much more aware of the sensation your clothes provide when you first get dressed in the morning than you are later in the day. This is because the nerves in your skin have adapted to the constant stimulation provided by your clothes and send a much weaker message to your brain, making you less aware of the sensation along your skin.

Treatment with GnRH agonists is frequently accompanied by what is known as a flare reaction: the cancer initially experiences a temporary growth spurt because of the high concentration of LH and FSH that is released at the beginning of treatment. Once the pituitary is desensitized by the GnRH agonist, the tumor begins to shrink because of the lack of sex hormones.

Androgen and estrogen antagonists function on target cells to prevent their response to the steroid sex hormones

An alternative to reducing androgen and estrogen levels is to prevent those that are produced from reaching their target cells. Antiandrogens and antiestrogens are antagonists that bind in place of the natural sex hormones to the receptors in the cells of the prostate and breast.[7.9] The net effect is that the cancer cells would no longer receive the stimulus to grow. Tamoxifen is an estrogen antagonist that has been used in the treatment of metastatic breast cancer for more than 30 years. It may also be taken as a form of chemoprevention by women who are at high risk of developing breast cancer. Additional information about tamoxifen can be found in Chapter 8.

> [7.9] The mechanism by which antiandrogens and antiestrogens work is similar to that of GnRH antagonists, which is illustrated in Figure 7.11.

Aromatase inhibitors achieve hormone ablation in postmenopausal women

The majority of a woman's estrogens is synthesized by her ovaries, yet a small amount of the hormone is produced in other tissues – principally,

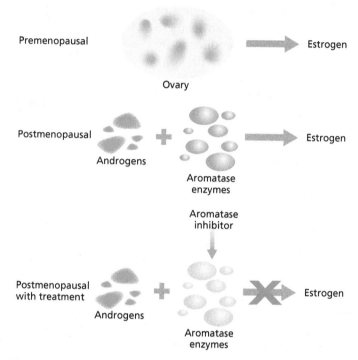

Figure 7.14 Aromatase inhibitors prevent the conversion of androgens into estrogen in peripheral tissues.

adipose (fat) tissue. In these tissues, aromatase enzymes convert androgens, produced by the adrenal glands, into estrogens. Drugs known as aromatase inhibitors block the activities of the key enzymes involved in the conversion process, thereby eliminating the synthesis of the small amount of estrogen usually produced outside of the ovaries (Figure 7.14). The ovaries of a postmenopausal woman are no longer producing estrogens, yet a small amount is still being produced by her adipose tissue.[7.10] Even the lowest estrogen levels can stimulate the growth of breast cancer cells. Aromatase inhibitors are used to treat breast cancer in postmenopausal women in order to eliminate the small amount of estrogens being produced by their bodies.

> 7.10 A woman enters menopause when her ovaries stop producing estrogens.

CAN CANCER GROWTH BE CONTROLLED BY INHIBITING ANGIOGENESIS?

Angiogenesis, the development of new blood vessels, must occur if a tumor is to grow larger than the size of the head of a pin (Chapter 5).

There has been a great deal of research into the isolation and development of molecules that regulate blood vessel growth, particularly angiogenesis inhibitor molecules. The pioneer of this avenue of research was Dr Judah Folkman, a visionary who saw the role angiogenesis inhibitors could play in the treatment of cancer. He theorized that preventing microscopic groups of metastatic cancer cells from stimulating angiogenesis may effectively limit their ability to grow into sizable tumors and cause harm. Furthermore, the inability of an already established tumor to stimulate additional blood vessel growth would halt its growth. It is also possible that the angiogenesis inhibitor drugs could trigger the cells of recently formed blood vessels in a tumor to undergo apoptosis. This could result in a decrease in tumor size due to a reduction in the vascularization that was sustaining its previous growth.

In 2004, for the first time, the US FDA approved two angiogenesis inhibitor drugs, Avastin and Erbitux, for the treatment of advanced stage metastatic colorectal cancer. Avastin is a monoclonal antibody that binds to vascular endothelial growth factor (VEGF), an angiogenesis promoter that stimulates the proliferation of the endothelial cells of blood vessels.[7.11] Avastin binds VEGF in the area that would normally make contact with the VEGF receptor (VEGFR) that is on the surfaces of the endothelial cells of the blood vessels (Figure 7.15a) This binding neutralizes the growth factor because it prevents the growth factor from triggering the intracellular signals necessary to promote the cells through the cell cycle and divide.

> [7.11] Monoclonal antibodies are proteins produced by immune system cells that are grown in a lab. These proteins are valuable because they bind very strongly and selectively to particular areas on specific molecules.

Erbitux is also a monoclonal antibody, but rather than binding to and neutralizing a growth factor, it binds to the epidermal growth factor receptor (EGFR). The presence of Erbitux on the EGFR prevents the signaling molecule epidermal growth factor (EGF) from binding to the receptor (Figure 7.15b). The cells will not be stimulated to grow as long as the ability of EGF to bind to its receptor is blocked by Erbitux.

In clinical trials, Avastin and Erbitux, when administered in combination with one or more chemotherapy drugs, extended both the period of time before the tumor began to grow again once chemotherapy began, and the survival of the patient by 4−5 months. These results are in comparison to those who received solely the angiogenic inhibitor or other form of chemotherapy. There are a number of angiogenesis inhibitors currently being tested in clinical trials. There are still issues regarding the use of anti-angiogenesis therapy that need to be clarified. For example, to maintain a constant concentration of the drugs in the body, it may be necessary to create a dosing schedule. This would allow the drugs to

Box 7.3

Judah Folkman, MD: the pioneer of antiangiogenesis cancer therapy

Dr Judah Folkman devoted his entire professional life to the study of cancer and rose very quickly to positions of great prominence. He was appointed Surgeon-in-Chief at Boston's renowned Childrens' Hospital at the age of 34 and was promoted to Professor of Surgery at Harvard Medical School when he was 35 years old. Scientists today regard Dr Folkman as a visionary in the field of cancer research, although that was certainly not always the case. His 1971 hypothesis that tumors played an active role in stimulating the angiogenesis needed to support their growth was, at the time, regarded by others as far-fetched and ill-conceived.

Dr Folkman advocated that antiangiogenesis should be a primary area of research as a potential cancer treatment method. This focus had received little attention previously and his results were initially difficult to independently verify, so there was much skepticism about his ideas. With time, refinement of the research, and incontrovertible data, Dr Folkman's theories gained more and more acceptance. His work is now considered one of the vital components of current day cancer research.

Three antiangiogenic compounds – angiostatin, endostatin, and vasculostatin – have demonstrated the ability to arrest tumor growth in laboratory mice and are currently being tested in human clinical trials in the USA. The related drug Avastin has been approved for use since 2004 in the US and in 27 other countries for the treatment of colon cancer while also being studied for use in kidney, breast and ovarian cancers. In September 2005, the government of China approved the use of endostatin along with traditional chemotherapy for 48 patients with one particular form of lung cancer. To date, endostatin has been shown to delay progression of the disease.

Folkman used to say, "We have dozens of mice downstairs that are getting old and gray because they had cancer and we treated them. Some have been on endostatin their whole lives, and they're fine." Thirty to forty similar drugs are undergoing studies in patients with varied types of cancers. Further applications of these drugs in the fields of heart disease, stroke prevention, and in the treatment of macular degeneration, an eye disease, are on-going.

Many scientists believe that once a less toxic drug treatment for human cancers is successful, it will be predicated on Dr Folkman's research. His former associates are currently in the process of identifying angiogenesis-based biomarkers in the blood and urine of mice. Detection of such molecules as part of a routine screening test may serve as an early warning indicator of cancer cells growing somewhere in the body. The goal is to detect and treat cancers months or possibly even years before symptoms develop, thereby improving the potential for early and successful treatment of most, if not all, cancers.

Dr Folkman's unexpected death in January 2008 at the age of 74 is a tremendous loss to the field of cancer research.

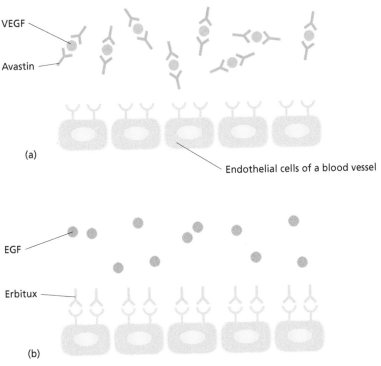

Figure 7.15 Angiogenesis inhibitors. (a) Avastin binds to and neutralizes VEGF preventing it from binding to VEGF receptors on endothelial cells of blood vessels and stimulating their growth. (b) Erbitux binds to EGF receptors thereby blocking the binding of EGF to stimulate the growth of the endothelial cells of blood vessels.

continually counteract or neutralize the signaling molecules that promote angiogeneisis. It may also be necessary to determine appropriate dosages on a patient-by-patient basis in order to accommodate for individual differences in the concentrations of angiogenesis promoter molecules.

ADDITIONAL ENZYMES TARGETED FOR INHIBITION

Rapidly dividing cells have a very high demand for protein and so require a large quantity of amino acids (Chapter 2).[7.12] Since the vast majority of proteins contain at least one of each of the 20 amino acids, a cell that lacks even a single type of amino acid would be in a dire state, not able to synthesize those proteins that require that particular amino acid. A cell without functional proteins would cease to function.

> [7.12] All proteins are composed of amino acids linked one to another in an order specified by the nucleotide sequence of a gene.

Eight of the 20 amino acids used to make proteins are termed essential. The word essential implies that our cells are unable to synthesize them from other molecules and, therefore, it is *essential* that they be supplied in the food a person eats. The remaining 12 amino acids can be synthesized by our cells and are termed nonessential. Asparagine is classified as a nonessential amino acid. There are certain types of white blood cells, however, that are unable to synthesize it and they satisfy their need for asparagine by obtaining it from the blood in which they are being transported.

Acute lymphoblastic leukemia (ALL), a cancer of the white blood cells, responds very well to asparaginase therapy. This form of treatment requires the injection of large doses of the enzyme asparaginase into the circulatory system. This enzyme destroys the amino acid asparagine by chemically modifying it.[7.13] The high levels of injected asparaginase practically eliminate asparagine in the blood, effectively starving the cancerous white blood cells of their sole source of the amino acid. The inability of the cells to obtain asparagine results in their death since they are not able to synthesize the proteins needed to carry out their functions.

> [7.13] Asparaginase removes an amine, a nitrogen containing group, from the side chain of the amino acid asparagine to yield aspartic acid and ammonia.

> **What do you suppose are some side effects of using asparaginase therapy?**

> [7.14] The EM is the web-like framework of proteins that forms a structural support system or scaffold to which the cells of a tissue adhere.

In order for the cancer cells of a tumor to invade deeper into a tissue, they must acquire the ability to degrade the extracellular matrix (EM; Chapter 5).[7.14] Malignant cells accomplish this invasion by producing one or more types of matrix metalloproteinases (MMPs), a group of enzymes capable of destroying the types of proteins that make up the EM. MMP inhibitors are currently being tested in clinical trials as a potential means of controlling the ability of a tumor to invade deeper into a tissue. This form of therapy would likely minimize any damage to the surrounding tissue caused by invasion.

The proteosome is a subcellular structure that degrades damaged or unneeded proteins, which is why it is oftentimes referred to as the trashcan or garbage disposal of the cell. This is the predominant way that cells eliminate unwanted proteins. The proper functioning of proteasomes is critical in maintaining the appropriate concentration of cell cycle inhibitor or promoter proteins that control the rate of proliferation. The high division rate and impaired ability to repair DNA damage common to malignant cells make it more likely that they have a higher turnover of cell cycle proteins and a higher concentration of damaged proteins in comparison to healthy cells. This would require a greater dependence on their proteasomes, creating a vulnerability that may be taken advantage of when treating cancer.

Cells exposed to compounds that function as proteasome inhibitors undergo cell cycle arrest and death by apoptosis. There are several proteasome inhibitors currently being tested in clinical trials. Bortezomib has been approved for the treatment of multiple myeloma, a type of white blood cell cancer. The drug appears effective against the myeloma cells at a dose that does not significantly impair the functioning of the proteasomes in healthy white blood cells.

BIOLOGICAL THERAPY STIMULATES THE BODY'S ABILITY TO FIGHT CANCER

The body has an ability to protect itself against infectious microbes and cancer as well as repair the ravages associated with various treatment methods. The objective of biological therapy, also known as immunotherapy, is to use drugs or methods that enhance the body's ability to detect and destroy cancer cells and repair or replace cells damaged from the use of treatment methods. Biological therapy attempts to boost the functioning of the immune system to help keep a cancer patient healthy, since the disease typically increases the susceptibility to infection. Biological therapy enhances the body's ability to fight cancer and protect against infection by utilizing:

• Interferons, interleukins, and cytokines – molecules cells of the immune system use to communicate with one another. The use of these compounds can increase the numbers of specific types of the body's defense cells as well as stimulate their activity levels.

• Colony-stimulating factors – proteins that promote the bone marrow's production of red and white blood cells and platelets, which often times allows a patient to withstand higher dosages of cytotoxic chemotherapeutics. An increase in red blood cells can improve the delivery of oxygen to the body's tissues, which in turn can improve a person's alertness and overall stamina and strength. A greater number of white blood cells will improve the performance of the immune system.

• Monoclonal antibodies – proteins produced by certain white blood cells that assist the immune system in targeting of infectious microbes and cancer cells for destruction. Certain monoclonal antibodies have the ability to bind specifically to proteins present on the surface of tumor cells. Cells with antibodies bound to them are essentially flagged as dangerous and are more easily detected and destroyed by cells of the immune system.

• Cancer vaccines – particular proteins from specific types of cancer cells that are used to jumpstart an immune response against the cancer cells. Traditionally, vaccines are prophylactic (protective) measures used to stimulate an immune response against a modified form of a microbial pathogen or toxin. If and when a person encounters the actual pathogen or toxin in the future, the immune system has a pre-established defense against it. Prophylactic cancer vaccines are currently being studied for their effectiveness in preventing cervical (Chapter 10) and liver cancers. A therapeutic cancer vaccine is administered when the disease is at an early stage; the immune system is deliberately exposed to proteins that are unique to the cancer cells in a manner that is intended to stimulate an immune response stronger than that which the body has initiated on its own. Numerous studies are being conducted to determine the efficacy of therapeutic vaccines against a wide array of cancers.

• Gene therapy – the introduction into cells of a DNA molecule that possesses a gene sequence which codes for a protein that will carry out a specific function. Ongoing research and clinical trials are determining the feasibility and effectiveness of introducing genes to immune system cells that will allow the cells to make proteins that will improve their performance. Genes are also being distributed into cancer cells to disrupt their functions or make them more susceptible to detection and destruction by immune system cells.

EXPAND YOUR KNOWLEDGE

1 Explain why "slash", "burn", and "poison" are terms used to refer to surgery, radiation, and chemotherapy.

2 Surgery, radiation therapy, and chemotherapy are the most commonly used methods in the treatment of cancer. Explain how it is possible that each of the *treatment* methods actually carries a small risk of *causing* cancer.

3 Provide the names of three antibiotics that are used in chemotherapy.

4 Why would a 36-year-old woman who is being treated with a GnRH agonist for breast cancer experience symptoms common to menopause?

5 (a) A flare reaction is experienced with what form of cancer therapy?
   (b) Explain what a flare reaction is and what is typically done to minimize its occurrence.

6 Would testosterone levels be reduced more quickly with GnRH antagonists or agonists? Explain.

ADDITIONAL READINGS

Alberts, B., Bray, D., Hopkin, K., Johnson, A., Lewis, J., Raff, M., Roberts, K., & Walter, P. (2004) *Essential Cell Biology*, 2nd edn. New York, NY: Garland Science.

Campbell, N. A. & Reece, J. B. (2005) *Biology*, 7th edn. San Francisco, CA: Pearson Benjamin Cummings.

Cooper, G. M. & Hausman, R. E. (2007) *The Cell: A Molecular Approach*, 4th edn. Sunderland, MA: Sinauer Associates.

Haskell, C. M. (2001) *Cancer Treatment*, 5th edn. W. B. Saunders.

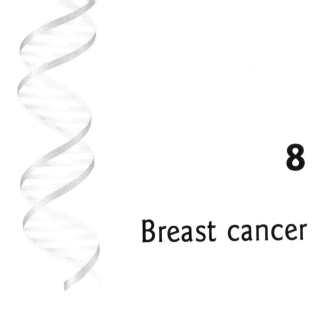

8

Breast cancer

Either you push forward and do the things you did yesterday or you start dying.
I don't want to do that.

Elizabeth Edwards, breast cancer patient

A woman's breasts undergo many changes in her lifetime. As she matures, she develops more awareness of them and they often become a symbol of her femininity. Thus a concern for many women is the development of a possible disfiguring, problematic, or fatal disease that would threaten her female identity. As a result, breast cancer has become a major topic of conversation among women of all ages despite the decreased likelihood of actually developing the disease in comparison to other health conditions (Table 8.1). Breast cancer patients and advocates, along with their families and friends, have engaged in a highly promoted fight against this disease. Countless dollars have been raised to improve breast cancer awareness and detection, research, and treatments.

Cancer: Basic Science and Clinical Aspects, 1st edition.
By C. A. Almeida and S. A. Barry. Published 2010
by Blackwell Publishing, ISBN 978-1-4051-5606-6.

Table 8.1 Women's lifetime risks table for common diseases

Disease	Lifetime risk
Heart disease	1 in 2
Diabetes	1 in 2.5
Stroke	1 in 3
Alzheimer's disease	1 in 6
Breast cancer	1 in 8

BREAST CANCER STATISTICS

Breast cancer is the most common form of malignancy experienced by women; approximately 182,000 women in the US were diagnosed with invasive breast cancer in 2008 and another 68,000 early, noninvasive forms of breast cancer were also detected. About one in eight women (12.5%) will experience this disease in one form or another in her lifetime (Table 8.2). This often quoted statistic does not take into consideration various risk factors such as the age of diagnosis. This and other contributing factors will be discussed in this chapter. Breast cancer is second only to lung cancer as the most common cause of cancer deaths in the US. It is encouraging to know, however, that the mortality rate has decreased in recent years as a result of early detection and improved treatment methods. Men are also at risk for developing breast cancer, albeit at a significantly lower rate. It is estimated that in 2008, approximately 2000 new cases of breast cancer will have occurred in men. The prognosis and treatment

Table 8.2 Probability of a woman developing breast cancer by age

Age	Risk
20	1 in 1837
30	1 in 234
40	1 in 70
50	1 in 40
60	1 in 28
70	1 in 26
Lifetime risk	1 in 8

modalities are essentially the same for both sexes and depend on the stage of the disease at the time of diagnosis.

WOMEN'S BREAST TISSUE: UNIQUE IN STRUCTURE AND FUNCTION

Both men and women have breast tissue. Its size in men and nonlactating women, however, is determined to a large extent by the amount of adipose (fat) tissue and connective tissue that is present (Figure 8.1).

The mammary glands are modified sweat glands that develop fully at the onset of puberty in girls and function to produce milk as a food source for a developing child. Each gland is composed of a highly branched lactiferous duct system that radiates from the nipple. Milk-producing sacs known as alveoli are clustered into lobules, which are arranged into lobes. There are typically 15–20 lobes in a single breast, organized in the shape of a cone with all of the lactiferous ducts being funneled to where the nipple is located. Delivery stimulates the mother's production of the hormone prolactin which activates the milk-producing glands. Milk is secreted from the alveoli into the ducts and travels to the nipple for excretion.

The growth of the epithelial cells that form the inner lining of the ducts is controlled by hormones that also determine sex differences in mammary growth development. In girls, the ductal epithelial cells proliferate during puberty in response to the increased concentrations of estrogens and progesterone. These hormones promote growth at the ends of the

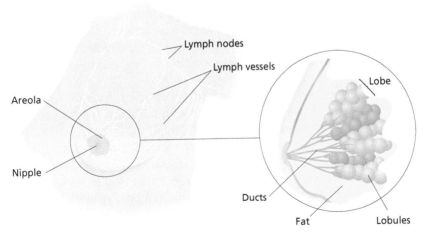

Figure 8.1 Breast tissue and lymphatic system.
Source: http://www.medem.com/medlib

ducts, known as buds, farthest from the nipple. During pubertal growth, the buds split, forming a more highly branched duct system. Expansion of the duct system and the accumulation of the surrounding adipose and connective tissue are usually complete by the end of puberty.

If a woman becomes pregnant, the lactiferous ducts expand again in response to the hormonal changes that occur during this time. Proliferation of the epithelial cells at the buds leads to more highly branched ducts. These newly formed epithelial cells of the ducts, rather than undergo their default response of apoptosis, remain intact by the hormonal signals that persist throughout lactation, the time a woman breast feeds. The cessation of suckling that occurs upon weaning a child off of breast milk removes not only the stimulus for lactation but also the apoptosis inhibition signals that have been maintaining the complex duct system. The production of milk by the alveoli is diminished as the epithelial cells that were formed during pregnancy to expand the ducts now undergo programmed cell death. Following lactation, approximately 90% of the branches that were formed during pregnancy are destroyed.

There are several types of breast cancer

Cancer may arise in any area of a breast in both females and males. The cell types which most commonly undergo malignant transformation are the epithelial cells of the lactiferous ducts or lobules. Ductal carcinoma *in situ* (DCIS) is viewed as an early form of breast cancer, as the tumor is found strictly in the ductal cells and has not invaded surrounding normal tissue. DCIS is most often discovered by a mammogram since it is too small to be detected by a self or clinical breast examination. The most common form of breast cancer, invasive ductal carcinoma (IDC), is a more advanced stage of DCIS. At the time of IDC detection, the malignant cells have broken through the initial epithelial cell layer and are present in the surrounding connective tissue. Lobular carcinoma *in situ* (LCIS) and invasive lobular carcinoma (ILC), which develop in the epithelial cells of the milk-producing lobules, are less common than ductal carcinomas.

Some rarer forms of breast cancers may also occur, such as inflammatory breast carcinoma (IBC) which accounts for about 1–5% of all breast cancers. Unlike other breast cancers, this highly aggressive form of the disease is typically diagnosed in younger women. Unfortunately, IBC is often not detected in routine screening methods such as a breast examination. A routine mammogram may only show a skin thickening rather than a lump. The unique symptoms may include redness, swelling, dimpling, warmth of the breast, an inverted nipple, or itchiness in response to what is believed to be a mosquito bite (Figure 8.2). More women are becoming aware of this form of the disease and healthcare providers are responding in positive ways.

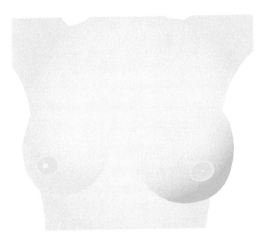

Figure 8.2 Inflammatory breast cancer may cause the breast to appear red or swollen.

CAUSES OF BREAST CANCER

More than 70% of women with breast cancer have *no* known factors that place them at higher risk of developing the disease. Even the healthiest of lifestyles provides no guarantee of immunity against breast cancer, although ongoing research may alter that theory. The two most common risk factors, gender and age, are beyond anyone's control.

Estrogen exposure: a significant factor contributing to the development of breast cancer

As mentioned previously, women are much more highly prone to developing breast cancer than men, and women over the age of 45 are one hundred times more likely to develop breast cancer than women 25 years old or younger. Combined, these risk factors implicate female sex hormones – primarily estrogens, the feminizing steroid hormones that promote proliferation of the epithelial duct cells of the mammary glands – as the prime contributors to the development of breast cancer.[8.1]

> [8.1] There is no single hormone named estrogen; rather, it is a term that applies to the hormones estradiol, estriol, and estrone that are produced by the ovaries.

The most crucial determinant of a woman's risk for breast cancer is her lifetime estrogen exposure. This exposure level is determined by the age at which she first menstruates, whether or not she has been on a form of birth control that contains estrogen, the age of her first full-term pregnancy, the number of times she was pregnant, her age at the onset of menopause, and whether or not she is or was on hormone or estrogen replacement therapy. For instance, girls who first menstruate before the age of 12 have a 10–20% higher risk of breast cancer than those who experience their first menstrual period after the age of 14. This increased risk may be attributable to the fact that girls who have an early menarche (first menstrual period) have higher estrogen concentrations during adolescence, which results in a larger amount of the hormone available to the body tissues. A higher bioactive estrogen concentration will provide a strong signal favoring the progression of the epithelial duct cells of the breast to progress through the cell cycle and divide.

The number of times a woman is pregnant and if and how long she lactates factor into her risk for breast cancer. Since women do not go through their menstrual cycle while pregnant or while lactating, they produce less estrogen. A woman who has been pregnant multiple times and had her first pregnancy before the age of 20 has half the risk of a woman who has never been pregnant, but a woman who has her first full-term pregnancy after 30 years of age is at an elevated risk for the disease.

As a woman approaches her fiftieth birthday, there is a decline in the function of her ovaries and therefore a decline in estrogen production. For every year that menopause is delayed after 50, there is a 3% increase in a woman's risk of developing breast cancer, most likely due to the increase in the number of menstrual cycles (estrogen exposure). An oophorectomy, the removal of both ovaries, will induce menopause. A woman younger than 35 who has had an oophorectomy and subsequent early menopause has a 60% lower risk of developing breast cancer than one who experiences normal (later) menopause. This is because the only source of estrogen in postmenopausal women is adipose tissue, which produces low levels of the hormone. One indication of the role estrogens play in the development of the disease is that, on average, postmenopausal breast cancer patients' estrogen levels are 15% higher than those in healthy women of comparable age and condition.

As a result of reduced estrogen levels, postmenopausal women experience a variety of unpleasant side effects. Atrophy (reduction in size) of the breast tissue, hot flashes, cold sweats, tachycardia (increased heart rate), weight gain, osteoporosis, growth of facial hair, hair loss, loss of libido (sex drive), and vaginal dryness can occur during menopause, lasting from months to years in some women. In an attempt to diminish the number and extent of side effects they experience, women have utilized estrogen replacement therapy (ERT). An alternative to ERT is hormone replacement therapy (HRT), which is estrogen used in combination with progestins, steroid hormones naturally produced by the placenta during pregnancy. There is an increase in breast cancer risk for every year a woman is on ERT or HRT, but studies are ongoing to confirm the extent of the risk.

Box 8.1

The Women's Health Initiative and breast cancer

The Women's Health Initiative (WHI) is a ground-breaking research program that was established in 1991 by the National Institutes of Health and the National Heart, Lung and Blood Institute. It includes nearly 162,000 postmenopausal women with the mission of determining the most common causes of death, disability and poor quality of life in this cohort. In particular, it is examining cardiovascular disease, cancer, and osteoporosis in women from 50 to 79 years old. One finding from the WHI announced in 2002 was that the long-term use of HRT (estrogen plus progestin) in women between the ages of 50 and 70 years old correlated with an increase in the number of breast cancers as well as an increase in heart disease, strokes, and blood clots in women. These early results clearly indicated risks associated with HRT; in fact, a particular part of the overall study was stopped due to the concern for all of its participating women. Following that change, it was announced in 2004 that ERT (estrogen alone) caused an increased risk of stroke and ovarian cancer in a small number of women, again leading to the early conclusion of this study as well. Controversy still exists over these outcomes as ERT and HRT was also found to contribute to a decrease in fractures and colorectal cancer, requiring women with a family history of these conditions to make a choice as to whether the benefits of remaining on HRT or ERT outweigh the risks. Data is still being collected and further announcements will be made in the next few years.

Several studies have indicated that a woman's breast cancer risk is greater with HRT than with ERT. In fact, after 5 years of HRT, a woman's risk for developing breast cancer is 30%, compared to the 10% risk for a woman on ERT. Because of this, many women seek alternatives to these therapies or have adjusted to living with the menopausal symptoms. It has been demonstrated that lifestyle choices such as exercise and dietary changes can lower the potential for developing osteoporosis and heart disease, which are more common in older women, decreasing the need for hormonal supplements.

Women who take an oral contraceptive (birth control pill) that contains estrogen and progestins have a 24% increase in breast cancer risk. After stopping the birth control pill, the risk decreases every year for 10 years, at which time the risk is equal to that of the general population. From 1940 to 1971, a synthetic estrogen drug called diethylstilbestrol (DES) was prescribed to some pregnant women to prevent miscarriage. Use of DES declined in the 1960s after studies showed that not only was it ineffective in preventing pregnancy complications but when given during the first

5 months of a pregnancy, DES interfered with the development of the reproductive system in a fetus. For this reason, although DES and other estrogens may be prescribed for some medical problems, they are no longer used during pregnancy. It was also found that women who were prescribed DES during their pregnancies are at a slightly higher risk of developing breast cancer. The children of DES mothers are currently being studied for health changes.

Genetics may influence the development of breast cancer

> **Give three examples of first-degree relatives.**

A family history of breast cancer puts a woman at higher risk for the disease if her first-degree relatives (i.e., mother, etc.) were affected and especially if they were younger than 40 at the time of diagnosis.[8.2] Breast cancer in any relative on the mother *or* father's side may also increase the risk. Furthermore, having a prior history of breast cancer in one or both breasts substantially increases one's risk of reoccurrence. A family history of the disease indicates that particular alleles of certain genes that predispose a person to the development of breast cancer are inherited rather than spontaneously formed during a person's life. Breast cancers that form as a result of these inherited mutations tend to have certain hallmarks, such as early onset, tumor development in both breasts, and multiple tumor locations within the breasts.

[8.2] The Sister Study, which began in 2004, is a long-term study of 50,000 women aged 35–74 who have one sister that has had breast cancer but whose other siblings have not. It is a national study sponsored by the National Institutes of Health to learn how environment and genetics affect the chances of getting breast cancer. Participants will have their blood sugar and urine tested, as well as toenail clippings and house dust collected and measured for other assays. Still ongoing, the study holds great promise to answer these significant questions. Information is available at http://www.sisterstudy.org.

The alleles of genes associated with the development of breast cancer fall into one of two classes: high- and low-penetrance. **Penetrance** is the proportion of individuals who possess a certain genotype (collection of alleles) and express its particular trait (phenotype). The cancer-causing alleles of the low-penetrance genes are more common in the population and, in combination with other risk factors (e.g., estrogen exposure, diet, physical activity, etc.), account for 90–95% of breast cancer cases. While the cancer-causing alleles within high-penetrance genes carry a very high risk for developing the disease, they are responsible for only 5–10% of breast cancer cases because they are so rare in the population. Cancer-causing alleles of the high-penetrance gene *p53* (Chapter 4) have been linked with approximately 50% of all human cancers, making it the most commonly mutated gene associated with cancer.

Cancer-causing alleles within the *BRCA1* and *BRCA2* high-penetrance genes are most commonly associated with inherited breast cancer. *BRCA1* and *BRCA2* are tumor suppressor genes; *BRCA1* maintains the structural integrity of the chromosomes while *BRCA2* functions in repairing damaged DNA. A woman possessing a cancer-causing allele within *BRCA1* or *BRCA2* has up to an 80% risk of developing breast cancer – much higher than the 12.5% risk for the general population.

Other factors associated with an increased risk in the development of breast cancer

While Caucasian women are at a slightly higher risk for developing breast cancer than others, breast cancer is more likely to be fatal in African-American women. Theories regarding this disparity range from reduced access to medical care to genetic differences that predispose these women to the development of more aggressive tumors. Additional studies may or may not confirm these possibilities.

> **Explain how reduced access to medical care can result in a higher death rate from breast or any other form of cancer.**

There is a well-documented association between alcohol consumption and breast cancer risk, regardless of age. The risk for breast cancer increases by 9% for every 10 g (approximately 1 drink) of alcohol ingested for up to 60 g per day. What is not as clear is the mechanism by which alcohol increases the risk of developing a breast tumor. Tumor formation may be due to acetaldehyde, the first metabolite produced when alcohol is broken down in the body. Acetaldehyde is a known mutagen and carcinogen that could promote the development of cancerous tissue. Drinking alcohol has also been demonstrated to raise estrogen levels in women regardless of whether they are pre- or postmenopausal. Additionally, vitamin deficiencies, a possible consequence of alcohol consumption, can have a wide range of negative effects. For example, the immune system may not function properly following alcohol consumption, allowing developing cancer cells to escape detection and increase in number. Cells may not be able to synthesize or repair their DNA properly, or levels of highly reactive and damaging oxygen radicals could be inappropriately increased or not reduced due to a lack of antioxidants.

REGULAR EXAMINATIONS OF THE BREAST ARE IMPORTANT FOR EARLY DIAGNOSIS

A great deal of fear and a number of myths surround breast cancer. Contrary to popular opinion, *most lumps in the breast are not cancerous!* In many cases, breast lumps are the result of a condition known as fibrocystic breast in which cysts (pockets of fluid) or scar-like connective tissue create areas of lumpiness. More than 50% of women between the ages of 30 and 50 experience this noncancerous condition; only about 5% of the fibrocystic changes indicate a high risk for the development of breast cancer. Another common nonmalignant condition encountered predominantly in young women is a fibroadenoma, a movable lump of fibrous or glandular tissue measuring less than 4 cm wide. Despite the fact that these are normally benign, a healthcare provider should be consulted as soon as possible when an abnormal area within a breast is detected.

Both women and men should be aware of the symptoms of breast cancer. Though a lump or swelling in the breast/underarm area are warning signs, everyone should be knowledgeable of other significant symptoms: nipple discharge that is not breast milk and may be clear or bloody; nipple crustiness or scaling that does not go away; inverted nipples; nipple pain; breast redness; and dimples on the breast. Having one or more of these symptoms in no way constitutes a diagnosis of breast cancer. In fact, more often than not they are *not* indicative of the disease. A healthcare provider should be contacted to discuss any of the preceding conditions as soon as they are discovered.[8.3]

> **Why is a lump or swelling in the underarm area a warning sign of breast cancer?**

> [8.3] Thanks to the Internet, people often receive inaccurate information about breast cancer. Some suggested causes are antiperspirants or deodorants, underwire bras, abortions, and breast implants. None of these rumors have been proven to be accurate! More information on these topics is available at the Centers for Disease Control and Prevention website: http://www.cdc.gov/hoaxes_rumors.html.

Women need to become familiar with changes in their breast

Women should perform a periodic breast self-examination (BSE) beginning in their twenties. At the same time, they should have a clinical breast examination (CBE) included in a physical examination by their healthcare provider at least every 3 years. After 40, this examination should be performed annually as its value in detecting breast cancer at an early stage cannot be overstated. The advantage of a BSE is that it allows women to be aware of how their breasts normally feel so that they can report any changes to a healthcare professional. While research has shown that BSE plays a small role in detecting breast cancer, often women who do detect their own breast cancer discover a lump while performing daily activities such as showering or getting dressed.[8.4]

> [8.4] The American Cancer Society no longer recommends that BSE be performed on a monthly basis. Studies have not yet shown that BSE reduces *deaths* from breast cancer. Most healthcare professionals agree that women should become familiar with their own breasts to be aware of changes.

Box 8.2

How to perform a breast self-examination

STEP 1: VISUAL EXAMINATION

Remove clothes from the waist up and stand in front of a mirror with your arms at your sides. Visually inspect your breasts for any changes in size, shape, color, dimpling, or symmetry. Repeat the inspection by bending forward and by holding your arms over your head.

STEP 2: PHYSICAL EXAMINATION

Lie down on your back and place your right arm behind your head. Using the finger pads of the three middle fingers on your left hand, examine the right breast. Keep the fingers flat and together without lifting them off the skin. Use three levels of pressure – light, medium, and firm – to feel all of the breast tissue. Cover the entire breast from top to bottom and side to side. Some prefer the up and down or vertical pattern while others begin at the nipple, moving outward in larger and larger circles until the whole breast has been examined (Figure 8.3). Repeat the process for the left breast using the finger pads of the right hand. Gently squeeze both nipples to see if there is any discharge. Sit up or stand and check the tissues under the arm (lymph nodes) by slightly raising it. Do not raise it straight up or it will tighten the tissue in that area and make it more difficult to examine.

WHEN TO PERFORM BSE

The best time to perform this examination is about a week after the first day of menstruation, so that the breasts are less likely to be tender or swollen. Those who no longer have a period should pick a day that is easy to remember, such as the first day of each month.

Perform a breast self-examination tonight.

Figure 8.3 Breast self-examination: using the pads of the first three fingers, examine each part of the breast.

If pain, tenderness, or a lump occurs in the breast, a diagnostic mammogram may be ordered to rule out breast cancer or to determine the need for further testing such as ultrasound or biopsy. Mammography is a low-dose X-ray of the breast tissue. The standard X-ray film or digital image is examined to detect small lumps or growths that may or may not be felt during a manual examination. If the mammogram detects an irregularity, it does not mean cancer is present – additional testing is required to rule out or confirm a diagnosis. Recently, computer generated digital mammography has been introduced, producing a higher quality image for the radiologist to analyze. While digital equipment is more expensive initially, one advantage is that the technologist can immediately review the image to assure exposure quality, reducing unnecessary recalls for the patient.

The mammogram procedure is simple and brief. A patient stands in front of an X-ray machine and the technician places the breasts, one at a time, on a plastic plate (Figure 8.4). Another plate is lowered on top of the breast, compressing the tissue. Although some women are uncomfortable with this pressure, the procedure lasts only a few seconds while the technician takes the image. A radiologist then examines and interprets the films.

The American Cancer Society and the National Cancer Institute recommend that women over the age of 40 should have a mammogram performed every 1–2 years if they have no history of breast cancer. Women who are at higher risk for breast cancer should consult their healthcare provider to discuss an earlier or a more frequent testing schedule. Most states have laws requiring health insurance companies to reimburse all or part of the cost of screening mammograms.

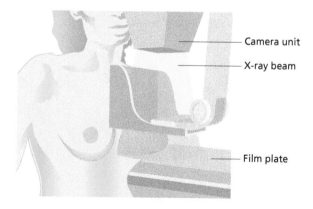

Camera unit

X-ray beam

Film plate

Figure 8.4 Mammography is an X-ray of the breast. Each breast is compressed horizontally, then obliquely and an x-ray is taken of each position.

While mammograms are a very significant tool in the fight against breast cancer, they are not infallible. Problems can occur, including false positive or false negative readings (Chapter 6). The former are characteristics that are reported by a radiologist as abnormal, but upon further testing prove to be benign. Mammograms may also miss up to 20% of breast cancers that are present at the time of the test (false negatives). False negatives are more common in younger women than older because their breast tissue is denser thereby making it more difficult to spot abnormalities. When a woman grows older, her breasts usually become fattier and less dense; this is one of the reasons why mammograms are not routinely ordered on young women unless there is reason to suspect an abnormality.

Should MRI scans be used routinely for breast cancer?

Researchers have attempted to establish better, superior techniques in finding early breast cancer, particularly by the use of magnetic resonance imaging (MRI; Chapter 6). Recent studies strongly suggest that women at moderate or high risk for breast cancer (genetic, familial predisposition, cancer in one breast, etc.) should have both a mammogram and an MRI performed on a regular basis.[8.5] Since breast density is not a problem with MRIs, they can be performed on younger women with better results than a mammogram. Unfortunately, since MRI scans are more sensitive than mammograms, this testing can lead to more false positives, and more unnecessary follow-up examinations and tests. It is worth remembering that most American women are *not* at high risk, that MRI scans are expensive, and that they are not always covered by health insurance.

> [8.5] The best quality MRI images are those that are run with equipment designed specifically for breasts, where the woman lies facedown on the bed, placing her breasts in a cut-out area.

> **Compare and contrast mammograms vs. MRIs. for breast examinations.**

Ultrasound testing (Chapter 6) may be used to analyze a suspicious lump or spot detected upon palpation or on a mammogram. This painless procedure involves passing a sensor "wand" emitting sound waves over the area of interest and does not involve exposure to radiation. A breast ultrasound is used to determine whether a lump is a cyst (filled with fluid) or a solid mass. If it is a cyst, a needle is used to withdraw the fluid in a process called aspiration. This can cause the mass to disintegrate, and no further treatment is needed. Ultrasound can also be used to confirm correct needle placement for a biopsy. Two examples of breast biopsies are fine needle aspiration and core biopsy. A fine needle aspiration uses a narrow needle to withdraw a sample of cells from the lump in the breast. The needle used in a core biopsy has a larger diameter that removes more tissue but not an entire lump.

Box 8.3

Research is ongoing to diagnose breast cancer *before* a tumor is detected

Breast surgeon Susan Love MD, has developed a procedure to hopefully diagnose breast cancer at the cellular level before it progresses to a tumor. Called ductal lavage, the procedure has been nicknamed a "breast Pap smear" because, like the test for cervical cancer, it is a nonsurgical approach to identifying abnormal cells. According to *Dr Susan Love's Breast Book*, this minimally invasive procedure has the potential to enable detection of cells when they are "just thinking about becoming cancer." Since research has indicated that breast cancer originates in the lining of the milk ducts, the procedure involves the identification of the location of the ducts and the collection of cells from the ducts for analysis. The first step utilizes a breast pump, or "aspirator," that suctions a small amount of fluid from the nipple. An anesthetic is then applied to the nipple and a hair-thin catheter is threaded into the duct. A salt-water solution is infused through the catheter and into the duct, and then sucked back out to allow the cells that have been washed out to be collected and analyzed (Figure 8.5). They may then be determined to be normal, atypical, or malignant.

Ductal lavage is not intended to replace mammography or clinical breast examinations. It has the potential to help women who are at high risk for breast cancer to assess their probability and make risk-reduction treatment decisions. This procedure is currently available in clinical trials at selected centers around the country.

Figure 8.5 Ductal lavage procedure. This minimally invasive procedure allows for the removal of fluid from the breast ducts, which are then microscopically examined for cellular changes consistent with breast cancer. Reproduced by permission of Susan M. Love, MD.

WHAT FOLLOWS A POSITIVE DIAGNOSIS?

Pioneered by Dr Armando Giuliano of the John Wayne Cancer Institute in California, the sentinel node biopsy/dissection has radically altered the way breast cancer surgery is performed. Lymph nodes are small bean-shaped tissues that help fight infections and are found under the arm. A sentinel node is the first lymph node into which a tumor cell is likely to enter if or when it has broken free of the tumor mass and gained access to the lymphatic system. In order to detect that specific node, a blue dye or radioactive tracer is injected into the tumor area, taken up by the lymphatic vessels, and transported to the sentinel node (Figure 8.6). A surgeon removes only those nodes that have a blue color or a small amount of radioactivity. The excised nodes are analyzed for the presence of tumor cells. The absence of cancer cells in any of the nodes is an indication that the cancer may not have metastasized. In the past, women had multiple lymph nodes removed, leading to some permanent complications such as lymphedema, accumulation of lymphatic tissue fluid due to inadequate draining of the area.

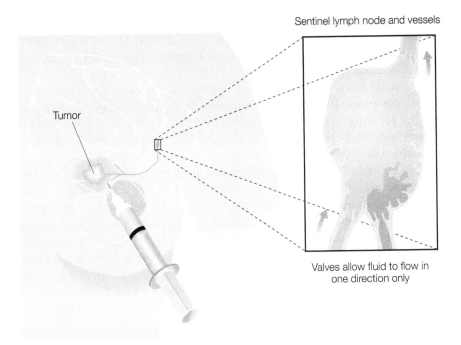

Sentinel lymph node and vessels

Tumor

Valves allow fluid to flow in
one direction only

Figure 8.6 The sentinel lymph node biopsy. Either dye or radioactive molecules are injected into the area around the tumor. The first lymph node to take up the dye or radioactive molecules is the sentinel lymph node. Source: http://apps.uwhealth.org/health/adam

New tests can determine the odds of metastasis and/or recurrence

There is always a lingering question as to whether a treatment option is necessary or the most appropriate for a patient's particular form of cancer. Newly available tests enable breast cancer patients and their physicians to be much more confident in their choice of treatment and prognosis. The Food and Drug Administration (FDA) has approved several tests that can determine whether after surgery a breast cancer patient has a low or high risk of a recurrence of the disease within the next 10 years. Since the tests have not been used regularly in the past, limited information is available on the success rate, but they have the potential to dramatic-ally improve breast cancer treatment by giving additional information about the tumor. The decisions made by physicians and patients will not be based solely on the results of these new tests but also on other clinical information and laboratory tests. The following paragraph has examples of tests that are currently being utilized and this list should expand in the future.

The *MammaPrint* test, developed by Agendia, has been used in the Netherlands since 2005 and was approved for use in the US in 2007. The test examines the gene signature of 70 cancer related genes in a sur-gically removed breast tumor by using microarray analysis, a DNA-based technology, to determine genetic activity. A specific formula devised by computer analysis is used to generate a score that will determine the risk of metastasis. Also in 2007, the FDA approved the *GeneSearch Breast Lymph Node Assay*, manufactured by Johnson & Johnson. This fairly rapid molecular-based laboratory test analyzes the sentinel lymph node tissue for the presence of cancerous growths as small as 0.2 mm in size. The test determines the activity level of two genes, Mammaglobin and Cytokeratin 19, which are typically expressed at low levels in lymph node tissue but overexpressed in malignant cells. These genes are known to cause a breast cancer to metastasize. Results from this assay enable a surgeon to know within 40 minutes whether breast cancer has spread. This is valuable informa-tion that can be used to determine whether or not more lymph nodes should be removed. The *Oncotype Dx* assay quantifies the RNA produced by a specific set of genes within cells of a tumor, thereby providing a snap-shot of the activity of the tumor cells. The result is a Recurrence Score, a number between 0 and 100 that corresponds to a specific likelihood of breast cancer recurrence within 10 years of the initial diagnosis in those with early stage disease.

Staging the tumor

Once a diagnosis of breast cancer has been made, further tests must be performed to determine if the tumor is contained within the breast or has spread to other parts of the body. This staging process is beneficial in determining the course and selection of treatment options as well as the prognosis. Staging classification depends on the size of the tumor, determined visually and/or by imaging tests such as mammography, ultrasound, CT or MRI scans which are also utilized to give information on the lymph nodes and distant organs. The TNM system, described in Chapter 6, is utilized for the staging of breast cancer (Tables 8.3, 8.4).

Table 8.3 Breast tumor classification

Size of tumor	Tis	T1	T2	T3	T4
	Carcinoma *in situ* or noninvasive breast cancer	<2 cm (<0.78 in)	2–5 cm (0.78–2 in)	>5 cm (>2 in)	Tumor may be any size but has ulcerated through the skin and/or spread to the chest wall
Tumor size comparison		Raisin	Strawberry	Lemon	
Lymph nodes	N0	N1	N2	N3	
	Nodes are negative	Cancer has spread to underarm nodes on the same side as the affected breast	Cancer has spread to 4–9 underarm nodes on same side as affected breast	Cancer has spread to 10+ underarm nodes or lymph nodes in other breast areas	
Metastasis	M0	M1			
	Cancer is contained, has not spread to distant tissues and organs	Cancer has spread to distant tissues and organs			

Table 8.4 Breast cancer staging

Stage 0	Carcinoma *in situ*
Stage I	0–2 cm tumor, no metastasis
Stage II	IIA: no tumor found, lymph nodes are positive OR tumor <2 cm, spread to lymph (underarm) nodes OR tumor 2–5 cm, no spread to lymph nodes IIB: 2–5 cm tumor has spread to lymph nodes OR 5+ cm, no spread to lymph nodes
Stage III	IIIA: no tumor found, cancer spread to lymph nodes, or nodes near breastbone OR tumor <2 cm, spread to lymph nodes or nodes near breastbone OR 2–5 cm tumor, spread to lymph nodes or lymph nodes near breastbone OR 5+ cm, spread to lymph nodes or lymph nodes near breastbone IIIB: any size tumor AND metastasis to chest wall and/or breast skin OR metastasis to lymph nodes or lymph nodes near breastbone IIIC: no tumor or tumor may be any size, may have metastasized to chest wall and/or breast skin. Has spread to lymph nodes near breastbone or lymph nodes. Further classified into operable and inoperable stage IIIC depending on location and number of positive lymph nodes
Stage IV	Metastasis to distant organs: usually bones, lungs, liver, or brain

TREATMENT OPTIONS ARE UNIQUE FOR EACH INDIVIDUAL

When a cancer diagnosis is made, there are various options for treatment. Surgery is generally the most effective and radiation and/or chemotherapy are also frequently prescribed.

Surgery is commonly used as one of the forms of treatment

A lumpectomy represents the least invasive of all surgical treatments for breast cancer. In this procedure, only the tumor and some surrounding tissue are removed. In order to avoid recurrence, a clear, cancer-free margin of normal breast tissue must be established. Lumpectomies are almost always followed by a regimen of radiation therapy and sometimes chemotherapy. A lumpectomy, however, may not be an option for all patients with breast cancer; it is usually done for DCIS and smaller invasive tumors. Therefore, this procedure should be discussed with a health-care provider to determine if it is a viable choice. A mastectomy[8.6] involves the surgical removal of the entire breast or

8.6 There are three kinds of mastectomies: Total – removal of the breast only; Modified Radical – removal of the breast and some underarm lymph nodes; and Radical – removal of breast, lymph nodes, and muscle underneath the breast.

both breasts. Sometimes the lymph nodes are also removed to determine if the cancer has metastasized.

The Halsted radical mastectomy was once the procedure of choice for breast cancer patients but is rarely used today due to disfigurement and side effects such as lymphedema (swelling).[8.7] A variation of the radical mastectomy allows for breast reconstruction and less swelling. Researchers have concluded that there are no significant differences in survival rates between women who opt for a lumpectomy (breast conserving surgery) versus those treated with a mastectomy. Since many factors enter into the decision of the choice of surgery, it is one that is made together by the patient and his/her surgeon.

[8.7] William Smith Halsted introduced the radical mastectomy in 1882 at Johns Hopkins University when he performed a procedure that removed the breast, lymph nodes, and chest muscles. This traumatic and painful breast cancer treatment was the preferred surgical procedure for breast cancer until the 1960s.

Some women at very high risk for breast cancer (i.e., 36–85% for those with *BRCA1* or *BRCA2* mutations) may consider the option of a prophylactic mastectomy – the removal of the breast tissue *prior* to the development of cancer. This controversial procedure is reserved for those who have received appropriate genetic and psychological counseling. Recent studies have indicated that those who have had a prophylactic mastectomy have a reduced risk of developing breast cancer. There is still a very small possibility for a woman to develop breast cancer following prophylactic surgery because it is not possible to surgically remove all of the breast tissue.

Hormone therapy is a major treatment for some breast cancers

The role of estrogen in the growth and differentiation of the mammary glands was discussed earlier in this chapter. The epithelial cells of the mammary glands contain estrogen receptors (ER), members of the nuclear receptor family, which trigger the transcriptional changes that cause the cells to proliferate upon estrogen binding. ERs are expressed in 60–80% of breast tumors, which are termed estrogen positive (ER+) tumors.

The determination of estrogen receptor status at the time of biopsy and/or surgery is critical because it is linked with prognosis and treatment choice. Patients with estrogen positive tumors have longer remissions and overall survival rates. Treatment of estrogen positive tumors is based on the premise that if the binding of estrogen to the receptor is blocked or the production of estrogen itself is inhibited, the tumor's growth would be controlled. Just because tumor cells produce ERs, however, does not necessarily mean that they will respond to estrogen. Only 70% of breast tumors that are estrogen positive respond to estrogen, and are therefore hormone-responsive. Interestingly, 5–15% of estrogen *negative* cells are also estrogen-responsive. The most commonly used drugs to control estrogen-dependent tumors are presented in the next section of this chapter.

The Her-2/*neu* proto-oncogene, which encodes for a transmembrane tyrosine kinase receptor, is a member of the epidermal growth factor receptor family.[8.8] An alteration in the DNA sequence that duplicates, rather than mutates, the Her-2/*neu* gene results in its transformation from a proto-oncogene to an oncogene. This increase in the number of copies of the Her-2/*neu* gene results in an overexpression of the encoded protein. The amplification of the Her-2/*neu* gene, represented as Her-2/*neu+*, is the most common oncogene amplification found in breast tumors. The overexpression of the Her-2/*neu* receptor, which occurs in approximately 1 out of every 5 breast tumors, results in an abnormally strong proliferation response to external growth signals. As a consequence, breast tumors with Her-2/*neu+* status grow at a characteristically rapid rate and are usually large and at an advanced stage when diagnosed, which is typically at a young age. These tumors are also often nonhormone-responsive, adding to the aggressiveness of, and poor prognosis associated with, the Her-2/*neu+* tumors. The method of choice for treatment of Her-2/*neu+* tumors is presented in a later section of this chapter.

> [8.8] In Chapter 4 it was mentioned that a kinase modifies proteins by phosphorylation – the addition of a phosphate group (PO_4^{2-}). The Her-2/*neu* protein is a tyrosine kinase, which means that the phosphate group is added to the side chain of the amino acid tyrosine in the target protein during phosphorylation.

The fact that approximately 70% of breast cancers are estrogen sensitive provides another option for breast cancer treatment. Drugs designed to treat estrogen positive breast tumors function in one of two ways: (a) they bind to the ER to effectively inhibit the binding of estrogen and thereby maintain the inactive state of the receptor; or (b) they inhibit the synthesis of estrogen. The choice of drug used to combat breast cancer is made subsequent to the determination of the tumor's estrogen receptor and Her-2/*neu* status and whether the woman is pre- or post-menopausal.

A drug used in the treatment of Her-2/*neu+* breast cancers is Herceptin (trastuzumab), a monoclonal antibody that binds specifically to the Her-2/*neu* receptor. Herceptin binds to the receptor in a way that effectively prevents the epidermal growth factor from binding. Even though the cancer cells are overexpressing the receptor for the growth factor, the signaling molecules are not able to reach their target. This markedly reduces the growth signal the malignant cells had been receiving prior to the presence of Herceptin. The objective with this form of treatment is to slow the rate of cell division by minimizing one of the factors that promotes the progression through the cell cycle.

Drugs that act through the estrogen receptor

Tamoxifen (Nolvadex), developed in 1967 as a fertility drug, was approved by the Food and Drug Administration (FDA) in the 1970s to treat breast cancer. The drug, a nonsteroidal anti-estrogen, competes with estrogen for binding to the estrogen receptor (ER). When tamoxifen, rather than estrogen, is bound to the receptor, activation does not occur; the changes in gene expression in epithelial cells of the mammary glands that are necessary to undergo proliferation will not occur, and the growth of the breast cancer cells is slowed. Tamoxifen was the most commonly prescribed drug for the treatment of breast cancer for many years but its side effects, such as hot flashes and stomach upset, rendered it hard to tolerate by all who used it. Scientists soon realized, however, that women who took it had fewer recurrences of breast cancer and a lower chance of developing it in the other breast. Consequently, in 1998 tamoxifen became the first drug approved by the FDA as a *chemopreventative* drug for women at high risk of developing breast cancer.

Although now prescribed routinely, tamoxifen has potential serious side effects such as possible blood clots, strokes, and an increased risk of endometrial cancer. Other side effects, which may be temporary, include vaginal discharge and depression. A newer drug, raloxifene (Evista™), has been prescribed to prevent and treat osteoporosis in postmenopausal women. It binds to and inactivates the ER in a tissue-specific manner similar to tamoxifen and is now also in use for those at high risk for developing breast cancer.

An alternative means of treating estrogen-sensitive breast cancers is to inhibit the synthesis of estrogen. Members of a newer class of drugs called aromatase inhibitors bind to and inhibit the enzyme aromatase, which catalyzes the conversion of androgens to estrogens in peripheral tissues (Chapter 7). Arimidex™ (anastrozole) and Femara™ (letrozole) are nonsteroidal compounds which bind reversibly to the active site of aromatase so as to compete with estrogen. Exemestane, a steroidal compound, also competes with estrogen for binding to the active site of aromatase, but does so irreversibly (i.e., a suicide inactivator) and permanently inhibits the enzyme from working. When any of these drugs are taken by postmenopausal women, whose only source of estrogen is from the peripheral tissues, estrogen levels are nearly depleted after 2–4 days. Due to positive results obtained from randomized trials, aromatase inhibitors have become the drugs of choice for the treatment of postmenopausal women with estrogen-sensitive metastatic breast tumors.

Chemotherapy is an important tool for breast cancer patients

Once diagnosed with breast cancer, the patient will often, but not always, be treated with some form of chemotherapy. Local therapy, such as surgery or radiation, will treat the tumor but not affect the rest of the body. Systemic therapy is delivered through the bloodstream either by mouth or injections to target cancer cells located throughout the body. Examples of systemic therapy are chemotherapy, hormone therapy, or immunotherapy. Some patients will receive systemic therapy to shrink the tumor before surgery; this is called neoadjuvant therapy. Breast cancer patients may be given systemic therapy after surgery even when there is no sign of cancer; this is known as adjuvant therapy. The goal of adjuvant therapy is to target cancer cells that may have metastasized and are undetectable.

A variety of chemotherapeutic drugs have been developed

A class of drugs called taxanes consists of mitotic inhibitors that interfere with microtubules, cytoskeletal protein filaments that provide structural support and are involved in the transport of materials around the cell (Chapter 2). During cell division, microtubules are rearranged to form the spindle apparatus involved in the separation of the replicated chromosomes to the daughter cells. The interference of the functionality of the microtubules by taxanes results in the improper segregation of the chromosomes to the daughter cells and leads to cell death.

The drug paclitaxel, commonly called Taxol, is a taxane that was originally isolated from bark of the Pacific yew tree. The extraction and purification of Taxol from the bark is a difficult and expensive process that yields only small quantities. Also, the Pacific yew tree is found in forests that are home to the endangered spotted owl, which limits the ability to acquire the starting material. A synthetic version of Taxol has

Box 8.4

Making Strides Against Breast Cancer

Since 1993, Making Strides Against Breast Cancer has been the American Cancer Society's premier event to raise awareness and money towards fighting breast cancer. In that time, nearly 3.5 million walkers have raised more than $230 million through this walkathon. Making Strides is a noncompetitive 3–5 mile walk supporting the American Cancer Society's unique mission to fight cancer on four fronts: research, education, advocacy, and patient services. It is a chance to honor breast cancer survivors as well as to educate others about prevention and early detection. Involvement in Making Strides helps ensure that progress against breast cancer continues and thus more lives are saved. More than 100 such events take place throughout the nation annually, with the majority of these events taking place during Breast Cancer Awareness month in October.

now been synthesized from a renewable resource – the needles and twigs of the Himalayan yew tree. Breast cancer patients may be treated with Taxol if there is a recurrence of cancer within 6 months of adjuvant chemotherapy or if the cancer has spread to the lymph nodes or has metastasized to other areas of the body.

MUCH HAS BEEN ACCOMPLISHED, MORE NEEDS TO BE DONE

Most breast cancer cases are diagnosed at an early stage. Furthermore, more than half of women diagnosed with breast cancer are diagnosed at a localized stage, where survival rates reach almost 98%. The overall 5-year survival rate is now 89%, the 10-year relative survival rate 81%, and the 15-year survival rate 73% according to the American Cancer Society's 2007–2008 report on breast cancer. The overall death rate from breast cancer has decreased annually since the 1990s and there has also been a decline in breast cancer incidence since 2000.

Although breast cancer is still the number one disease feared by many women, it is one in which much research has been successfully performed and is still ongoing. Thanks to early testing and diagnosis, women and men can make decisions based on scientific facts as well as family history. This has led to a new term – previvors – those who do not have breast cancer but are at a higher risk for its development based on results from genetic testing or the presence of abnormal microscopic cells in an examined breast. This is definitely an improvement from the past, when those with breast cancer were given one option – a radical mastectomy – with no choice of treatment decisions. Breast cancer research has paved the way for other cancers in the field of research and therapy. Hopefully, lessons learned will benefit all cancer patients.

EXPAND YOUR KNOWLEDGE

1 Explain why inherited mutations are associated with tumor development in both breasts whereas sporadic mutations are associated with tumor development in only one breast.

2 The pink ribbon is associated with breast cancer. Why, when and how did this originate?

3 How do the numbers for incidence and deaths from breast cancer of US patients compare to those of other countries? What factors would affect these numbers?

4 How does the US breast cancer research budget compare to that for the number 1 cancer that affects men? Is it less or more and why?

5 Women in North America have the highest rate of breast cancer in the world. Give some reasons for this fact.

ADDITIONAL READINGS

Dr Susan Love Research Foundation. Susan Love's website: http://www.susanlovemd.org

Geiger, A. M., Yu, O., Herrinton, L. J., Barlow, W. E., Harris, E. L., Rolnick, S., Barton, M. B., Elmore, J. G., and Fletcher, S. W. (2005) A population-based study of bilateral prophylactic mastectomy efficacy in women at elevated risk for breast cancer in community practices. *Archives of Internal Medicine*, **165**: 516–520.

Love, S. and Lindsey, K. (2005) *Dr. Susan Love's Breast Book*, 4th edn. New York, NY: DaCapo Press, Perseus Books Group.

Marieb, E. N. & Hoehn, K. (2007) *Human Anatomy and Physiology*, 7th edn. Upper Saddle River, NJ: Prentice Hall.

Martini, R. & Bartholomew, E. (2007) *Essentials of Anatomy & Physiology*, 4th edn. Upper Saddle River, NJ: Prentice Hall.

Rossouw, J. E., Anderson, G. L., Prentice, R. L., LaCroix, A. Z., Kooperberg, C., Stefanick, M. L., Jackson, R. D., Beresford, S. A., Howard, B. V., Johnson, K. C., Kotchen, J. M., and Ockene, J. (2002) Risks and benefits of estrogen plus progestin in healthy postmenopausal women: principal results from the Women's Health Initiative randomized controlled trial. *Journal of the American Medical Association*, **288**: 321–333.

Susan Komen for the Cure website: http://ww5.komen.org/

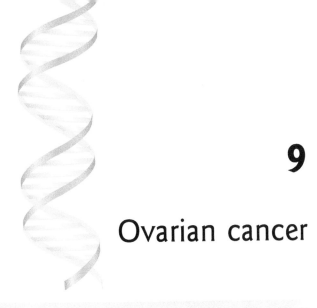

9

Ovarian cancer

Honey, there's nothing wrong with your reproductive organs.
Paula Williams' gynecologist, only days before she was diagnosed
with ovarian cancer

CHAPTER CONTENTS

Cancer: Basic Science and Clinical Aspects, 1st edition. By C. A. Almeida and S. A. Barry. Published 2010 by Blackwell Publishing, ISBN 978-1-4051-5606-6.

Ovarian cancer may not be as common as some other malignancies, yet it causes fear in many women because of its high mortality rate, the nonspecificity of symptoms, and the lack of early diagnostic tests. It is highly treatable if detected early, but the presence of vague and/or believed to be unrelated symptoms most often results in definitive diagnoses occurring at one of the later stages of the disease, when there is a lower survival rate. It is the deadliest of the several cancers of the female reproductive system so it is understandable why women are cencerned about this disease. The good news is that recent findings related to symptoms should bring about a dramatic change to the traditional thinking and/or fears.

OVARIAN CANCER STATISTICS

It is estimated that approximately 22,000 American women were diagnosed with ovarian cancer in 2008 and that about 15,500 died from it that same year.

Unfortunately, only one in five women with advanced ovarian cancer will have survived 5 years after diagnosis. Incidence and mortality rates are highest in Caucasian women compared to other racial and ethnic groups, and the rates are highest in industrialized countries, with the exception of Japan, for reasons not yet established. If it is detected before it has spread, there is a 90% survival rate, but early stage diagnosis only occurs in about 25% of the cases. Women below the age of 65 have a better survival rate than older women, largely because their general health overall is better.

According to a 2003 study by the CDC and the North American Association of Central Cancer Registries, there may be some cause for optimism. The researchers found that there has been a 1.4% decline in ovarian cancer incidence each year for all races combined, as well as particularly significant annual declines in Caucasian and Hispanic women. Studies have also demonstrated that the mortality rate from the disease has decreased by about 9% in the past few years. There are no clear cut answers to these changes, but one would hope that the latest treatments are contributing to this decline.

STRUCTURE AND FUNCTION OF OVARIES

The ovaries are a pair of almond-shaped glands about 4 cm (1½ inches) long located on either side of the uterus in the lower abdomen, and are a crucial part of the female reproductive system (Figure 9.1). The uterus, also known as the womb, is a pear-shaped hollow organ, separated from the vagina by the cervix, and is the site of fetal development.

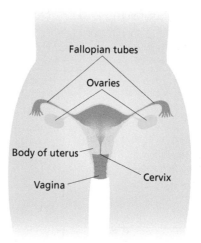

Figure 9.1 The female reproductive anatomy. The ovaries, fallopian tubes, uterus, cervix, and vagina comprise the female reproductive system.

The ovaries are the main source of female sex hormones (estrogen and progesterone) which influence breast development, body shape, and the presence of body hair in women. Approximately 2 weeks before a woman's monthly period, one of the two ovaries releases a single egg (ovulation). The egg travels through the fallopian tubes, which connect each ovary to the uterus. Sperm migrate through the cervix, up the uterus and into the fallopian tubes, where a single sperm can fertilize an egg, if present. In preparation for the potential of receiving a fertilized egg, the endometrial lining of the uterus becomes thickened and supplied with more blood vessels. A fertilized egg will migrate to the uterus and become embedded in the nourished endometrium, where it develops over the next 9 months into a fully developed fetus. If an egg is not fertilized, it does not implant in the endometrium but instead travels to the uterus and is expelled along with the endometrial lining in the process called menses (menstruation).

THERE ARE THREE TISSUE CATEGORIES OF OVARIAN CANCER

Tumors that develop in the ovaries are named according to one of three types of cells from which the cancer originates – germ, stromal, and epithelial cells. Germ cells produce eggs inside the ovary; about 5% of ovarian cancers are germ cell tumors, but most of these tumors are benign. Germ cell tumors, although rare and aggressive, are seen most often in young women or adolescent girls. They are generally curable by surgery and chemotherapy if found early enough. The stromal cells are located in between the germ cells and produce most of the female hormones. Ovarian stromal cell carcinoma also accounts for about 5% of the ovarian cancer tumors and is usually diagnosed in an early stage. It is treated by removing the ovary with the tumor.

The epithelial cells that cover the ovary are the source of most ovarian malignancies. The three subgroups of epithelial ovarian tumors are benign tumors that do not spread or cause serious illness. Tumors of low malignant potential (LMP tumors), also known as borderline tumors, are not clearly cancerous. Found in younger women, they grow slower, are less serious than most ovarian cancers, and occur in about 15% of all ovarian tumors. Even though they are not cancerous and have a high cure rate, they are still abnormal and are treated in the same fashion as ovarian cancer. The most common type, epithelial ovarian cancer (EOC), occurs in 65–90% of all cases and is usually diagnosed in women aged 40 years or older. It is significant to note that risk for all forms of the disease begins to increase at age 40 and peaks in the late seventies. Another rare type of cancer that resembles EOC in appearance but actually originates *outside* the ovaries in the peritoneum, which lines the pelvis or abdomen, is called primary peritoneal carcinoma. This appears and acts like ovarian cancer, mimicking many of the symptoms, and can occur in women who have had a hysterectomy (the removal of the reproductive organs) and who no longer have ovaries.

SYMPTOMS OF OVARIAN CANCER ARE VAGUE AND OFTEN MISSED

Much debate has ensued recently as to the symptoms of ovarian cancer. Experts in the field of cancer have often said that early ovarian cancer is asymptomatic or subtle, thus leading to the phrases "the disease that whispers" or the "silent cancer." Such symptoms are often gastrointestinal and may include abdominal or pelvic pain, swelling, bloating, pressure or discomfort, and/or a feeling of fullness even after a light meal; vague or mild gas, nausea or indigestion that is persistent; loss of appetite; lower backache or pain in the leg; unexplained change in bowel habits such as constipation or diarrhea and/or a sense that the bowel has not completely emptied; vaginal bleeding or prolonged, heavy menstrual bleeding; mid-cycle spotting or post-menopausal bleeding; ongoing unusual fatigue; frequent or urgent urination; shortness of breath; weight gain or loss with no known reason; or pain during intercourse. While all agree that the location of the ovaries deep within the abdominal cavity causes the symptoms to often times be vague and difficult to detect, many women who were later diagnosed felt that their doctors ignored their concerns when first presented. Many of these symptoms could and should trigger a visit to a gynecologist, internist, or gastroenterologist.

In 2007, the Gynecological Cancer Foundation (GCF) announced the first national consensus on ovarian cancer symptoms based on women who were ovarian cancer survivors. They believed that there were common symptoms that were not so "silent" in spite of what healthcare professionals had been saying for years. Researchers now agree that any of the following symptoms on an almost daily basis for more than a few weeks are significant: bloating, pelvic or abdominal pain, difficulty eating or feeling full quickly, and/or urinary symptoms (urgency or frequency).

Many of the preceding symptoms could occur in any woman at any time and it does not immediately suggest that she has cancer. These symptoms can easily be confused with perimenopause, menopause or common gastrointestinal illnesses.[9.1] If any of them *persist for more than 2–3 weeks*, however, they should be reported to a physician. Since the symptoms are so vague (a.k.a. "silent"), in the past, less

[9.1] Women with suspicious symptoms may be referred by their gynecologists to a *gynecologic oncologist*, a physician who is specially trained in treating cancers of the female reproductive system.

than 25% of ovarian cancers were found in the early stages. Advanced stages occur when the tumor grows and creates pressure on the bladder and rectum, leading to subsequent formation and build-up of abdominal fluid. This is often what brought most women to their physicians,

> **Name five other medical conditions that may have the same symptoms as ovarian cancer.**

whereas an awareness of earlier changes by both women and their physicians would have led to an increase in more timely diagnoses. One study found that 94% of women surveyed who were diagnosed with ovarian cancer had symptoms in the year prior to their diagnosis.

CERTAIN FACTORS HAVE BEEN ASSOCIATED WITH A HIGHER RISK

The exact causes of ovarian cancer have not been determined but some factors seem to be consistent among the women diagnosed. It is important to realize that benign ovarian cysts (fluid-filled sacs), which are especially common in young women and found less frequently after menopause, are not cancerous, nor are they associated with an increased risk of developing

---

Box 9.1

Gilda Radner's Tragic Story

Gilda Radner, who died in 1989 at the age of 43, was a wife, comedian, actress, and author of *It's Always Something*, her personal account of her ultimately unsuccessful struggle with ovarian cancer.

She jokingly called herself the "Queen of Neurosis" but actually had many symptoms of ovarian cancer for over a year prior to her diagnosis. She was seen by physicians regularly, yet her symptoms were not recognized for their true nature. It would have been highly beneficial for her and her providers to know that some family members had ovarian cancer previously, but even she was unaware of this history. In past generations, ovarian cancer might have been known as "stomach" cancer or mis-identified in some other way. This reinforces how important it is for both women and healthcare providers to be educated on the risk factors and symptoms of this and all cancers.

At the end of her book, she wrote: "I had wanted to wrap this book up in a neat little package. I wanted a perfect ending. Now I've learned, the hard way, that some poems don't rhyme, and some stories don't have a clear beginning, middle, and end. Life is about not knowing, having to change, taking the moment and making the best of it, without knowing what's going to happen next."

the disease. These cysts usually self-dissolve, but if they become excessively large will have to be removed surgically.

Increasing age and genetics are two significant factors

Aging increases the risk of developing cancer of the ovary, as is true with most other malignancies. Most cases occur in women over the age of 50, with an even higher risk after age 61 (the average age of diagnosis is 63), but it can and does affect younger women in their twenties and thirties as well. If a woman has a first-degree relative with ovarian cancer, this will put her at three times the risk of also developing the disease. Having two or more first-degree relatives with ovarian cancer sharply increases this possibility, more so than if a grandmother, aunt, or cousin has it.[9.2] Other factors that increase the risk are occurrences of breast or colon cancer in the family or a personal history of breast, ovarian, endometrial or colon cancer. Anyone with several cancers in their family, especially in those relatives who were diagnosed at a younger age, should discuss this trend with their physician because it is more likely that they could benefit from genetic testing and counseling.

The genetic predisposition of ovarian cancer is being studied at length. Not more than 5–10% of those diagnosed have a family history of the disease, but for those who do, it is vital information to have. Some doctors are now ordering the *BRCA1* or *BRCA2* (Chapter 8) test to assess the cancer risk in those patients and their families.[9.3]

Studies have shown that there are certain mutations of the *BRCA1* and *BRCA2* genes that are particularly common in some women of Eastern European Jewish (Ashkenazi) descent. A specialized blood test that will pinpoint these mutations is performed in conjunction with a meeting with a genetic counselor who can explain the significance of the results and their effects on future decisions.[9.4] Those women

> [9.2] The National Society of Genetic Counselors has excellent tips on creating a family medical tree or pedigree and it can be found at http://nsgc.org/consumer/familytree/index.cfm

> [9.3] Remember that while all of us are born with *BRCA1* and *BRCA2* genes, certain alterations in them can make a woman more susceptible to breast and ovarian cancer, but many with these mutations will never develop the diseases.

> [9.4] The Gilda Radner Familial Ovarian Cancer Registry at the Roswell Park Cancer Institute in New York State is an international registry of families that have two or more members with ovarian cancer, and its goal is to identify new genes for the disease. By collecting family histories, medical records and tissue samples from ovarian cancer patients, the registry hopes to acquire information that will lead to better methods for detecting or predicting ovarian cancer in order to help future generations of women at high risk and, in turn, all women. Research cannot be successfully undertaken without reliable data to guide it, and the Registry has the intent of providing that information.

Box 9.2

Johanna's Law

Johanna Silver Gordon died at the age of 58 from ovarian cancer after a 4-year battle with the disease (Figure 9.2). As happens with so many women, not knowing the common symptoms of ovarian cancer contributed to a delay in diagnosis. To reduce the frequency with which this occurs, her sister Sheryl Silver spearheaded legislation in her memory that is now known as *Johanna's Law*. Also known as the *Gynecological Cancer Education and Awareness Act*, it was signed by President Bush in early 2007. It funds programs that increase the awareness and knowledge of women and healthcare providers with respect to gynecologic cancers. This legislation seeks to narrow the life-threatening education gap that has led to so much suffering and so many deaths. Educating women on the symptoms and risk factors of gynecologic cancers will enable them to seek appropriate medical help more quickly than in the past. Adopting medical school curricula to increase the education to physicians in training will also lead to earlier diagnosis and ultimately contribute to a longer survival rate. The bill calls for a *National Public Awareness Campaign* through education strategies and outreach programs that would represent a significant step forward in the education of women about symptoms of this and other cancers. Possible means to do this may include public service announcements, billboard campaigns, making pamphlets available in doctor's offices, libraries, etc.

Figure 9.2 The tragic death of Johanna Silver Gordon led to the development and passage of the Gynecological Cancer Education and Awareness Act in 2007, also known as Johanna's Law.

who test positive for these mutations may choose to have a prophylactic oophorectomy (removal of healthy ovaries and fallopian tubes so that cancer cannot grow in them) or a prophylactic mastectomy (removal of the breasts). Obviously, either choice is a dramatic one with far reaching ramifications, which is why these options are chosen only after careful consideration and discussion.

Prolonged estrogen exposure is also a risk factor

An increased risk of estrogen exposure has been associated with the development of ovarian cancer. Oftentimes, the growth of cancer cells is stimulated by estrogen. Women who have never been pregnant, never taken birth control pills, or never had children (nulliparity) are more likely to develop this malignancy than women who have had children. Thus, the more children a woman bears, the lower her risk of developing ovarian cancer. There is a direct relationship between the frequency of ovulation and the risk of cancer, since having fewer pregnancies increases a woman's lifetime estrogen exposure (Chapter 8). Although it is not substantial enough to be worrisome to otherwise healthy childless women, it is yet another reason why awareness of symptoms is essential for all. Furthermore, some studies have demonstrated that those who have taken estrogen replacement therapy (ERT) for more than 10 years were at a slightly increased risk for developing ovarian cancer.

Other risk factors associated with the development of ovarian cancer

The once seemingly harmless use of talcum powder on the genital area has been associated with an increased risk. This is most likely attributable to the fact that talcum powder formerly contained particles of asbestos, a known carcinogen. Since its inclusion in the product was discontinued over 20 years ago, it is considered to be safe for current day use. Women who have had endometriosis (a painful condition in which the tissue of the uterine lining grows outside the uterus and may adhere to the ovaries) are at a higher risk of developing ovarian cancer, possibly due to the inflammation it causes.

It is critical to understand that most women with any of these risk factors, including family history, will never actually develop ovarian cancer, but they do have an increased statistical chance of it occurring, however slight. Similarly, most women with ovarian cancer may have had no prior risk factors for the disease. Women need to know the symptoms and risk factors, and feel comfortable enough to meet with and question their healthcare providers when they are concerned.

There are ways to lower the risk of developing the disease

Women who have taken the birth control pill for a number of years (3–5 straight) have a decreased risk of developing ovarian cancer, since oral contraceptives interfere with ovulation. This is only recommended for those under the age of 50 as there are risks as well as benefits for users. Some studies have demonstrated that breast feeding for more than a year might also lower the risk. Tubal ligation ("tying" the fallopian tubes) or having a hysterectomy (removing the uterus) or oophorectomy has been utilized to decrease the risk but they are obviously drastic measures not recommended for all women.

DIAGNOSTIC TOOLS ARE AVAILABLE BUT NOT ALWAYS USED OR RECOMMENDED

There is no accurate screening test for ovarian cancer. However, most women who report symptoms to their physicians should have pelvic and rectovaginal examinations, transvaginal ultrasounds, and possibly the blood test for the CA-125 antigen, a protein marker for ovarian cancer.

During a pelvic examination, the healthcare professional will feel for lumps or changes in the shape of the vagina, cervix, uterus, fallopian tubes, and ovaries. A speculum will be inserted into the vagina to look at the cervix and take samples for the Pap test. As a woman ages and undergoes menopause, her ovaries start to shrink to about half of their original size and gravitate towards the rear of the uterus. They become so small that they are often undetectable during a routine pelvic exam, so it is recommended that women over the age of 40 have a rectovaginal (bimanual pelvic) examination annually (Figure 9.3). During this procedure, the practitioner inserts a lubricated, gloved finger into the vagina and then simultaneously places a second finger into the rectum. By pressing on the lower abdomen with the other hand, she is able to feel the location and size of the pelvic organs. A skilled practitioner can do this procedure with minimal discomfort and it should also be performed on women of any age who present with any of the previously listed symptoms.

Since ovarian tumors and normal ovarian tissue reflect sound waves differently, an ultrasound not only enables a tumor to be detected, but it can help in the determination of whether the mass is solid or a fluid-filled cyst (Chapter 6). The procedure used to examine the vagina, uterus, fallopian tubes, ovaries, and bladder is called a transvaginal ultrasound (TVUS) (Figure 9.4). A probe is inserted into the vagina and sound waves are reflected off of the organs in the pelvic area, thereby creating echoes. A computer then translates the returned sound waves into a black and white, 2-D picture known as a sonogram.

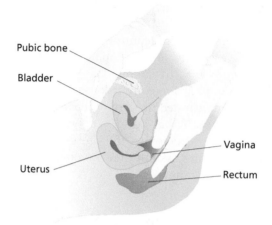

Figure 9.3 The rectovaginal exam. The practitioner inserts fingers into the vagina and rectum simultaneously to feel for any abnormalities.

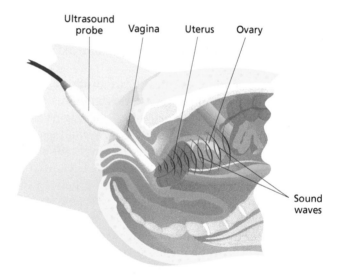

Figure 9.4 The transvaginal ultrasound. A woman's reproductive organs can be examined by ultrasound. Source: https:/stamhealth.org/services/healthlibrary/images

The CA-125 blood test is currently used for women at high risk or with questionable symptoms. A CA-125 level less than 35 U/ml (microns per milliliter) is usually considered to be normal. This test, however, is *not* recommended for general screening purposes because it has a tendency to generate too many false positives (Chapter 6). This inevitably leads to unnecessary testing and unfounded anxiety. Factors which cause the CA-125 level to be elevated in the absence of cancer include: endometriosis, pelvic inflammatory disease (PID), uterine fibroids, benign ovarian tumors, pancreatitis, kidney, liver or cardiac disease, early pregnancy, or menstruation. Furthermore, the test is unreliable for detecting early ovarian cancer because about half the women who have early stage ovarian cancer do not produce enough CA-125 to cause a positive result.[9.5] It also does not detect all types of ovarian cancer. The test is more useful in monitoring for reoccurrence after a woman has been diagnosed and treated for the disease.

> [9.5] The Prostate, Lung, Colorectal and Ovarian Cancer Screening Trial (PLCO Trial) (Chapter 6) is a large-scale study to determine if particular screening tests ultimately lead to a reduction in the number of deaths from these cancers. The women in the trial will have a physical examination of the ovaries, a transvaginal ultrasound, and the CA-125 test performed on an annual basis.

ADDITIONAL PROCEDURES ARE NECESSARY TO CONFIRM SUSPICIOUS RESULTS OR IF THERE IS METASTASIS

As with most cancers, imaging studies can confirm that a pelvic mass is present, but cannot determine whether it is cancerous. CT scans may show the size of the tumor and whether it is invading other organs, determine if lymph nodes are enlarged and the condition of the kidneys, bladder, or liver, which may be affected. Often, this procedure requires an injection of a dye or contrast agent to enable a more accurate picture. Positron emission tomography combined with CT scans (PET/CT) is considered to be a good tool to determine if the cancer has spread (Chapter 6). Chest X-rays may detect cancer that has metastasized to the lungs. A barium enema X-ray requires the patient to take laxatives to clean out her colon (Chapter 15). The next day, barium sulfate is inserted into the rectum and colon. This fluid prevents X-rays from penetrating, allowing the outline of the soft tissues of the colon and rectum to be observed more clearly for evidence of metastasis to these locations. A colonoscopy can be used to rule out colorectal cancer, which is significant in that it shares some of the same symptoms as ovarian cancer (Chapter 15). MRIs are not often used to diagnose ovarian cancer, but may be utilized to determine if it has spread to other organs.

If the woman has a swollen abdomen, the physician may perform a paracentesis, which is an aspiration of abdominal fluid (Figure 9.5). In this procedure, the doctor will inject a mild anesthetic to numb the area and then insert a needle to extract the fluid (ascites). The cells within the fluid are then microscopically examined to determine if they are cancerous. A distended abdomen may indicate that the cancer has spread, or, it could be caused by other conditions that are not cancer related.

A doctor may utilize laparoscopy to visually examine the ovaries and the surrounding area (Chapter 6). After administering anesthesia, this

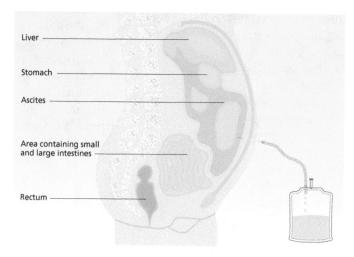

Liver

Stomach

Ascites

Area containing small
and large intestines

Rectum

Figure 9.5 Paracentesis. This procedure is used to drain excessive fluid from the abdomen to examine for the presence of cancerous cells. Source: www.cancerhelp.org.uk

procedure, less invasive than surgery, is performed by inserting a long tube with a fiberoptic camera into a ½ inch (1.25 cm) incision made in the abdomen. If the physician observes any suspicious growths, a biopsy will be performed to examine for cancerous cells. A CT guided needle biopsy, either fine or core, will remove the tissue sample for examination by a pathologist. If necessary, ovaries can also be removed by laparoscopy, thereby avoiding a major surgical incision and procedure. Afterwards, a few stitches are used to close the incision. One side effect of the procedure is some bloating or discomfort for a couple of days, which will gradually pass.

THE FIGO SYSTEM MAY BE USED TO STAGE OVARIAN CANCER

The Federation Internationale de Gynécologie et d'Obstétrique (International Federation of Gynaecological Oncologists) is known as the FIGO staging system (Table 9.1) and is one of the most commonly used worldwide for ovarian cancer. There are other systems in use but all are very similar in scope and design.

TREATMENT OPTIONS FOR OVARIAN CANCER

Treatment of ovarian cancer begins with surgery

Unlike many other cancers, if the decision is made to operate, a biopsy can be performed during the surgery to confirm the disease. Traditional

Table 9.1 Staging of ovarian cancer

Stage 1	The cancer is present *within* the ovary (or ovaries)
Stage 1a	The cancer is contained within one ovary
Stage 1b	The cancer is contained within both ovaries
Stage 1c	Includes 1a or 1b AND some cancer on the surface of one or both ovaries OR cancer cells have been identified in abdominal fluid removed during surgery OR the ovary ruptures (bursts) before or during surgery
Stage 2	The cancer is in one or both ovaries and has metastasized to the other organs *within* the pelvis, such as the bladder, rectum, or uterus
Stage 2a	The cancer has spread into the fallopian tubes or the uterus (womb)
Stage 2b	The cancer has grown into other tissues in the pelvis such as the bladder or rectum
Stage 2c	Includes 2a or 2b AND some cancer on the surface of one or both ovaries OR cancer cells identified in abdominal fluid removed during surgery OR the ovary ruptures (bursts) before or during surgery
Stage 3	The cancer is in one or both ovaries and has spread outside the pelvis to the lining of the abdomen or to the lymph nodes in the upper abdomen, groin or behind the womb
Stage 3a	Cancer is identified microscopically in tissue taken from the lining of the abdomen (peritoneum)
Stage 3b	Tumors <2 cm (0.78 in) appear on the lining of the abdomen
Stage 3c	Tumors >2 cm on the lining of the abdomen; or the lymph nodes in the upper abdomen, groin, or behind the womb are cancerous
Stage 4	The cancer has spread into distant organs such as the liver or lungs. If the cancer is only on the *surface* of the liver but not present within the liver, it is regarded as Stage 3 and not 4

surgical options may then include the removal of the cancerous tissue, an oophorectomy (removal of the affected ovary), or a hysterectomy. The choice of surgery will depend on the health of the woman and the extent of the metastasis. Women of childbearing age and intent obviously would want to keep at least one of their ovaries and uterus intact. The decision to do so would depend on the stage of the cancer and the kind of tumor found. Most women who have surgery will have their uterus and/or cervix (hysterectomy) and both ovaries and fallopian tubes (salpingo-oophorectomy) removed. In some cases of cancer, the upper part of the vagina and supporting tissues will also be excised. The omentum, a layer of fatty tissue that covers the abdominal contents, is typically removed to prevent the cancer from spreading there. Removal of the ovaries and/or the uterus renders the woman incapable of becoming pregnant and she will enter a state of medically induced menopause, if she has not naturally done so already.

During surgery, all visible cancerous tissue will be excised and samples of lymph nodes, tissues and fluids will be removed from different parts of the pelvis and abdomen. These tissues will be sent to the pathology laboratory for diagnosis and staging and examined for evidence of metastasis. If the cancer of the ovary is judged to be advanced, the surgeon may perform debulking or cytoreduction surgery, a procedure in which as much of the tumor as possible is excised, because removing the entire tumor would cause too much damage to an organ or surrounding areas. This will reduce the amount of cancer that is left to be treated later with radiation therapy or chemotherapy. Although ovarian cancer can spread to the brain, lungs, breast and lymph nodes, it usually stays within the abdomen and affects organs such as the intestines, liver and colon because of their close proximity.

Chemotherapy rather than radiation is used very often following surgery

The dosage and frequency of chemotherapy varies from once a day, to once a week, to once every few weeks, etc. depending on the type and stage of the cancer. Part of the treatment regimen will include drugs intended to reduce the unpleasant side effects of the chemotherapeutics. Usually more than one drug is used in order to minimize the chance of resistance, thus the term combination therapy. The most commonly used drugs for ovarian cancer are cisplatin or carboplatin, both platinum compounds, and a taxane such as paclitaxel or docetaxel (Chapter 8). All are given orally or intravenously in 3- or 4-week cycles. Although many patients have a complete clinical remission after primary therapy, most have some degree of relapse. If there is recurrence, the woman will undergo additional cycles of chemotherapy.

Many drugs used are given intravenously (injection into a vein, also known as IV) or given through a temporary catheter, a thin tube placed into a large vein. Often, ovarian cancer drugs are administered into a catheter placed through the skin and into the abdomen, allowing the drugs to be injected directly into the area of the cancer. This form of intraperitoneal (IP) therapy allows for most of the drugs to remain in the abdomen. In June 2006, The National Cancer Institute issued an announcement for a preferred method of treatment, one that utilizes both the IV and IP methods after surgery, for women with advanced ovarian cancer. According to a large clinical study, the combined forms of chemotherapy extended overall survival for these women by about a year. Certain chemotherapy drugs, such as cisplatin and paclitaxel, worked better when utilizing the IP method since they could be given more frequently and in higher concentrations. The IP method does have more side effects and is more complicated than IV, but it appears to be more effective in killing cancer cells in the peritoneal cavity, where ovarian cancer is likely to spread or first reoccur.

Although radiation therapy is not used routinely as a primary treatment for ovarian cancer, it may be used to kill cancerous cells that remain in the pelvic area or that have spread to another organ in the body, or as a localized treatment to improve symptoms. It may be given by external beam or internally using implants. Internal radiation therapy is given either by delivering a small amount of radioactive material directly into the tumor or by intraperitoneal injection of radioactive liquid through a catheter.

> **Why is radiation not a routine treatment for ovarian cancer patients?**

STEPS ARE TAKEN TO PREVENT RECURRENCE BUT DO NOT ALWAYS WORK WHEN THE CANCER IS ADVANCED

After the completion of primary treatment, second-look or look-back surgery by laparoscopy or laparotomy (a large incision made into the abdomen) was commonly performed in the past to allow the surgeon to see the effects of the chemotherapy directly inside the abdomen in order to determine the success or failure of the treatment. This practice, however, is not as common as it used to be because it does not appear to contribute to an increased survival rate. Consolidation therapies, which are chemotherapy treatments given after the initial therapy to help prevent recurrences, may be given. Currently, second-look surgeries and consolidation therapies are being studied to determine their overall effectiveness. Second-line (those that are given if the most common ones do not work) chemotherapy drugs are also available and patients with any stage of ovarian cancer are appropriate candidates for clinical trials.

MUCH NEEDS TO BE DONE IN THE FUTURE

Future hope for success in ovarian cancer prevention and treatment begins, as always, with an increased awareness of this disease's risk factors and symptoms. A recent poll done in the US showed that while more than one-half of women surveyed felt that they were at risk for developing a gynecologic cancer, almost the same number could not name any of the symptoms and/or were unaware of how to reduce their risks for developing these cancers. In contrast to breast cancer, which becomes obvious with a lump in the breast, women must become more aware of the various, and oftentimes vague, symptoms of ovarian cancer that are often attributed to something else.

Genetic counseling is currently being offered to select families in the hopes of reducing the numbers of cases both for these families and for future generations. At least 350 different mutations of the *BRCA1* gene and more than 100 mutations of the *BRCA2* gene have been identified. A woman with a positive test result will need to work closely with her healthcare provider to decide how she can reduce factors that may contribute to her risk of developing the disease.

Much attention is being given to proteomics, a method of analyzing proteins in the blood to detect a characteristic "signature" that may signal ovarian cancer at the earliest stages. Researchers hope to use blood tests to identify specific patterns as early biomarkers of the disease. They have compared blood samples from persons with and without ovarian cancer and will continue to research this technique.

The drug Combretastatin A4P (CA4P) has been evaluated in a Phase II clinical trial in combination with carboplatin and paclitaxel for the treatment of ovarian cancers that are resistant to one of the platinum drugs. Previously, Phase Ib clinical trial results showed a 67% response rate to this combination of treatment among women with all types of advanced ovarian cancer who failed to respond to previous cancer therapy. Manufactured by Oxigene Inc., the experimental drug has been granted orphan drug status by the US Food and Drug Administration, which will grant the company 7 years of exclusivity if it eventually wins approval. The drug works by selectively disrupting abnormal blood vessels associated with solid tumor growth. Clinical trials are ongoing in the UK.

Many other agents that work in novel ways are offered to selected patients through clinical trials. For example, there are agents, such as EMD 72000, that block growth receptors on cancer cells, causing the cells to slow their growth and ultimately die. Another is CTLA4, an antibody that enables the immune system's T cells to battle cancer cells more effectively. Additionally, HMFG-1, an antibody that will deliver a radioactive material directly into cancer cells lingering in the abdomen following surgery, holds promise.

Vitamins A and D as well as COX-2 inhibitors (anti-inflammatory drugs) are being studied for their possibility of reducing a woman's risk. Cancer vaccines are also currently being tested but are far from ready for general use. Unlike a typical vaccine to prevent measles or the flu, these vaccines are not designed to prevent cancer, but rather to "teach" the immune system how to recognize cancer cells and attack them before they spread.

Advancement in the treatment for ovarian cancer has improved dramatically over the decades. In the 1960s, there was only one drug available while today there are about 30 drugs in actual use with many more in clinical trials. Women are strongly encouraged to investigate these trials and to educate themselves and their families about the symptoms and risk factors. More "noise" is needed to eradicate this "silent" cancer.

EXPAND YOUR KNOWLEDGE

1 Symptoms of ovarian cancer can resemble other conditions. Give some specific examples of what women or their doctors may think is wrong with them.

2 Since there is no other test, would it make sense for the CA-125 test to be used as a screening test for ovarian cancer for all women once they get to a certain age? Why or why not?

3 Explore the latest accomplishments of Johanna's Law.

4 Why would women who have taken ERT be at an increased risk of developing ovarian cancer?

5 Investigate the field of proteomics. How is the blood analysis performed and for what other cancers has it been studied?

ADDITIONAL READINGS

Armstrong, D. K., Bundy, B., Wenzel, L., Huang. H. Q., Baergen, R., Lele, S., Copeland, L. J., Walker, J. L., and Burger, R. A. (2006) Intraperitoneal cisplatin and paclitaxel in ovarian cancer. *New England Journal of Medicine*, **354**: 34–43.

The Gilda Radner Familial Ovarian Cancer Registry, Roswell Park Cancer Institute, Buffalo, New York. www.ovariancancer.com

Goff, B. A., Mandel, L. S., Melancon, C. H., and Muntz, H. G. (2004) Frequency of symptoms of ovarian cancer in women presenting to primary care clinics. *Journal of the American Medical Association*, **291**: 2702–279.

Johanna's Law's website: www.johannaslaw.org/

Marieb, E. N. and Hoehn, K. (2007) *Human Anatomy and Physiology*, 7th edn. Upper Saddle River, NJ: Prentice Hall.

Martini, R. and Bartholomew, E. (2007) *Essentials of Anatomy & Physiology*, 4th edn. Upper Saddle River, NJ: Prentice Hall.

Ovarian Cancer National Alliance's website: www.ovariancancer.org

Radner, G. (1989) *It's Always Something*. London: Harper Collins.

Ryberg, M. ed. (2003) *Warnings, Signs and Whispers: True Stories that Could Save Lives*. 1st Books.

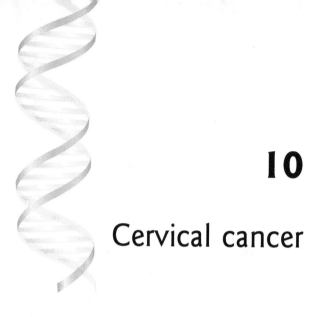

10

Cervical cancer

> You'd think that if they invented a vaccine that protects you from cancer, everyone would want it.
>
> Christine Gorman, "A Shot Against Cancer" *Time Magazine*, June 11, 2006

Cancer: Basic Science and Clinical Aspects, 1st edition. By C. A. Almeida and S. A. Barry. Published 2010 by Blackwell Publishing, ISBN 978-1-4051-5606-6.

Cervical cancer is one of the most common cancers to affect a female's reproductive organs. In the past, it had a high fatality rate because there was no means of early diagnosis or successful treatments. That has changed radically thanks to excellent basic testing that enables an early diagnosis as well as a vaccine to prevent most infections. The change in incidence and mortality has not been worldwide, however, because many countries still do not have access to these services.

CERVICAL CANCER STATISTICS

The American Cancer Society estimated that there were approximately 11,000 new cases of cervical cancer and 4,000 deaths from the disease in the US in 2008. The numbers are disproportionally higher outside this country. Almost half a million women develop cervical cancer and more than 288,000 die of the disease annually worldwide, making it the second most common cancer in women and the leading cause of death from cancer in developing countries, disproportionately affecting the poorest and most vulnerable. At least 80% of cervical cancer deaths occur in developing countries, with most being found in the poorest regions – south and southeastern Asia, sub-Saharan Africa, the Caribbean, and parts of Latin America. In many of these regions, it is the *most* common cancer among women.

The primary reason for such a high mortality associated with the disease in these areas is the lack of knowledge of or access to the simple, inexpensive **Pap test** (named after Dr Papanicolaou, the physician who developed it), the most reliable detector of this cancer at an early stage. Thanks to this procedure and improved treatments, cervical cancer mortality has decreased by more than 70% over the past 50 years in the US. The 5-year survival rate for early detected invasive cervical cancer is now above 90% and the overall 5-year survival rate for all stages combined is about 73%.

STRUCTURE AND FUNCTION OF THE CERVIX

Cervical cancer is usually slow growing and begins in the lining of the **cervix**, which is the lower, narrow portion of the uterus or womb, the pear-shaped tissue where a fetus grows (Figure 10.1). The cervix connects the body of the uterus to the vagina, the birth canal leading to the outside of the body. This donut-shaped structure, about the size of a thumb, stays closed during pregnancy, keeping the baby inside the womb. During childbirth, it stretches to an opening large enough for the baby to pass through.

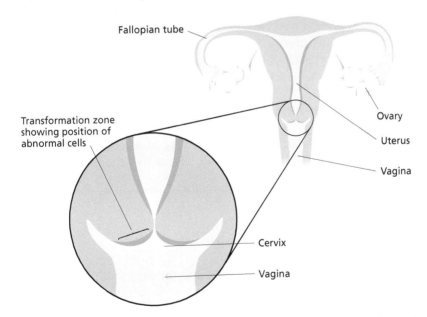

Figure 10.1 The cervix. The junction between the vagina and the uterus is the cervix.

SYMPTOMS OF CERVICAL CANCER

Early cervical cancer may have no noticeable signs

Since cervical cancer usually grows slowly, it can remain at an early stage for 2–10 years. Generally, these early cervical pre-cancers or cancers will *not* cause any noticeable signs or symptoms. Some examples of symptoms that *may* occur and cause concern are: vaginal bleeding or unusual discharge not associated with menstruation; pelvic pain; and/or bleeding or pain during or after intercourse, between periods or after menopause, douching, or after a pelvic examination. Since these symptoms aren't unique to cervical cancer and can occur with many female reproductive tract disorders, a healthcare provider should be seen in order to perform diagnostic tests. If cervical cancer is detected early, there is an excellent prognosis.

PELVIC EXAMINATIONS AND PAP TESTS ENABLE EARLY DETECTION

Caught in its early stages, cancer of the cervix can be completely cured. Thanks to the development and regular use of the Pap test, it is possible for physicians to detect the disease years before any symptoms are present and before any damage is done. The American College of Obstetrics and Gynecology recommends an annual pelvic examination and Pap test for all women by the age of 21 or within 3 years after they begin sexual intercourse, whichever comes first. Pelvic examinations, an internal examination of the uterus, vagina, ovaries, fallopian tubes, bladder, and rectum, should be performed on an annual basis for all women but not all women need an annual Pap test.

Pap test guidelines have been established

Routine Pap tests are the most effective way to detect cervical cancer in the earliest stages. While some women of any age choose to have them

---

Box 10.1

How is a pelvic examination and Pap test performed?

The healthcare provider will examine the cervix and the skin around and inside a woman's vagina for any irritation, sores, swelling, or rashes. The provider will then insert a speculum, a metal or plastic device shaped like a duck's bill. The

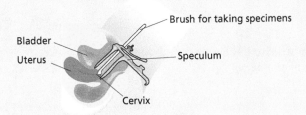

Figure 10.2 The Pap test. During a pelvic examination, the cervix is swabbed with a brush to obtain cells that will be analyzed.

speculum is closed upon entry and then gently spread open once inside the vagina allowing the provider to look inside the vagina and cervix for signs of inflammation or disease (Figure 10.2). A sample will then be collected for the Pap test, a simple, inexpensive screening procedure used to identify precancerous or cancerous cells.[10.1]

To perform this test, a small tool is gently rubbed against the cervix to collect cells from the surface of the cervix and vagina.[10.2] The collection instrument is then suspended in a special liquid, which preserves the cells for later examination. Two methods used are ThinPrep® and SurePath®.[10.3] The specimen is sent to a clinical laboratory where it is processed and examined by a cytotechnologist (CT), a skilled professional who has been trained to look for microscopic signs of cancer and other diseases by observing the smallest changes in color, shape, and size of the nuclei and cytoplasm of the cells.

After the speculum is gently removed, the clinician will feel the position, size and shape of the uterus, cervix and ovaries, and check for cysts or abnormal areas or any signs of infection or irritation such as tenderness or pain.

[10.1] The Pap test was introduced as the Pap smear since the cells were spread on a microscopic slide. The newer liquid-based cytology specimen is now the predominant method used in the US.

[10.2] The tool used for Pap tests could be a soft brush, a narrow wooden or plastic spatula, or a "cervical broom," a miniature broom-shaped tool with rubber fringes.

[10.3] ThinPrep was the first liquid based Pap test and is the one that people think of first, much as the way someone thinks of "band-aids" as both a brand and a product. Sure-Path is a newer product that is now used in many cytology laboratories.

Table 10.1 Pap test guidelines

Age	Recommendation
Three years after a woman has become sexually active or by the age of 21, whichever comes first	Initial Pap test
21–29 or Single sex partner or no sex partner Usage of the birth control pill	Annual regular Pap test or liquid-based test every years
30–69	If woman has had three normal Pap tests in a row, she should have a regular Pap test every 2 years or a liquid-based test every 3 years
70+	If a woman has had three or more normal tests in a row and no abnormal results in the last 10 years, she may stop having Pap tests
Women at high risk of cervical cancer, or who have certain risk factors such as multiple sex partners, HIV infection, a weakened immune system, a history of HPV (genital warts), an abnormal Pap smear, or women who's mothers took DES during pregnancy	Frequent Pap tests based on discussion with healthcare provider
Women who have had a total hysterectomy (removal of the uterus and cervix) for reasons other than the presence of cancer or a precancerous lesion	May choose to discontinue Pap test

performed annually, others may follow a schedule suggested by certain health agencies (Table 10.1). Currently, in the US, the decision on when to have a Pap test is usually governed by risk of occurrence and scope of insurance. Because cervical cancer is a very slow growing disease, if women have had a history of normal smears, it is assumed that the probability of an abnormal smear within the next year or two is quite low and there may be restrictions applied by the insurer.

For the most accurate results, women should try *not* to have the test performed during menstruation, nor should they douche or have sexual intercourse for 48–72 hours before the test. Contraceptive foam or jelly or spermicide should be avoided for a few days before the test as they have the potential to affect test results. The best time for a woman to have a Pap test is around 2 weeks into her cycle, days 10–20.[10.4] Many women confuse the need for pelvic examinations with Pap tests, especially since they are usually done at the same time. Since the pelvic examination is a part of a woman's regular healthcare examina-tion and will help identify diseases of the female organs, it should never be discontinued, even if Pap tests are no longer performed.

> [10.4] The first day of menses is day 1 of the menstrual cycle.

> **What is the difference between a pelvic examination and a Pap test?**

Pap test classifications

Pap test results are often reported in one of three categories using The Bethesda System (TBS), the most recent means of classification developed by the United States National Cancer Institute in 2001. The system of reporting consists of three parts: whether or not the sample was satisfactory for evaluation, an indication of whether the collected cells were normal or abnormal, and a descriptive diagnosis of abnormal cells when necessary. Examples of descriptive normal or abnormal results may include:

Negative for intraepithelial lesion or malignancy indicates that no signs of cancer or precancerous changes or significant abnormalities were found. Although the test may be negative for cancer, it is helpful in iden-tifying infections such as chlamydia, yeast, trichomonas, herpes, or other infections *excluding* cell changes caused by human papillomavirus (HPV).

Epithilial cell abnormalities reveal that the cells of the inner lining of the cervix may have changes, either dysplasia (a precancerous con-dition) or cancer itself. This category is further divided into groupings for squamous cell changes (80–90% of cervical cancers) and glandular cell abnormalities (10–20%) (Figure 10.3).

The most common abnormal Pap test result is atypical squamous cells of undetermined significance (ASC-US), used when the microscopic identification does not differentiate between an infection, irritation, or a precancerous condition. Results may be reported as negative or atypical when infections such as yeast or herpes are identified. It is estimated that approximately one in 20 Pap tests are reported with atypical findings and almost one in four women will probably have an ASC-US result at some point in her lifetime. Reasons for these common findings are unknown but very often the woman does not have a significant cervical lesion or cancer. Some young women may have an atypical result which will be cleared up by their own immune systems, but this is less likely to happen to older women for reasons not yet identified. This result *does* place a

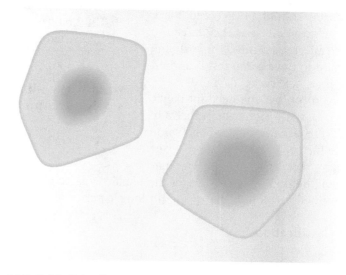

Figure 10.3 Epithelial cell abnormality. The cell on the left is normal and the one on the right is abnormal, evident by the large size of the nucleus in the abnormal cell.

woman into a higher risk category for cervical cancer, however, and requires them to have follow-up testing.

Squamous intraepithelial lesions (SILs) are further subdivided into low grade SIL, high grade SIL, and squamous cell carcinoma. Low grade SILs may go away without being treated, whereas high grade SILs are more likely to develop into cancer if not treated. About 80–90% of cervical cancers are squamous cell carcinomas composed of cells that resemble flat, thin squamous cells that cover the surface of the endocervix. There are procedures that can cure all SILs and prevent cancer from developing; some are performed in a physician's office and some need further surgical intervention. Once a diagnosis of squamous cell carcinoma is made, it means that the cancer is likely to be invasive and will need further testing and traditional oncology treatments.

Cancers of the glandular cells are reported as adenocarcinomas, which may originate in the cervix, uterus, or another part of the body. About 10–20% of the cervical cancer cases are adenocarcinomas and they seem to be found more often in younger women between 20 and 30 years old. Occasionally, cervical cancer cells resemble both squamous cell carcinomas and adenocarcinomas and are referred to as adenosquamous or mixed carcinomas. If it is unclear whether or not the cells are cancerous, they are termed atypical glandular cells and require further tests to confirm a diagnosis.

Other malignant neoplasms would include malignancies such as malignant melanoma, sarcomas and lymphoma that may rarely occur in the cervix.

Box 10.2

Dr George N. Papanicolaou, founder of the Pap smear

Dr Papanicolaou (1883–1962) was born in Greece. He moved to the US in 1913 where the pathologist and his wife Mary, a technician, worked in the Department of Anatomy in New York's Cornell Medical School. His early work involved examining vaginal smears of guinea pigs to determine the existence of a menstrual cycle. His valuable contribution to the basic understanding of endocrinology of reproductive organs was made by observing the changes in the female genital tract. His 1933 monograph "The Sexual Cycle of the Human Female as Revealed by the Vaginal Smear" led to his observation that malignant tumors of the cervix shed cancerous cells into the vaginal fluid, a revolutionary concept which was not accepted by fellow experts. At that time, no one believed that individual cancer cells could be identified and the fields of cell biology and cytology were nonexistent. In 1943, in collaboration with Dr Herbert Traut of Cornell University, they published a monograph entitled "Diagnosis of Uterine Cancer by the Vaginal Smear" that demonstrated a quick and simple test to detect the presence of abnormal cells. It was named the Pap smear in his honor. He published the *Atlas of Exfoliative Cytology*, a compendium of cytological findings for multiple human organ systems in 1954, and authored over 150 publications. In 1960, the American Cancer Society's members lobbied for the Pap smear to be part of their campaign to educate the public on cancer. Unfortunately, his death came at a time when he had just begun receiving the recognition he deserved. His contributions have without question saved millions of women from dying from cervical and uterine cancer.

Box 10.3

Bill and Melinda Gates Foundation

Bill and Melinda Gates have established a foundation to help reduce inequities in the US and around the world. The core of the foundation is based on two simple values: "All lives – no matter where they are being led – have equal value" and "To whom much has been given, much is expected." Their goal is to bring innovations in health and learning around the world. One such example is the formation of the Alliance for Cervical Cancer Prevention (ACCP), a major new effort to prevent cervical cancer worldwide by clarifying, promoting, and implementing strategies for preventing new cases in developing countries. ACCP works with developing-country partners to assess new approaches to screening and treatment, improve service delivery systems, ensure that community perspectives and needs are incorporated into program design, increase the awareness of cervical cancer, and develop effective prevention strategies. Their efforts are yielding very encouraging results.

RISK FACTORS FOR CERVICAL CANCER

Human papillomavirus (HPV) is the major risk factor for the development of cervical cancer

Human papillomavirus, or HPV, is the most common sexually transmitted infection in the US, with more than 20 million American men and women currently infected and over 6 million new cases annually. HPV includes a group of more than 100 types of viruses containing over 30 different strains or types that could infect the genital area of men and women (Table 10.2). It is most common in women and men in their late teens and early twenties and can spread through skin-to-skin contact with any infected part of the body. It is considered to be so common that it is estimated that 80% of sexually active women have been exposed to it at some point in their lives and will have acquired the infection by the age of 50. The majority of those who have it will not show any symptoms and will clear the infection without any treatment. Only a small number of women who carry it will develop cervical cancer, yet it causes the disease in over 90% of the women diagnosed.

These infectious agents are called papillomaviruses because they often cause papillomas (warts).[10.5] Genital warts, which are caused by certain types of HPV, may appear as soft, moist, pink or flesh-colored swellings, be raised or flat, single or multiple, small or large, or cauliflower shaped and can be found on the inside or outside of the genitals. It may take months or years after infection for warts to appear, if they appear at all,

[10.5] The types of HPV that cause warts on the hands or feet are different from those that are sexually transmitted.

Table 10.2 Selected HPV types and associated disease

HPV type	Disease
16, 18, 31, 45, 51, 52, 59, 68	Cervical cancer
16, 18	Penile, vulvar cancers
16, 18, 53, 58, 61	Anal cancer
16	Head and neck cancer
5, 8	Squamous cell carcinoma
1	Plantar wart
6, 11	Genital wart
2, 4	Common wart

and HPV can be "silent" or not demonstrate any symptoms for many years before it is found on a test. The appearance of genital warts does not necessarily predict an increased risk of cervical cancer. Since HPV is a sexually transmitted infection, it can be avoided by delaying intercourse until one is mature enough to make good choices, by using condoms, or by having fewer sexual partners. There is no guarantee, however, that even one chosen partner will not have an HPV infection already. In 2005, the Centers for Disease Control and Prevention estimated that at least 50% of sexually active people are infected with HPV during their lifetimes.[10.6]

> 10.6 Women who smoke and are infected with HIV are more likely to develop cervical abnormalities if they also have HPV.

Other factors may put women at risk

Since the development of cervical cancer is often many years removed from the initial incidence of HPV infection, a direct cause and effect correlation cannot always be made. Therefore, it is suspected that additional factors contribute to the development of the disease. Some possible risk factors cited include giving birth to many children (parity), having many sexual partners, having sexual intercourse before the age of 18 (immature cells seem to be more susceptible to the precancerous changes), smoking cigarettes (the exact mechanism linking cigarette smoking to cervical cancer is not known, but tobacco use increases the risk of both precancerous changes and cervical cancer), long-time use of oral contraceptives, exposure to DES (Chapter 8), family history of cervical cancer, having other sexually transmitted diseases such as chlamydia, gonorrhea, syphilis or HIV/AIDS, and having a weakened immune system. A healthy woman's immune system typically prevents the virus from doing harm, but for a small group of women, the virus may survive for years before converting the healthy cells to cancerous ones. Half of the cervical cancer cases occur in women between the ages of 35 and 55.

THE HPV TEST IS A VITAL DIAGNOSTIC TOOL

Human papilloma virus (HPV) infection is the most common cause of cervical cancer, responsible for over 90% of all cases. The HPV test, which also involves the collection of cervical cells, is performed during the pelvic examination. There are currently two types of analysis: the HPV DNA test will identify 13 of the high-risk types of HPV that are most likely to lead to cervical cancer even before changes to the cells of the cervix can be observed. The HPV PCR test, now being used in many hospitals and

10.7 The polymerase chain reaction (PCR) is a technique that enables many copies of a specific section of DNA to be made; literally over a billion copies can be made from a single molecule of DNA. The technique allows for the determination of whether a particular sequence of DNA is present or absent in a sample (e.g., is HPV DNA present or absent in a sample), or produces an amount of DNA that is needed to carry out a form of analysis (e.g., determination of what version of the *BRCA1* or *BRCA2* genes a woman possesses).

laboratories, will identify additional high-risk HPV types as well as the 13 found in the DNA test.[10.7] *Neither* of the HPV tests is a *substitute* for the Pap test, but are given *in addition* to it. They are not used to screen women younger than 30 with normal Pap test results, since women in this age group who have HPV infections usually do not have cervical cancer and their infections typically will be cleared up by their own immune system.[10.8] Some of them, however, may have high grade severe dysplasia that could progress to cervical cancer if left untreated, so they need to be carefully observed.

10.8 Currently, insurance providers may not be willing to pay for young women to have HPV testing if they have a negative Pap test, because traditionally most atypical results will clear up on their own.

ADDITIONAL TESTS ARE NECESSARY TO EXAMINE THE CERVIX

If irregularly shaped cells (dysplasia), potentially cancerous cells, are found on a Pap test, a colposcopy, performed in a physician's office (Figure 10.4), is usually the next step. A gynecologist applies a 3–5% acetic acid solution (like vinegar) on the cervix, then examines it and the vagina with a colposcope, a thin lighted tube painlessly inserted through the vagina into the cervix. The vinegar solution allows abnormal areas to temporarily turn white, making them easier to locate. Sometimes iodine is used instead, which will stain normal cervical tissues and not abnormal ones. Once abnormal areas are located, treatment can be performed immediately and locally by using a metal probe cooled with nitrogen or a heated wire to remove the abnormal tissue. Some clinicians may use laser surgery, which almost always destroys any precancerous cells and prevents them from developing into actual cancer. During a colposcopy, a clinician can also perform a cervical punch biopsy (Figure 10.5), the removal of a sample of tissue to see whether cancer cells are present. An endocervical curettage may be done at the same time, which is a scraping of cells from inside the endocervical canal with a curette, a small spoon-shaped instrument with a sharp edge or a cytobrush. The excised tissue is then examined under a microscope for malignant characteristics.

If any of the above demonstrates potentially cancerous cells, further options are available and may be performed in a doctor's office or in a

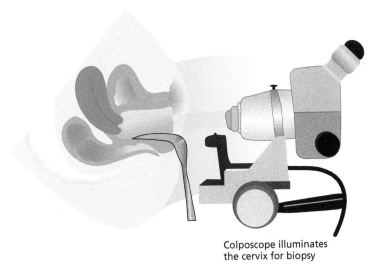

Colposcope illuminates
the cervix for biopsy

Figure 10.4 Colposcopy. A colposcope is used to examine the cervix. Adapted from: http://www.mdconsult.com/das/patient/body/

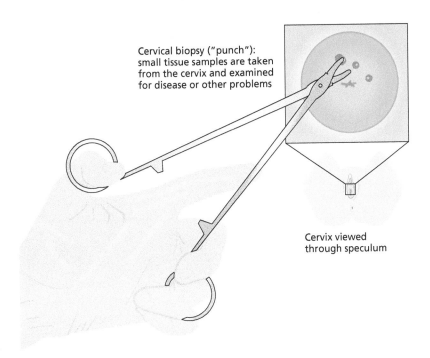

Cervical biopsy ("punch"):
small tissue samples are taken
from the cervix and examined
for disease or other problems

Cervix viewed
through speculum

Figure 10.5 Cervical punch biopsy. Tissue samples are removed from the cervix for analysis. Adapted from: http://www.nmh.org/nmh/adam/adamencyclopedia/Imagepages

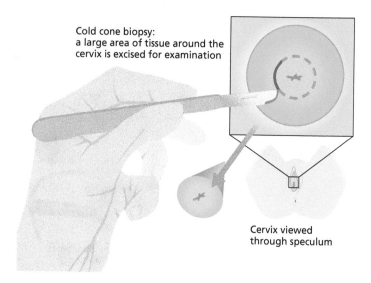

Cold cone biopsy:
a large area of tissue around the
cervix is excised for examination

Cervix viewed
through speculum

Figure 10.6 Conization. A cold cone biopsy is the removal of a large area of tissue from the cervix. Adapted from: http://www.mdconsult.com/das/patient/body/

hospital under general anesthesia. Conization (Figure 10.6) is the surgical removal of a larger, cone-shaped sample of tissue from the cervix and cervical canal (the spindle-shaped canal extending from the uterus to the vagina), areas where precancers and cancers are most likely to develop. This is done by using either a knife (cold knife cone biopsy), laser (narrow beam of light), or the loop electrosurgical excision procedure (LEEP) (a thin wire heated by electricity). When cervical cancer is found in the earliest stages, conization is the only form of surgical therapy necessary. If the cancer has spread to tissue deeper in the cervix or to neighboring organs, further treatment is needed. The FIGO (International Federation of Gynecology and Obstetrics) staging system is utilized based on clinical results rather than on surgical procedures (Table 10.3). Imaging results such as X-rays, CT and MRI scans, and ultrasounds can determine if the cancer has spread.

TREATMENT DEPENDS ON THE STAGE

Cells taken from the surface of the cervix can appear abnormal or dysplastic but may not be cancerous. They have the potential, however, to undergo changes that will eventually lead to cancer. The prognosis for cervical dysplasia is excellent, with most of the cases either resolving on their own or being cured with proper treatment and follow-up.

Cryosurgery, a procedure performed with an instrument that freezes and destroys abnormal tissues, will control most cases of this pre-invasive cervical cancer. Infrequently, it may still remain unclear as to whether the abnormal cells are confined to the cervix or have arisen from inside the

Table 10.3 Cervical cancer staging

Stage	Description	Survival prognosis
Stage 0 (cervical dysplasia)	Superficial cancer affecting only the surface layer of cells lining the cervix	99+%
Stage 1	The cancer has invaded the cervix but has not spread	85–95%
Stage 1A	Cancer can only be seen under a microscope	
Stage 1B	Cancer can be seen without a microscope or can be seen with a microscope if it has spread deeper than 5 mm (about 1/5 inch) into connective tissue of the cervix or is wider than 7 mm	
Stage 2	Cancer has spread beyond the cervix, but is still inside the pelvic area	65–80%
Stage 2A	Cancer has spread beyond the cervix to the upper part of the vagina	
Stage 2B	Cancer has spread to the tissue next to the cervix (parametrial tissue)	
Stage 3	Cancer has spread to the lower part of the vagina or the pelvic wall. It may block the ureters (tubes that carry urine from the kidneys to the bladder)	40–50%
Stage 3A	Cancer has spread to the lower third of the vagina but not to the pelvic wall	
Stage 3B	Cancer extends to the pelvic wall and/or blocks urine flow to the bladder, or the cancer has spread to pelvic lymph nodes	
Stage 4	Cancer has spread to nearby organs or other parts of the body	
Stage 4A	Cancer has spread to the bladder or rectum	40–50%
Stage 4B	Cancer has spread to distant organs beyond the pelvic area such as the lungs	<10%

> **Give other examples of when a D and C would be performed.**

uterus. In this situation, a **D and C**, Dilatation (the cervical opening is stretched) and Curettage (curette is inserted to scrape and clean out cells from the lining of the uterus and cervical canal), is performed while the woman is under anesthesia.

Surgical treatment is necessary for some cervical cancers

If the cancer is found at a later stage, but only in the cervix, doctors will perform a hysterectomy, either simple (the removal of the uterus and cervix) for very minimal cancers, or radical (the removal of the uterus, cervix, the upper parts of the vagina, and the connective tissues that keep the uterus in place). The fallopian tubes and/or ovaries may or may not be removed, but either surgery's outcome results in the woman's inability to become pregnant. Before performing a radical hysterectomy a physician may perform a pelvic lymph node dissection of nearby lymph nodes to determine if they contain cancercus cells. Positive lymph nodes may require other forms of treatment such as chemotherapy and/or radiation, treatments that are also recommended for large tumors and cancer that has spread outside the cervix.

Radiation and/or chemotherapy is recommended for most cervical cancers

Radiation therapy is also an effective treatment option for some Stage 1 and 2A cervical cancers. It may be given externally or internally by brachytherapy, where a radioactive source is placed inside the vagina. Besides the traditional side effects of radiation, exposing the ovaries to radiation may cause them to stop functioning. The combination of chemotherapy with radiation is the current standard therapy for women with Stage 2B to Stage 4 disease. One common option is to give cisplatin weekly during the radiation. Recent clinical studies suggest that the combination of cisplatin plus another drug would benefit these women with advanced or recurrent cervical cancer.

In 2006, the US Food and Drug Administration approved a combination of Hycamtin (topotecan hydrochloride) with cisplatin to be used as the first drug treatment for those with Stage 4B, recurrent or persistent cervical cancer which has spread to other organs and has been determined to not likely respond to surgery or radiation therapy. Hycamtin has also been used in the treatment of ovarian cancer and small cell lung cancer. Side effects for this drug include neutropenia (a lowering of white blood cells which could increase the risk of infections) and/or thrombocytopenia (a decrease in platelets, which could cause excessive bleeding) and/or anemia (drop in red blood cells due to excessive bleeding).

As is typical for most cancers, prognosis depends on the stage, size, type of tumor, and if it has metastasized to other organs. Treatment is also based on these same factors, but the patient's age and desire to have children will also be considered. If the woman is already pregnant, the treatment may be delayed until after the baby is born, but this depends on in which trimester the cancer is found.

A VACCINE WILL PREVENT MANY CASES OF CERVICAL CANCER

Unlike many cancers, whose origins are suspect at best, cervical cancer can be prevented. Prevention is based on the discovery that infection with the human papillomavirus (HPV) is the major risk factor for the development of cervical cancer. An HPV vaccine is now available that will protect against the viral strains that are responsible for about 70% of cervical cancers (HPV types 16 and 18) and 90% of genital warts (HPV types 6 and 11). Gardasil, made by Merck & Co., is the first vaccine specifically designed to prevent cancer, precancerous genital lesions, and genital warts due to HPV.

The CDC's Advisory Committee on Immunization Practices (ACIP) has recommended that this three dose vaccine be given routinely to girls when they are 11–12 years old. The ACIP also allows for vaccination of girls beginning at 9 years old at the discretion of the healthcare provider and of girls and women 13–26 years old. It is not effective against an existing HPV infection, which is why it is recommended for females *before* they become sexually active. Women who have been vaccinated must still be continually tested because the vaccine *does not protect* against the viral strains responsible for the remaining 30% of cervical cancers, *nor* does it protect against other sexually transmitted diseases. Drug companies are studying Gardasil and other compounds as a means of protection against other strains of the virus as well.[10.9] Since this new vaccine is targeted toward young girls to prevent a sexually transmitted infection, it is sometimes very controversial. Some parents question the need for this vaccine and worry that it may lead to promiscuity. Today, magazine and television ads attempt to convince parents of the value of the vaccine.

> **What other vaccinations are given to 12 year old girls?**

[10.9] Cervarix, made by GlaxoSmith-Kline, is another HPV vaccine for women. It has been approved in 67 countries, but not yet in the USA. The company hopes to have it approved by the FDA by the end of 2009.

What are some recent
accomplishments of PATH?

Unfortunately, while vaccinations for many childhood diseases are common in the US, they are far less so in the developing world where the majority of HPV cases exist. The Bill and Melinda Gates Foundation has donated millions of dollars to the Seattle-based Program for Appropriate Technology in Health (PATH). This organization works with drug makers and public health organizations to help plan for and pilot introduction of the HPV vaccine in India, Peru, Uganda, and Vietnam, countries chosen because of their commitment to projects aimed at preventing cervical cancer, and because of their effective childhood vaccination programs.

Thankfully, many advances have been made in the battle against cervical cancer, especially in the field of prevention and testing. The HPV vaccine has the potential to greatly decrease and eliminate a woman's chances of ever developing this disease. Hopefully time, research and more education will result in future successes towards the eradication of this cancer.

EXPAND YOUR KNOWLEDGE

1 Give two reasons why the HPV vaccine is recommended to be given when a girl is 12.
2 Which cancer(s) can possibly be seen on a Pap smear? Which gynecological cancers will *not* be found on a Pap smear?
3 Has Cervarix been approved for use in the USA or any other countries? What strains of HPV will it protect against?
4 How does being HPV positive affect men?
5 What progress has been made on an HPV vaccine for men?

ADDITIONAL READINGS

American College of Obstetricians and Gynecologists' website: www.acog.org/
American Social Health Association's website: www.ashastd.org
Colgrove, J. (2006) The ethics and politics of compulsory HPV vaccination. *New England Journal of Medicine,* **355**: 2389–2391.
Journal of Women's Health website: http://www.liebertpub.com/Products/Product.aspx?pid=42
Kim, S.-W. and Yang, J.-S. (2006) Human papillomavirus type 16 E6 protein as a therapeutic target. *Yonsei Medical Journal,* **47**: 1–14.
Lowy, D. R. and Schiller, J. T. (2006) Prophylactic human papillomavirus vaccines. *Journal of Clinical Investigation* **116**: 1167–1173.
Mahdavi, A. and Monk, B. J. (2006) Vaccines against human papillomavirus and cervical cancer: promises and challenges. *The Oncologist,* **10**: 528–538.

Marieb, E. N. and Hoehn, K. (2007) *Human Anatomy and Physiology*, 7th edn. Upper Saddle River, NJ: Prentice Hall.

Martini, R. and Bartholomew, E. (2007) *Essentials of Anatomy & Physiology*, 4th edn. Upper Saddle River, NJ: Prentice Hall.

Merck & Co., Inc.'s website: www.gardasil.com

Zangwill, M. (2004) Cervical cancer: the hidden risk of HPV. *CURE*, **3**: 29.

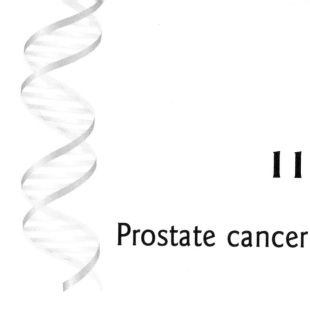

11

Prostate cancer

I know what the doctors say, but [cancer] never met someone like me.
Michael J. Tenaglia, founder of the Miracle Foundation

CHAPTER CONTENTS

- Prostate cancer statistics
- Function of the prostate gland
- Certain factors influence the development of prostate cancer
- Symptoms of an enlarged prostate
- Screening and diagnostic tests for prostate cancer
- There are traditional and unique treatment options available
- Prostate cancer mortality rates have decreased
- Much attention is being given to the number one cancer affecting men
- Expand your knowledge
- Additional readings

Cancer: Basic Science and Clinical Aspects, 1st edition.
By C. A. Almeida and S. A. Barry. Published 2010
by Blackwell Publishing, ISBN 978-1-4051-5606-6.

Among the greatest concerns of middle-age and older men is the possibility of developing prostate cancer. There are psychological as well as physical elements associated with the advent of this disease that understandably strike at the heart of a man's masculinity.

PROSTATE CANCER STATISTICS

Discounting skin cancers, the most common cancer occurring in men is that of the prostate. Approximately one out of every six American men will experience the disease during his lifetime and in 2008, about 186,000 men were diagnosed and 29,000 died from it. The encouraging news is that it is one of the few cancers that has a good prognosis – only one in 35 will die from it. It is considered to be so common that it has been estimated that the majority of men will eventually develop prostate cancer if they live into their seventies or eighties and, since this is a slow growing cancer, they may never even know they have it.

FUNCTION OF THE PROSTATE GLAND

Sperm are produced in the seminiferous tubules of each testicle and are stored within the epididymis, a coiled tube that sits on the top and back of each testicle. During sexual arousal, contractions cause the sperm to travel from the epididymis through the ductus or vas deferens to the prostate. The prostate is a male gland that produces a thin, milky, alkaline fluid that functions to protect the sperm by neutralizing the acidic environment of the vagina once the sperm have been ejaculated into it. The gland is positioned just below the bladder at the base of the pubic bone and in front of the rectum (Figure 11.1).

Within the prostate, each of the vas deferens merges with a seminal vesicle duct to form an ejaculatory duct. It is here that the sperm are mixed with the seminal fluid, a fructose-rich liquid produced by the seminal vesicles that nourishes sperm, and the prostate fluid. The semen travels in the right and left ejaculatory ducts that merge within the prostate to form the urethra, the tube that also transports urine from the bladder to the outside of the body through the penis. Approximately 60% of semen is seminal fluid, another 30% is prostatic fluid, and the remainder consists of approximately 200–500 million sperm.

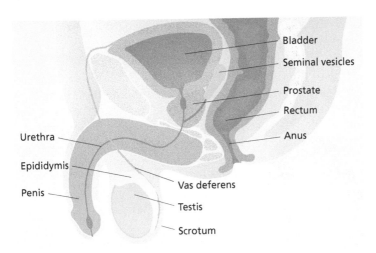

Figure 11.1 The male reproductive tract. Sperm are produced in the testicles, stored in the epididymis, and during ejaculation travel through the vas deferens and the prostate, and then exit through the urethra.

CERTAIN FACTORS INFLUENCE THE DEVELOPMENT OF PROSTATE CANCER

Although a definitive cause for the development of prostate cancer has not yet been determined, there have been a number of factors that have shown to be associated with an increased risk. Unfortunately, the associations between many of the risk factors and the disease are currently not well understood. Since many of the causes cannot be controlled, prevention is difficult.

Age and family history are the principal risk factors

The likelihood of developing prostate cancer, like that of most other cancers, increases with age, particularly after the age of 50. Nearly half of all men exhibit some degree of neoplasia among the epithelial cells of the prostate by the age of 50. Published studies on men that have died of prostate cancer as well as those who have died of other causes have provided an estimate of the disease's prevalence in the general population. The results of the studies indicate that 40%, 60%, and 80% of men in their fifties, sixties, and seventies, respectively, have prostate cancer. Again, although the majority of men will develop the disease in their lifetimes, many will never know or be negatively affected by it.

An understanding of the genetic role is still being constructed. A man's risk of developing prostate cancer is doubled if he has a first-degree male relative who has been diagnosed with the disease. In fact, his risk is slightly higher if his brother rather than his father is afflicted. A man has a higher risk for developing prostate cancer if he has any first-degree relatives diagnosed with the disease at a young age. So far research has made progress toward identifying several genes that play a role, but it has failed to identify a single gene that plays a predominant one. It is estimated that only 5–10% of prostate cancers are due primarily to the inheritance of high-risk prostate cancer susceptibility genes. The *HPC1* gene (*h*ereditary *p*rostate *c*ancer gene 1) was the first to be associated with an increased risk of developing the disease. Men who possess mutations within *BRCA1* or *BRCA2* genes (Chapter 8) may have a slight to moderate increased risk. Unfortunately, genetic testing to determine a man's predisposition for the disease is not currently available.

Certain races and nationalities are at higher risk

The highest incidence of prostate cancer occurs in African-American men and they are statistically more likely to be diagnosed at a more advanced stage than men of any other race. Compared to Caucasian-American men, African-American men are 60% more likely to develop it and are more than twice as likely to die from it. By contrast, east Asian men are least likely to be affected by the disease. Caucasian and Hispanic men develop prostate cancer at the same rates. The reasons for differences in risk according to race are unknown and are the focus of ongoing research studies.

The locales with the greatest number of prostate cancer diagnoses are North America and northwestern Europe, whereas the lowest numbers of cases occur in Asia, Africa, and Central and South Americas. It *has* been established that the discrepancies in case number are not solely due to differences in screening rates or diagnostic methods.

A person's diet affects his risk

Diet may factor into the prostate cancer rates when analyzed by nationality. A diet high in red meat or low in fruits and vegetables, which often times goes hand in hand, puts men at a slightly higher risk of developing the disease. There is also some evidence that indicates that a high consumption of dairy fat (milk, cream, cheese, ice cream, etc.) and calcium carries an increased risk.

There have been some studies that have investigated an association between prostate cancer and lycopenes, plant pigments that function as antioxidants (molecules that prevent cell damage). The results are unclear, however, because some studies have demonstrated that lycopenes reduce the risk of developing prostate cancer while others have not. Lycopenes are present in fruits and vegetables with a red color, such as tomatoes, pink grapefruit, and watermelon.

A large clinical study, titled The *Sel*enium and Vitamin *E Cancer Prevention Trial* (SELECT), was initiated in 2001 with the objective to determine if either vitamin E or the mineral selenium reduces the risk of prostate, lung, and colon cancer as well as to investigate their suspected contribution to a longer healthier life. Vitamin E and selenium, a nonmetallic trace element obtained from rice, wheat, seafood, meat, and Brazil nuts, are also antioxidants. More than 35,000 men from the US, Puerto Rico, and Canada were enrolled in the study between 2001 and 2004 and will be followed until 2011.

Prostate cancer is hormone dependent

The androgens are the male sex hormones and principal among them is testosterone. Just as a woman's breast cells are growth responsive to estrogens, a similar effect exists between androgens and prostate cells. Androgens, specifically testosterone, not only play a critical role in stimulating the growth of prostate cells during normal development, but they also increase the rate of growth of cancerous prostate cells. What is not as clear is whether testosterone plays as significant a role in the development of prostate cancer as estrogens do in the development of breast cancer.

Finasteride, a drug taken by men to treat a benign enlargement of the prostate, blocks the production of the androgen dihydrotestosterone (DHT) from testosterone. In 1993 the Prostate Cancer Prevention Trial was initiated with the objective to determine if finasteride could prevent prostate cancer in men age 55 and older. A total of 18,882 men that exhibited no evidence of prostate cancer at the beginning of the trial were enrolled. Half of the men were randomly assigned to take the drug while the other half took a placebo. The trial was halted prematurely in 2003 because there was a clear finding that the drug reduced the incidence of prostate cancer by 25%. Interestingly, those men taking finasteride that did develop prostate cancer had a slightly higher incidence of developing high-grade tumors. An analysis is being done in order to determine whether finasteride played a role in the development of the aggressive prostate cancers.

SYMPTOMS OF AN ENLARGED PROSTATE

Growth of the prostate begins during fetal development and continues through to adulthood. Once adulthood is reached, the prostate will maintain its size as long as testosterone is being produced. The portion of the prostate that surrounds the urethra has a tendency to exhibit a form of growth termed benign prostatic hypertrophy or benign prostatic hyperplasia (BPH). When the growth of the prostate is large enough, it will begin to produce symptoms regardless of whether it is benign or malignant. These can include difficulty starting or holding back the flow of urine, a weak or interrupted urine flow, pain or burning during urination, blood in the urine, an increased need to urinate, painful urination, or impotency (difficulty in getting or maintaining an erection). Each is the result of a compression upon the urethra by the increased size of the prostate. These symptoms, however, may reflect other noncancerous conditions as well, such as prostate or bladder infection, and are generally associated with the inevitable aspects of the aging process, so a healthcare provider should be contacted to discuss diagnostic testing.

> What is the difference between hypertrophy and hyperplasia?

SCREENING AND DIAGNOSTIC TESTS FOR PROSTATE CANCER

In an attempt to detect prostate cancer at its earliest stage, it is recommended that the digital rectal examination (DRE) and prostate specific antigen (PSA) tests be performed during routine annual examinations starting at the age of 50. However, if there is a family history of prostate cancer or if the man is African-American, then annual testing should begin at age 45. Screening can also be performed in response to a patient exhibiting one or more of the symptoms mentioned previously.

Much controversy exists as to the value of screening tests, as they often produce false positive results that lead to unnecessary invasive procedures.[11.1]

During a DRE, the healthcare provider inserts a gloved and lubricated index finger into the rectum and presses against the prostate to assess its size, shape, and texture (Figure 11.2). While some discomfort may be felt when the finger is inserted and the pressure on the prostate may result in a desire to urinate, the procedure is brief and not painful. It is important to note that not all anomalies of the prostate can be detected through a direct tactile examination.

[11.1] An objective of the *Prostate, Lung, Colorectal, and Ovarian* (PLCO) Cancer Screening Trial, a large randomized study of 155,000 men and women between the ages of 55 and 74, is to determine whether screening reduces the number of fatalities associated with these cancers by enabling treatment to occur during the pre-symptomatic stage.

A more sensitive means of assessing the state of the prostate is the PSA test. The prostate produces PSA, a protein component of prostate fluid, and a small amount of it is found in the bloodstream. As the size of the prostate increases, so does the amount of PSA it produces. Blood is drawn from the arm and analyzed for PSA levels; an amount equal to or less than 4 nanograms per milliliter (ng/mL) is considered normal. If the PSA value is between 4 and 10 ng/mL, it is considered intermediate or borderline and there is a 25% chance of cancer being present. A high PSA value, one that is greater than 10 ng/mL, is associated with a greater than 50% chance of having prostate cancer. The DRE and PSA test complement one another. The DRE may detect an abnormality in approximately

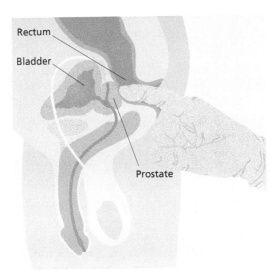

Figure 11.2 Digital rectal examination. This procedure allows the practitioner to examine the prostate.

15% of men with prostate cancer that does not produce an elevated PSA level. Conversely, the PSA test may indicate the presence of the disease when it cannot be palpated.

It should be kept in mind that an elevated PSA level is not synonymous with prostate cancer. There are a number of conditions other than cancer that can result in this elevation. For example, PSA level normally increases with age or with an enlarged prostate that is either the result of normal growth as a man ages (benign prostatic hyperplasia) or due to infection or inflammation (prostatitis), and for a day or two after ejaculation. It is also possible for certain prescribed medications (e.g., finasteride) and herbal supplements (e.g., Saw Palmetto) that are advertised for prostate health to *decrease* PSA levels, potentially masking any elevation that may occur in the PSA level. A patient should notify his healthcare provider of all medications, vitamins, and dietary supplements being taken because they may have an effect on test results.

Subsequent tests should be performed to confirm suspicious screening results

There are a few indications that should be considered when determining whether or not a prostate biopsy is necessary following either an abnormal DRE or elevated PSA test. First, a PSA level in the borderline range for a 50-year-old man versus an 80-year-old man is likely to have different implications because PSA levels normally increase with age. The PSA velocity, the rate at which the PSA level rises, should be determined through regular PSA tests. If the PSA level is obtained every year, sharp rises (>0.4–0.75 ng/mL/yr) can be detected, even if the values are below 4 ng/mL. The PSA in blood is present in two forms, bound and unbound (free) to protein. The percent-free PSA, the amount of free PSA relative to the total amount of PSA, is also an important diagnostic value. Although there is not unanimous consensus among physicians, it is recommended that a biopsy should be performed when the percent-free PSA is 25% or less. A study determined that using this cut-off value would reduce the number of unnecessary biopsies performed by 20%, and approximately 95% of prostate cancers would still be detected.

A procedure that allows a physician to obtain an image of the prostate is called transrectal ultrasound (TRUS) (Figure 11.3). It is carried out by placing a probe, about the size of a cigar, into the rectum. This probe produces sound waves that bounce off of the prostate, creating an echo pattern that is detected by the probe and converted into an image on a video screen. On its own, TRUS is not a procedure that is used to detect prostate cancer at an early stage, although it can be useful in determining the size/volume of the prostate. An enlarged prostate will produce a greater amount of PSA, and TRUS enables a correlation between the size of the prostate and the PSA level. The higher the PSA density, which is

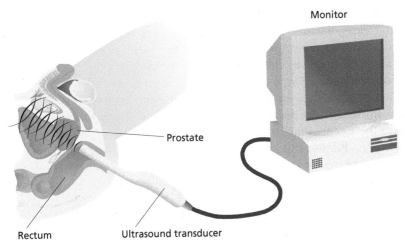

Figure 11.3 Transrectal ultrasound. This procedure is performed by the insertion of a probe into a man's rectum to check for abnormalities in the prostate. Sound waves bounce off the prostate and a computer uses the echoes to create a picture called a sonogram.

determined by dividing the PSA value by the volume of the prostate, the greater the likelihood of cancer.

The results of a core needle biopsy are essential in diagnosis

The principal way to definitively diagnose prostate cancer is by TRUS-guided core needle biopsy. The procedure is typically done in the urologist's office (i.e., a physician who specializes in the male and female urinary tracts and men's reproductive systems) under local anesthesia. This technique uses an ultrasound probe with a guide that holds the tip of a biopsy gun. A biopsy gun is a hand-held device that uses a very narrow spring-loaded needle to obtain a sample of prostate tissue through the wall of the rectum (Figure 11.4). The probe with the biopsy needle is inserted into the rectum and placed against its wall. The needle is positioned in the desired

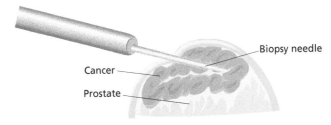

Figure 11.4 Prostate biopsy gun. This instrument is used for a prostate core needle biopsy. Guided by the TRUS, a small needle is inserted into the prostate gland to remove tissue from it.

location with the aid of the TRUS, and then the gun is activated. The needle extends from the gun's tip and passes through the wall of the rectum and into the prostate. Prior to the needle being retracted into the gun's tip, a sheath closes at the tip of the needle, trapping a small core of prostate tissue. The extraction of each core takes only a fraction of a second, but some urologists may inject the prostate with an anesthetic since up to 25% of patients may experience a sharp stinging sensation when the needle pierces the prostate.

During a biopsy, the urologist wants to sample the area of the prostate that is suspected to be cancerous but also wants to determine if the cancerous cells have spread throughout the gland, and if so, to what extent. At the minimum, a sextant biopsy is performed in which six core samples, one from the top, middle, and bottom of both the right and left sides of the prostate, are taken to obtain a representative sample of the prostate. Up to 24 core samples may be necessary depending on the specifics of the patient's condition and to increase the likelihood of detecting cancerous tissue.

A typical biopsy takes approximately 15–30 minutes and the patient may experience soreness or light rectal bleeding during the next 2 or 3 days. Blood may be present in the urine for the next few days or in the semen for up to a month after the biopsy. Given the effectiveness of the examination as a diagnostic tool, most patients consider these side effects to be a slight price to pay.

> **Why is blood likely to be found in semen following a core needle biopsy of the prostate?**

The Gleason grading system is utilized to stage prostate cancer

> 11.2 An adenocarcinoma is a malignancy that originates in the epithelial cells of glandular tissue.

In nearly all cases of prostate cancer it is the glandular cells, which produce prostatic fluid, that are cancerous. Therefore, a commonality between prostate and breast cancers is that the vast majority of them are adenocarcinomas.[11.2] It is theorized that the initial stages in the development of prostate cancer may occur in a man's twenties with a condition called prostatic intraepithelial neoplasia (PIN). PIN is determined following a microscopic analysis of a tissue sample taken from the prostate during a biopsy. As the term PIN implies, the epithelial cells have altered characteristics, such as changes in size and shape, but are not considered to be cancerous at this point. The degree to which the cells look abnormal can be categorized as low or high grade. A high-grade PIN diagnosis is associated with a 30–50% chance of cancer being present in the gland and necessitates the need for close monitoring and frequent biopsies.

The biopsy samples are sent to a lab for microscopic analysis by a pathologist. The tissue will be examined to determine whether or not cancer cells are present. Their appearance provides insight into how quickly the cancer cells are likely to grow and spread. The Gleason grading system, named after Dr Donald F. Gleason, the physician who developed it, is a 5 point scale that is most commonly used and is based on the appearance or differentiation of the cells that make up the glands of the prostate. The more the cancer cells resemble normal prostate cells (i.e., well differentiated), the more likely it is that their biological function will be normal and the cancer will be less aggressive.

Tissues with Gleason grades of 1 and 2 possess well-defined cells and glands that closely mimic normal tissue. The most common Gleason grade is 3, which defines tumors possessing moderately differentiated cells but whose boundaries between glands are less well-defined. The complete lack of a uniform arrangement of cells and distinctive boundaries between glands is indicative of poorly differentiated tissues and so these tumors are given a Gleason grade of 4 or 5. Cancers of this grade are typically aggressive.

In order to deal with the distinct possibility that different regions within a tumor have different grades, the pathologist is likely to report a Gleason score upon examination of a prostate biopsy. The two most common grades from the biopsy samples are added together to obtain the Gleason score which can range from 2 to 10. For example, if a pathologist had examined 13 core samples from a single biopsy and found that the most common grade was 3 and the second most common grade was 2, then the Gleason score would be 3 + 2 = 5. The higher the score, the more aggressive the tumor and the greater the likelihood that the cancer has metastasized, leading to a poorer prognosis. Cancers with Gleason scores of 2–4 are typically slow growing, whereas scores of 5–7 indicate moderate aggressiveness, and tumors with scores of 8–10 are fast growing and likely to metastasize.

> **What is the difference between Gleason grading and score?**

A variety of tests are available to determine the stage of the disease

The diagnosis of cancer by a prostate biopsy is typically followed up by one or more tests in order to stage the disease within the gland but also to determine whether it has spread to surrounding tissues. A localized stage of prostate cancer is confined to the prostate whereas a regional stage has spread to nearby tissues such as the seminal vesicles or bladder, but not to distant sites. Staging plays an important role in determining which form of therapy would be best for a patient. Greater than 90% of prostate cancers are diagnosed when they are at the local and regional stages, providing a very favorable prognosis.

Prostate cancer has a tendency to metastasize to the bones (Chapter 5), accounting for persistent pain or stiffness. A particular type of radionuclide bone scan can be used to locate areas in the skeleton that contain very active cells which may be sites where metastatic cancer cells reside. This procedure entails the injection of a small amount of low-level radioactive material into the patient's bloodstream that travels through-out the body. Cells that are very metabolically active, such as cancer cells, will take up more radioactive material than will less active cells. A camera that scans across the body will detect the radiation being given off. Those cells that have more radioactivity than others will appear as "hot spots" on a video screen. The presence of "hot spots" doesn't necessarily imply the presence of cancer cells, as they will also appear at sites of arthritis or other forms of bone disease, or at the location of an old injury.

It is possible, with the very detailed 2-D images or 3-D models created by magnetic resonance imaging (Chapter 5), to distinguish between healthy and diseased tissue. The prostate can be clearly observed by MRI, and can therefore be used to determine the precise location of cancer within the gland or whether it has spread to the seminal vesicles or bladder.

11.3 Antibodies are proteins that have high degrees of binding specificity, meaning that a given antibody is capable of binding to a specific protein. A monoclonal antibody is a collection of a single type of antibody with an ability to bind specifically to a single type of protein. A polyclonal antibody is a mixture of antibodies with each antibody having an ability to bind to specific type of protein. The protein that antibodies bind to is referred to as an antigen. Therefore, the prostate-specific membrane glycoprotein is also known as the prostate-specific membrane antigen (PSMA).

Benign hypertrophic and cancerous pro-state cells produce a prostate-specific membrane glycoprotein (PSMG), which is a carbohydrate-linked protein located only in the cellular membranes of these types of cells. Normal prostate cells do not produce the PSMG or produce very small quantities of it. During a ProstaScint scan, a radioactive monoclonal antibody with a binding specificity for PSMG is injected into the patient and a body scan is taken to detect where the radioactive antibody is located.[11.3] A ProstaScint scan is useful in determining the exact location of cancer cells within the prostate as well as in local lymph nodes or nearby tissues.

The detection of cancer cells in lymph nodes that have been removed from the area around the prostate is an indicator of metastasis. A lymph node biopsy or dissection can be performed in one of two ways. During a surgical biopsy, an incision is made in the lower abdomen and the lymph nodes are excised. Based on prior test results, the surgeon may have a strong suspicion that metastasis is likely to have occurred (e.g., PSA value of 15 ng/mL or a Gleason score of 8). Therefore, the surgically removed lymph nodes would be subjected to an immediate frozen section examination by a pathologist prior to the removal of the prostate in the next stage of the operation. If the patholo-gist determines that cancer cells are present in the biopsied lymph nodes, then the surgeon will usually not proceed with the removal of the pro-state. This is because it is not likely to cure the cancer but will probably result in undesirable complications or side effects. A less invasive means of removing lymph nodes is by laparoscopy. A small incision is made in the abdomen through which the surgeon inserts a laparoscope, a slender tube with a video camera at the end, to visualize the pelvic cavity. The laparoscope is used to guide the surgical instruments (inserted through additional small incisions) that remove the lymph nodes. Laparoscopy is performed when the lymph nodes are to be biopsied without the concomitant removal of the prostate.

The staging of a patient's prostate cancer is based on the categorization of the tumor using the TNM system (Chapter 6) and Gleason score in a process called stage grouping. The Roman numerals I through IV are used to represent the least to most advanced stages of cancer, respectively. Table 11.1 provides a breakdown of the T, N, and M categories and subcategories for prostate cancer and Table 11.2 provides the overall stage groupings. A simplified view of the four stages of prostate cancer is illustrated in Figure 11.5.

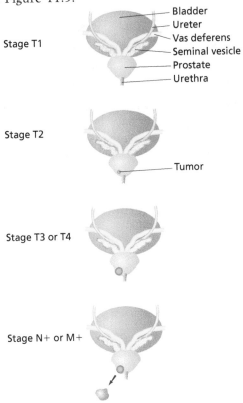

Stage T1

Bladder
Ureter
Vas deferens
Seminal vesicle
Prostate
Urethra

Stage T2

Tumor

Stage T3 or T4

Stage N+ or M+

Figure 11.5 Stages of prostate cancer. (a) Stage T1 – usually detected by DRE or ultrasound. (b) Stage T2 – confined to prostate, detected by DRE or ultrasound. (c) Stage T3 – metastasized to adjacent tissue or seminal vesicles; Stage T4 – metastasized to nearby organs such as the bladder. (d) Stage N+ or M+ – metastasized to pelvic lymph nodes, lymph nodes, distant organs, or bones. Adapted from: http://www.fda.gov/FDAC/features/1998/prosdiag.html

Table 11.1 TNM staging for prostate cancer

Category	Subcategory	Explanation
TX		Tumor cannot be assessed
T0		No evidence of tumor
T1		Tumor cannot be detected by DRE or an imaging method
	T1a	Cancer is detected upon examination of prostate tissue surgically removed during the treatment of benign prostatic hypertrophy; cancer represents <5% of removed tissue
	T1b	Cancer is detected upon examination of prostate tissue surgically removed during the treatment of benign prostatic hypertrophy; cancer represents >5% of removed tissue
	T1c	Cancer detected upon needle biopsy in response to elevated PSA
T2		Cancer detected by DRE but appears confined to prostate
	T2a	Cancer occupies half or less of one side of the prostate
	T2b	Cancer occupies more than half of one side of the prostate
	T2c	Cancer occupies both halves of the prostate
T3		Cancer extends beyond the capsule of the prostate and may be present in one or more of the seminal vesicles
	T3a	Cancer extends beyond the capsule of the prostate but is not present in one or more of the seminal vesicles
	T3b	Cancer extends beyond the capsule of the prostate and is present in one or more of the seminal vesicles
T4		Cancer in tissues adjacent to the prostate but not one or more of the seminal vesicles
NX		Lymph nodes were not assessed
N0		No regional lymph node metastasis
N1		Regional lymph node metastasis
MX		Distant metastasis was not evaluated
M0		Metastasis beyond one or more regional lymph nodes has not been detected
M1		Metastasis beyond one or more regional lymph nodes
	M1a	Metastasis to one or more distant (outside of the pelvis) lymph nodes
	M1b	Metastasis to one or more bones
	M1c	Metastasis to one or more sites with or without bone disease

Table 11.2 Stage grouping for prostate cancer

Stage	Stage grouping
I	T1a, N0, M0, Gleason score of 2
II	T1a, N0, M0, Gleason score of 2–4
	T1b, N0, M0, Gleason score of 2–10
	T1c, N0, M0, Gleason score of 2–10
	T2, N0, M0, Gleason score of 2–10
III	T3, N0, M0, Gleason score of 2–10
IV	T4, N0, M0, Gleason score of 2–10
	Any T, N1, M0, Gleason score of 2–10
	Any T, any N, M1, Gleason score of 2–10

THERE ARE TRADITIONAL AND UNIQUE TREATMENT OPTIONS AVAILABLE

Following the diagnosis of prostate cancer a man will usually have more than one treatment option available. There are several factors that go into deciding which method will afford the best possible prognosis given the patient's condition, including age and life expectancy, stage of the cancer, growth rate of the tumor, potential risks, complications and side effects of each treatment, the likelihood that a treatment will be curative, and other health conditions. Currently the methods that are available are radical prostatectomy, external beam radiation therapy, brachytherapy, hormonal therapy, chemotherapy, and watchful waiting. Often, a combination of treatments is preferred.

Surgical options vary for prostate cancer

Surgery on the prostate requires general or spinal anesthesia. With general anesthesia the patient is asleep whereas with spinal or epidural anesthesia the patient remains awake but sedated and numb from the waist down. Surgery can be used to remove all or a portion of the prostate.

The surgical removal of the entire prostate, a radical prostatectomy, is an appropriate treatment option when the cancer has not spread outside of the prostate (Figure 11.6). The operation is performed in one of three ways. Two of these procedures enable the prostate to be accessed through incisions in the lower abdomen. In a radical retropubic prostatectomy an incision in the shape of an upside down T is made above the pubic bone. In a radical laproscopic prostatectomy several small incisions

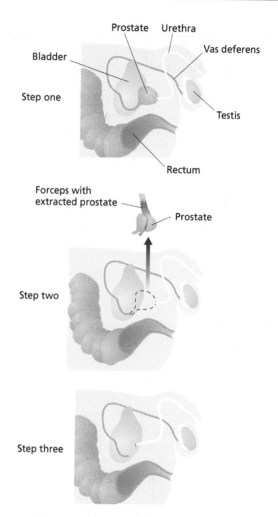

Figure 11.6 Radical prostatectomy. In this procedure the entire prostate and seminal vesicles are removed. It allows the complete removal of the tumor when it is confined to the organ.

are made just above the pubic bone for the insertion of a laparoscope and surgical instruments. The third procedure, a radical perineal prostatectomy, entails gaining access to the prostate through an incision in the perineum, the tissue between the anus and scrotum.

Each procedure has its benefits and drawbacks. The prostate and lymph nodes can be accessed through abdominal, but not perineal, incisions. The minimally invasive laparoscopic procedure is relatively new and requires a great deal of skill by a highly experienced surgeon, but it enables patients to go home the day after the procedure while the retropubic and perineal procedures require a 3-day hospital stay. A prostatectomy is a delicate operation because of the potential to damage or remove two small bundles of nerves, one on either side of the prostate, that are involved in achieving an erection. If the nerves are not within the cancerous region, then the surgeon will try and remove the prostate while sparing the nerves. This is very difficult to do when approaching the prostate from the perineum rather than from the abdomen. During the first 3–12 months after a radical prostatectomy nearly all men will need assistance to achieve an erection in the form of medications (e.g., Viagra, Cialis, Levitra) or vacuum devices. While the ability to achieve an erection will forever be diminished, the degree of erectile dysfunction increases with the age of the patient at the time of the operation.

An alternative surgical procedure that is performed for palliative rather than curative means is transurethral resection of the prostate (TURP) (Figure 11.7). The procedure is typically performed when a man with prostate cancer is not a candidate for a radical prostatectomy but is experiencing difficulty urinating. It is also conducted on a man with a benign enlargement of the prostate. During the procedure, a surgeon extends a thin resectoscope through the end of the penis and into the urethra. When the resectoscope is in the portion of the urethra that passes

Another common side effect of a radical prostatectomy is the loss of bladder control to varying degrees for several weeks to months following surgery. A man typically has a catheter, a thin tube inserted through the urethra and into the bladder, for a period of 10–21 days to drain urine while the bladder recovers. Approximately a third of men that have a radical prostatectomy will forever experience a small degree of incontinence when they sneeze, cough, laugh, or exercise.

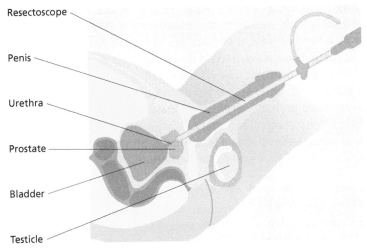

Figure 11.7 Transurethral resection of the prostate. This is a surgical procedure to remove tissue from the prostate by using a resectoscope inserted into the urethra.

through the prostate, the instrument heats up and cauterizes (burns) away the excess tissue. A catheter remains in place for 1–3 days to enable the patient to void urine while the urethra heals.

An option for the treatment of prostate cancer confined to the gland is cryosurgery, which is also referred to as cryosurgical therapy or cryosurgical ablation. During the procedure a surgeon uses TRUS to guide several hollow tubes inserted through the perineum to the prostate. TRUS is continuously used during the procedure to monitor the placement of the tubes, through which very cold liquid nitrogen flows, and the freezing or ablation (destruction) of the tissue. A catheter that contains warm circulating saltwater is placed into the urethra to protect it from freezing.

Cryosurgery is less common than a radical prostatectomy or radiation therapy because of the difficulty of precisely controlling the temperature during the freezing process. Recent advances, however, have improved the capabilities of the procedure, and data from long-term studies needs to be collected to determine its effectiveness compared to the more traditional surgical and radiation therapies. Side effects are more severe when cryosurgery is performed as a secondary method of treatment, that is, when cryosurgery is performed after radiation therapy was unsuccessful. The freezing may affect the functioning of the bladder and rectum but only temporarily. The principal side effect is permanent impotence because of nerve damage that occurs during the freezing of the tissue.

Radiation therapy can be either external or internal

The use of radiation therapy (Chapter 7) is an option when the cancer is localized to the prostate or has spread to adjacent tissue. If the cancer is advanced, radiation can be used to try and reduce the size of the tumor or to kill metastatic cancer in the bones where it is causing pain. It is a noninvasive form of treatment that does not require anesthesia and is done on an outpatient basis. The two forms of radiation treatment that target the prostate and nearby tissues are external beam radiation therapy (EBRT) and brachytherapy. A third form of therapy, systemic radiation therapy, is used to treat metastatic prostate cancer in many bone locations.

A patient may receive external beam radiation therapy by laying on a table and having the diseased area of the prostate subjected to multiple beams of radiation that enter from different angles and converge in the diseased area. This method requires the prior use of imaging techniques to precisely map out the location of the cancer in the prostate.

Common side effects following external beam radiation therapy are associated with the rectum and bladder because they are located adjacent to the prostate. During

> **Why are multiple beams of radiation at various angles or a single beam that rotates with a fixed focal point used rather than a single beam aimed directly at the cancerous area?**

treatment a man may experience diarrhea, colitis (irritated colon), rectal bleeding or leakage, frequent urination, a burning sensation when urinating, and/or blood in the urine. Occasionally, the problems with the colon persist even after treatment is complete. In various studies, anywhere from 30% to 77% of men experienced some degree of impotence following the completion of EBRT. The impotence does not occur during radiation therapy but rather after its completion. EBRT can also result in fatigue that may last for months during and after treatment.

The prostate can also be treated with internal radiation therapy or brachytherapy (Chapter 7) by the insertion of radioactive pellets or "seeds" the size of grains of rice directly into the gland (Figure 11.8). This option is typically available to men who do not have enlarged prostates but have early stage, slow growing prostate cancer. The precise placement of the pellets within the gland is accomplished with the aid of images from TRUS, MRI, or CT scans and the use of a needle inserted through the perineum. In low-dose rate (LDR) brachytherapy approximately 80–120 of the pellets may be implanted permanently, depending on the size of the prostate, and will decay to undetectable levels within 3–6 months.[8.4] Alternatively, pellets that emit larger dosages of radioactivity may be placed into the prostate for 5–15

> [8.4] The levels of radiation become undetectable because the amount of radioactivity will diminish daily.

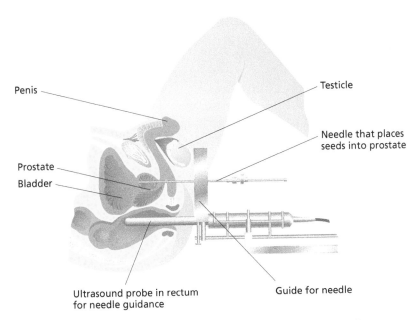

Penis	Testicle
	Needle that places seeds into prostate
Prostate	
Bladder	
Ultrasound probe in rectum for needle guidance	Guide for needle

Figure 11.8 Brachytherapy. During this procedure radioactive seeds are implanted into the prostate by using a needle guided by the use of an ultrasound image. Adapted from: www.stuarturology.com

minutes before removing them. The procedure is repeated up to three times over a series of consecutive days. This method is referred to as high-dose rate (HDR) brachytherapy.

When brachytherapy is used in combination with external beam radiation it is possible to deliver higher dosages of radiation to the diseased prostate tissue. Side effects of brachytherapy can cause incontinence, bowel problems, and impotence but on a much smaller scale than external beam radiation.

Prostate cancer tends to metastasize to bone which can be very painful. EBRT may be used to target the destruction of metastatic prostate cancer in one or more bone locations. When the cancer has metastasized to multiple locations, however, EBRT is a less practical treatment option. Systemic radiation therapy utilizes the radioactive isotopes strontium-99 and samarium-153 which, after being injected, travel throughout the body and accumulate within cancer cells present in the bones. Pain is usually diminished once the radiation has killed the cancer cells.

Hormone deprivation is a factor in treatment for prostate cancer

Prostate cancer cells grow in response to androgens, the male hormones. The vast majority of testosterone, the principal androgen, is produced by the testes while the remainder is produced by the adrenal glands, which reside on top of the kidneys. The objective of androgen deprivation or androgen suppression therapy in the treatment of prostate cancer is to either prevent the synthesis of testosterone or block target cells from responding to testosterone so as to slow the growth or reduce the size of prostate tumors. This form of hormone therapy is used when the cancer has spread beyond the prostate and cannot be treated by surgery or radiation, or as a follow up treatment when surgery or radiation failed to be effective. It can also be used as a neoadjuvant form of therapy to reduce the size of the prostate prior to surgery or radiation, or as an adjuvant in addition to radiation therapy. Typically, the smaller a tumor is, the more easily and effectively surgery or radiation therapy can be performed.

Androgen deprivation is not curative but works for a number of years in controlling the growth of the cancer or maintaining a state of remission. Typically, the cancer will reach a point where it is no longer responsive to hormone therapy and will become androgen independent. Rarely are prostate cancers androgen independent upon diagnosis and those that are independent are likely to not be carcinomas and are best treated with chemotherapy.

There are several methods used to reduce the level of testosterone in the body and all but one are reversible. The surgical removal of both testes, a double radical orchiectomy or castration, removes the source of more than 90% of the testosterone that a man produces. Since this procedure is not reversible or curative, most men opt to try other methods of hormone therapy rather than have their testicles permanently and irreversibly removed, a process that can be psychologically devastating.

The pituitary is a small gland on the underside of the brain. It produces several types of hormones, two of which, luteinizing hormone (LH) and follicle stimulating hormone (FSH), are termed gonadotropins because they are involved in promoting the gonadal tissue (testes and ovaries) to synthesize androgens in men and estrogens in women. The pituitary releases the gonatropins in response to gonadotropin-releasing hormone (GnRH) which is secreted from the hypothalamus, an adjacent structure in the brain. Treatment of prostate cancer with GnRH agonists (Chapter 7) results in an initial release of LH and FSH from the pituitary in high concentrations, but the gland's cells eventually become desensitized (produce fewer receptors) to the agonist signal and no longer secrete the gonadotropins. Once a man has ceased to produce LH and FSH he will also no longer synthesize androgens. During the initial phase of treatment, the cancer is likely to grow in response to the surge in LH and FSH; this is known as a flare reaction. A man with metastatic prostate cancer in the bones may experience an increase in bone pain during the flare reaction. If the prostate cancer has metastasized to the spine, the flare reaction may result in not only pain but also paralysis as a result of the cancer's growth compressing the spine.

Whereas agonists stimulate the pituitary to the point that it becomes desensitized to GnRH and no longer secretes LH or FSH, GnRH antagonists (Chapter 7) bind to the same receptors but do not stimulate the cells. This effectively prevents the gland from receiving the signal necessary for it to produce and release the gonadotropins required to stimulate the testes to produce androgens. While men taking GnRH antagonists do not experience a flare reaction, up to 5% of patients have undergone severe allergic reactions to the drug. For this reason, GnRH antagonist therapy is reserved for those men with advanced prostate cancer or for those who are unable to have or have refused other forms of hormone therapy.

While GnRH agonists and antagonists function to reduce the levels of LH and FSH, an alternative therapeutic strategy is to prevent already synthesized androgens from binding to their receptors on hormone responsive tissue cells. Antiandrogens are testosterone antagonists that bind in place of testosterone to the androgen receptors on cell membranes, effectively blocking the cells' ability to respond to androgens. Prostate cancer cells would no longer be receiving the stimulus to grow that they normally receive from testosterone. Antiandrogens can be used alone but are usually a part of combination hormone therapy or combined androgen blockage (CAB). They are commonly used in combination with an orchiectomy or during the early stage of treatment with GnRH agonists in an attempt to prevent the flare reaction.

> **How would antiandrogens prevent the flare reaction caused by GnRH agonists?**

The treatment to decrease testosterone levels or prevent cells from responding to androgens has a number of side effects, such as a loss of libido (sexual desire), hot flashes, tenderness and enlargement of the breasts, loss of muscle and bone mass, a greater chance of bone fracture, anemia, weight gain, fatigue, depression, and decrease in mental acuity and HDL (the so-called "good" cholesterol). A man experiencing some of these side effects may perceive that he is being emasculated or feminized, even if he did not undergo a double radical orchiectomy. Fortunately, many of the side effects can be readily controlled through medication and exercise.

Chemotherapy is used in advanced stages of prostate cancer

Chemotherapy (Chapter 7) is the use of drugs to damage cellular DNA or to interfere with the functions of cells in an attempt to kill them or to slow their rate of division. Cancer cells typically have a reduced ability to repair DNA damage, making them more susceptible than healthy cells to chemotherapeutic drugs. Chemotherapy is usually reserved for the treatment of men with prostate cancer that has either metastasized beyond the prostate, has not responded well to hormone therapy, or has returned following previous treatment methods.

The objective of treating prostate cancer with chemotherapy is not to obtain a cure but rather to slow the growth or reduce the size of the cancer. Typically, a patient takes two or more drugs in order to minimize the likelihood that the cancer cells will become resistant to either of the drugs. Regardless of the drugs used, many of the side effects are the same and are a consequence of their effects on healthy cells, especially those that divide rapidly, as well as on cancer cells. Many of the side effects can be treated with medications and all are temporary, ranging from weeks to months in duration.

Watchful waiting is unique to prostate cancer

In its early stages, prostate cancer is often slow growing and slow to metastasize. As a result, one treatment option may be watchful waiting, also known as expectant management, observation, or deferred therapy, in which no active form of treatment is taken, but rather the patient undergoes more frequent testing, such as a DRE and PSA test every 6 months, and has a yearly TRUS-guided biopsy. A man may opt for watchful waiting if the cancer is confined to a single area of the gland and is expected to grow very slowly, or if he is currently not experiencing any symptoms, has other serious health problems, or is older and expected to live out his normal lifespan without any further problems. The benefit of watchful waiting is that the man would not be at risk for the complications associated with other treatments.

A large study demonstrated that men who were diagnosed with early stage prostate cancer but not treated for it were asymptomatic for 10–15 years. It is not known whether watchful waiting is any more effective than prostatectomy for treatment in the early stages. In an attempt to answer that question, the National Cancer Institute and the Department of Veterans Affairs are currently conducting the *prostate cancer versus observation trial* (PIVOT), a randomized trial of 2000 men below the age of 75 that have been diagnosed with clinically localized prostate cancer.

PROSTATE CANCER MORTALITY RATES HAVE DECREASED

Prostate cancer is the second leading cause of cancer death in men, second only to lung cancer. The encouraging news is that the mortality rate has fallen about 3.6% every year since 1992. Probable reasons for this decline include the successful early detection of prostate cancer by PSA tests and/or the advances in the effective treatment for both early and advanced occurrences. The prognosis for greater than 90% of men diagnosed is very good as long as the disease is detected at the local and regional stages. These men have nearly 100%, 93%, and 77% survival rates after 5, 10, and 15 years, respectively. The less than 10% who are diagnosed with metastasis to distant locations have a 5-year survival rate of 34%.

MUCH ATTENTION IS BEING GIVEN TO THE NUMBER ONE CANCER AFFECTING MEN

There is a great deal of research being conducted and a large number of studies being carried out to better understand the causes and determine the best diagnostic and treatment methods for prostate cancer. Efforts are being made to identify and characterize genes involved not only in the development of prostate cancer but also in its aggressiveness. The *EZH2* gene has recently been found to be expressed more often in aggressive prostate cancers than in nonaggressive ones. If further studies determine that *EZH2* can serve as a marker gene for aggressive cancers then it could also serve as a means of screening men with early stage cancer and deciding who would or would not be a good candidate for watchful waiting. A new technique being studied is Doppler ultrasound which produces a color image of the prostate gland based on the volume of blood flow within it. Cancerous lesions, which tend to have a greater number of blood vessels, are more likely to be detected by this method and allow a urologist to better pinpoint specific areas to biopsy.

Research in new treatments is ongoing. For instance, some angiogenesis inhibitor drugs have been approved by the FDA for the treatment of other cancers and are now being tested in clinical trials for their effectiveness in treating prostate cancer. Endothelin A is used by prostate cancer cells to grow, and is therefore a target of inhibition by a drug that has shown favorable results when used to treat advanced cases. An alternative approach to cryotherapy that is still in its infancy is to kill the cancer cells by heating (rather than cooling) them using either high intensity sound waves or magnetic fields. Under current review is the potential effectiveness of delivering radioactivity specifically to prostate cancer cells by using tagged monoclonal antibodies. Currently, there are several vaccines that are being investigated in clinical trials but none have been approved by the FDA.

Advances in medicine, treatment, and technology have been nothing short of startling in recent decades. An even greater promise can be foreseen in the not too distant future. If other cancers respond to research to the same extent as prostate cancer, then prospects for combating these diseases will be brightened considerably.

EXPAND YOUR KNOWLEDGE

1 The DRE and PSA test are both means of screening for prostate cancer. Why should they both be performed rather than one or the other?

2 Which of the following men is at a greater risk of having prostate cancer? Explain:
   (a) Man 1: PSA level of 5 ng/mL with a prostate volume of 4.3 mL^3.
   (b) Man 2: PSA level of 7 ng/mL with a prostate volume of 6.4 mL^3.

3 When should a biopsy of the prostate be performed?

4 What may be a serious side effect of the flare reaction that usually occurs with GnRH agonist treatment in a patient who's prostate cancer has metastasized to his spine?

5 In what ways are the treatment options for men with prostate cancer similar to those available to women with breast cancer?

6 Investigate the use of robotics in prostate cancer surgery.

ADDITIONAL READINGS

Marieb, E. N. and Hoehn, K. (2007) *Human Anatomy and Physiology*, 7th edn. Upper Saddle River, NJ: Prentice Hall.

Martini, R. and Bartholomew, E. (2007) *Essentials of Anatomy & Physiology*, 4th edn. Upper Saddle River, NJ: Prentice Hall.

Moore, G. E. (2005) *A Call to Action: The Michael Milken Story*. Santa Monica, CA: The Prostate Cancer Foundation.

The Prostate Cancer Foundation's website: www.prostatecancerfoundation.org

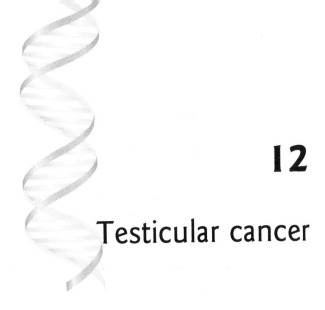

12

Testicular cancer

> I think about what I went through with cancer every single day, if not every single minute of every single day, because it's changed the person I am. It's made me a stronger, much better person.
>
> Chris Crichton, testicular cancer survivor

CHAPTER CONTENTS

- Testicular cancer statistics
- Structure and function of testicles
- There are three types of testicular tumors
- Risk factors for the disease
- Symptoms of testicular cancer
- Testicular self-examination (TSE) is recommended
- Blood and imaging tests are used to determine diagnosis and possible metastasis
- Testicular cancer treatment results in a high cure rate
- Causes and treatments are being studied
- Expand your knowledge
- Additional readings

Cancer: Basic Science and Clinical Aspects, 1st edition. By C. A. Almeida and S. A. Barry. Published 2010 by Blackwell Publishing, ISBN 978-1-4051-5606-6.

A diagnosis of testicular cancer, even if it has metastasized, need not be devastating as this particular cancer can be treated and most often cured, especially when detected early. Recent developments in the treatment of this disease make it one of the true success stories. Still, the incidence of this disease, which is rare compared with other types of cancers, has risen in recent decades and it is not yet known why.

TESTICULAR CANCER STATISTICS

Compared to other cancers discussed in this book, testicular cancer has a very low incidence and mortality rate. Approximately 8,000 of men developed this disease while almost 400 died from it in 2008. Although testicular cancer accounts for approximately 1% of all cancers in American men, it is the most commonly diagnosed and deadliest cancer of men between 15 and 34 years old, occurring most often between the ages of 20 and 39.

It is most prevalent in white men, especially those of Scandinavian descent. The testicular cancer rate has more than doubled among white men in the past 40 years, but has only recently begun to increase among black men. Hispanic, American Indian, and Asian men develop it at a higher rate than black men but at a lower rate than white men. The reason for the increased numbers and racial differences, although intriguing, is not known. The number of deaths has decreased dramatically since the 1960s due to the many improvements in treatment, technology, and early detection.

STRUCTURE AND FUNCTION OF TESTICLES

The testes (testicles) are the primary reproductive organs (gonads) of the male (Figure 12.1). Men usually have a pair of these oval-shaped glands, each somewhat smaller than a golf ball, hanging beneath the base of the penis in a sac of skin called the scrotum.

Testicles produce sperm and testosterone

Testes produce male hormones, particularly testosterone, that are essential for the development and maintenance of the male secondary sex characteristics including the growth of hair on the face and body, an increase in skeletal muscle mass, and the enlargement of the larynx that causes the deeper pitch of the male voice. The testes primarily serve as the site of production and maturation of male gametes (sperm). As they mature, sperm travel from the testicle to the epididymis, which is a short hollow tube on the outside of the testicle and may appear as a small "bump" on

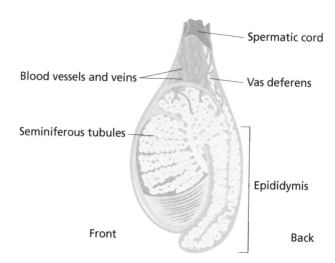

Figure 12.1 The testicle. Sperm are produced in the seminiferous tubules and stored in the epididymis.

the outer or middle sides of the testis. During sexual arousal, the sperm travel through tubes (vas deferens) where seminal fluid produced by accessory reproductive glands (seminal vesicles and prostate) is added to produce semen, which is extraordinarily concentrated and dense at approximately 100 million sperm per milliliter. This fluid travels through the paired vas deferens to a single tube (urethra) in the center of the penis, and is then expelled out of the body during ejaculation.

> **What is the difference between sperm and semen?**

THERE ARE THREE TYPES OF TESTICULAR TUMORS

The testicles are made up of several different kinds of cells, two of which, germ and stromal, have the potential to develop into testicular cancer. Ninety-five percent of testicular cancers are germ cell (Chapter 4) tumors, the two main types being seminomas and nonseminomas which are identified by their differences in appearance under the microscope.

> **What are germ cells? What other type of cancer covered in this book is a germ-cell cancer?**

Further classifications are necessary

Seminomas, the most common malignancies affecting the testes, are believed to arise from the epithelium or lining of the sperm-producing/ transporting tubules in the gland and tend to be more responsive to treatment. They are further classified as classical seminomas, which represent 95% of testicular cancers, and the rarer spermocytic seminomas. Classical seminomas usually occur in an older population of men than do nonseminomas. Men with the less common spermocytic seminomas are usually diagnosed around the age of 55, or about 10–15 years older than those with classical seminomas. Spermocytic seminomas are slow growing and usually not metastatic. Greater than 90% of patients with seminomas are cured regardless of the stage of the cancer.

Nonseminomas tend to develop earlier, grow faster, and spread more quickly than do seminomas. They usually occur in men between their late teens and early forties. Nonseminomas are further classified into embryonal carcinoma, yolk sac carcinoma, choriocarcinoma, and teratoma and various combinations of these types. The type is not of great significance, though, because all are treated using the same therapy. They do, however, have different patterns of metastasis. Some tumors are a mix of both seminomas and nonseminomas and are treated as nonseminomas.

Stromal tumors are a rarer type of cancer that may develop in the supportive tissues of the testicles where hormones are produced. There are two types of stromal tumors, Leydig cell tumors, occurring in the cells that produce male sex hormones, and Sertoli cell tumors, found in the cells that provide nourishment to germ cells. Stromal tumors that spread beyond the testicles are treated differently than those that have not metastasized. Although each may have a unique treatment, they are usually all treated successfully.

RISK FACTORS FOR THE DISEASE

Testicular cancer is more often found in younger than older men

Testicular cancer is typically, but not always, a disease of young men. It is the most common cancer to occur between the ages of 20 and 34, the second most common between the ages of 35 and 39, and the third most common cancer between the ages of 15 and 19. Other than age, most men have no common risk factors for developing testicular cancer, but some associations have been made.

Some factors associated with testicular cancer

One commonality among patients may be cryptorchidism, an undescended testicle (Figure 12.2). Most boys' testicles normally develop in the abdomen and then move down into the scrotum before birth. If a testicle does not descend by birth, it will often do so within the baby's

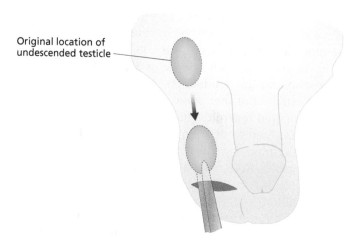

Original location of undescended testicle

Figure 12.2 Correction of cryptorchidism. An undescended testicle that had failed to move into the scrotum during development of a male fetus is surgically lowered into the scrotum.

first year. If this does not occur, surgery may be performed to correct the condition. About 14% of testicular cancer cases occur in men who have had cryptorchidism, arising particularly in the affected testicle. Up to 25% of the cases, however, have occurred in the normal testicle, leaving scientists to wonder if some other problem causes both conditions.

Family history also increases the risk in that if a man has the disease, his brothers or sons may also develop it, yet very few multiple cases occur in families. A team of scientists from the Institute of Cancer Research has discovered that one of the genes responsible, *TGCT1*, is on the X chromosome, which men inherit from their mothers. If mutant alleles of this gene are inherited, a man's risk may increase by as much as 50 times, but much more work needs to be done to clarify its role. If a man has had cancer in one testicle, he has about a 3–4% chance of eventually developing cancer in the other testicle. Congenital abnormalities of the testicles, penis, or kidney, or an inguinal hernia (hernia in the groin area) may be factors that place a man at an increased risk of developing this disease. One other risk factor is Klinefelter's syndrome (Figure 12.3). Men with

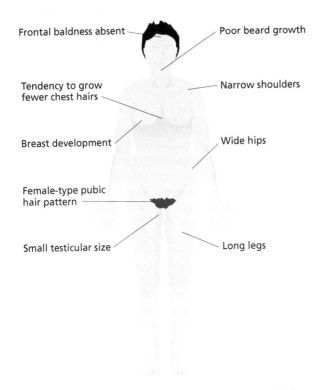

Frontal baldness absent

Poor beard growth

Tendency to grow
fewer chest hairs

Narrow shoulders

Breast development

Wide hips

Female-type pubic
hair pattern

Small testicular size

Long legs

Figure 12.3 Klinefelter's syndrome. Males who have an XXY chromosome pattern rather than the usual XY pattern found in their cells have Klinefelter's syndrome. Not every man with Klinefelter's syndrome has all of the symptoms above, as each case is unique.

this rare condition have an extra X chromosome, resulting in low levels of testosterone, small testes, sterility, and enlarged breasts.

Recent studies have determined that multiple pigmented skin spots or moles on the skin of the back, chest, abdomen, and face, also known as atypical nevi, increase both the risk of testicular and melanoma skin cancers (Chapter 13). Interestingly, men that have been cured of testicular cancer are at an increased risk of developing a melanoma later in life as well. A slightly higher risk of testicular cancer has been associated with men infected with human immunodeficiency virus (HIV), especially those with AIDS.

SYMPTOMS OF TESTICULAR CANCER

Most testicular cancers are detected by the patient unintentionally during self-examination or by a physician during a routine examination. The first sign may be a small, painless lump or swelling on either testicle. An early tumor may be about the size of a pea, but it can grow rapidly to the size of a marble or even larger. Some men may feel pain or discomfort with or without swelling or enlargement in the testicle or scrotum.[12.1]

> 12.1 Men may have similar symptoms when they have a bacterial infection in the epididymis, which is treatable by antibiotics.

There are a variety of conditions that can result in symptoms that mimic those of testicular cancer. Sometimes the testicle can become enlarged because fluid has collected around it resulting in a hydrocele, or the veins in the testicle can dilate, causing enlargement and lumpiness that is termed a varicocele. Orchitis is the painful inflammation of the testicle, causing swelling that can be a result of a bacterial or viral infection. Any kind of testicle injury may also produce similar symptoms. Only a healthcare provider can distinguish between these conditions and a testicular tumor, and virtually all of them are treatable by medication and surgery.

Symptoms may occur in areas away from the testicles

Other symptoms may be a feeling of heaviness, discomfort, pain or sudden collection of fluid in the scrotum, or a dull ache in the lower abdomen, back or groin. In some rare cases, men will experience breast tenderness or growth (gynecomastia), because some tumors produce hormones that affect breast development.[12.2] Other effects of

> 12.2 The normal increase in pectoral (chest) muscle mass that occurs in adolescent males is not the same as gynecomastia.

excessive hormone production may be a loss of sex drive or the growth of hair on the face or body at a younger age than usual (before puberty). A small number of men with testicular cancer have no symptoms, especially in the early stage. The condition may be discovered in later years during testing for infertility, which seems to have a connection to this disease. According to a 2005 study, men with infertility are 20 times more likely to have testicular cancer compared with the general population.

Even metastatic testicular cancer may not have any symptoms, but lower back pain is often an indicator of later stage disease. If it spreads to the lungs, a man may experience shortness of breath, chest pain, coughing, or the spitting up of blood. Advanced testicular cancer can also spread to the liver or brain and *still* be successfully treated, an indication of how successful medical science has been in harnessing this one particular form of cancer.

TESTICULAR SELF-EXAMINATION (TSE) IS RECOMMENDED

Examination of a man's testicles is traditionally part of a general annual physical examination. Men need to be aware of the symptoms of testicular cancer and should seek prompt medical evaluation if a lump, no matter its size, is found. Similar to breast cancer, there is no evidence that screening with clinical examination or testicular self-examination (TSE) is effective in reducing *mortality* from this disease, but by examining one's testicles on a regular basis, a man will become familiar with what is normal and what is different. As mentional previously, the current treatment interventions provide very favorable outcomes. TSE remains a method of early detection, however, and many physicians recommend that all men perform monthly testicular self-examinations after puberty.

How to perform a TSE

The best time to conduct the testicular self-examination is during or after a bath or shower, when the skin of the scrotum is relaxed. Suggested steps are as follows:
- Hold the penis out of the way and examine each testicle separately.
- Hold the testicle between the thumbs and fingers with both hands and roll it gently between the fingers.
- Look and feel for any hard lumps or nodules (smooth rounded masses) or any change in the size, shape, or consistency of each testicle. It is common to have one testicle slightly larger, or hang lower than the other, but any noticeable increase in size or weight is significant.

Box 12.1

Lance Armstrong LIVESTRONG™

On a daily basis, there is the likelihood that you will see someone wearing a yellow rubber LIVE**STRONG**™ bracelet. The wristbands represent unity in the battle against all cancers and hope for all those who are living with the disease (Figure 12.4).

Lance Armstrong's name and story are familiar to most of us in this information age. In 1996, this well-known competitive cyclist, at the age of 25, was diagnosed with an aggressive form of testicular cancer, having ignored the early warning signs. The subsequent spread of the disease led to two surgeries, resulting in the removal of a testicle and cancerous lesions on his brain. Chemotherapy complemented his treatment.

After a forced retirement to allow for his recovery, Armstrong returned to his sport with a new found determination born from his desire to make the public more aware of methods involved in cancer diagnosis and treatment. He became a household name for his masterful performances in the grueling Tour de France, the Super Bowl of cycling, winning the race a record seven consecutive times

Figure 12.4 The LIVE**STRONG**™ yellow wristband represents each wearer's commitment in the cancer battle. The cost per bracelet is only $1, but the total amount raised from their sale is now in the millions. Other charitable foundations have followed suit, and it is not uncommon to see pink, blue and other colored wristbands being sold as fundraisers.

Figure 12.5 Lance Armstrong, a competitive cyclist, established his foundation during his treatment, and before he knew he would recover from testicular cancer. His grassroots campaign has raised millions of dollars used for research and support for all cancers and has inspired many people to become involved.

between 1999 and 2005 (Figure 12.5). He subsequently retired from competitive riding and now devotes his energies to the Lance Armstrong Foundation, which is dedicated to cancer advocacy for those affected and to raising money for cancer research. This includes the LIVE**STRONG** Survivor*Care* network which provides services and counseling for all those affected by cancer including patients, survivors, and caregivers.

BLOOD AND IMAGING TESTS ARE USED TO DETERMINE DIAGNOSIS AND POSSIBLE METASTASIS

As with any disease, a complete medical history and general physical examination must be performed by a medical care provider to determine the general health of the patient. The testicles must be observed for any sign of swelling, lumps or tenderness and the abdomen must be palpated in order to see if the lymph nodes inside are enlarged, as this could be a sign of metastasis.

Ultrasounds and blood tests are important for determining the presence of the disease

If there is a question of testicular cancer, one of the first tests to be ordered will be an ultrasound, a painless imaging procedure that will visualize a testicular mass in order to help differentiate testicular cancer from an infection or other condition. Blood tests are also utilized, since there are tumor markers that may indicate the presence of testicular cancer. The two primary protein markers utilized are alpha-fetoprotein (AFP) and human chorionic gonadotropin (hCG). One or the other of these proteins is elevated in about 85% of nonseminoma cancers. These tests are not diagnostic but they are indicators to be used in conjunction with imaging studies to determine if cancer is present and/or to follow the progress of recovery after treatment. Lactate dehydrogenase (LDH) is an enzyme, found in many body tissues, that is released into the blood when cellular damage occurs. Although not specific for testicular cancer, if the level is elevated, it can give the physician additional information to enable a diagnosis.

Imaging tests are important, not in the initial diagnosis, but in determining if there is metastasis. A chest X-ray is a simple test that could demonstrate if the cancer has spread to the lungs, but the superior CT scan will help to determine if it has spread to the lungs, liver, or other organs, while MRI scans are typically used to examine the brain and spinal cord. PET scans can also determine whether enlarged lymph nodes contain scar tissue or active tumor. Table 12.1 provides the staging (Chapter 6) for testicular cancer.

> PET scans can determine whether a mass is an active tumor or simply scar tissue. Can other imaging techniques do this? Why or why not?

TESTICULAR CANCER TREATMENT RESULTS IN A HIGH CURE RATE

The prognosis and treatment options for testicular cancer depend on the stage of the cancer, levels of the protein markers, type of cancer, size of the tumor, and number and size of invaded retroperitoneal lymph nodes.

Surgical variations depend on the cell type

Testicular cancer is a highly treatable, often curable form of cancer that is treated according to cell type. Most cases are successfully treated with

Table 12.1 Staging of testicular cancer

Stage 0	(Carcinoma *in situ*): germ cell cancer that is contained and preinvasive
Stage I	No metastasis to lymph nodes or distant sites, and blood tests are normal
Stage II	Metastasis to regional lymph nodes only and has not spread to distant lymph nodes or organs
Nonbulky stage II	Metastasis to retroperitoneal (behind the abdominal cavity) lymph nodes; lymph nodes are not larger than 5 cm (2 inches).
Bulky stage II	Metastasis to 1 or more retroperitoneal lymph nodes; lymph nodes are larger than 5 cm
Stage III	Metastasis to distant lymph nodes and/or to distant organs, such as the lungs or liver
Nonbulky stage III	Metastasis confined to lymph nodes and lungs, and no mass is larger than 2 cm (about $^3/_4$ inch)
Bulky stage III	Large number of metastases, and metastases of the lymph node are larger than 2 cm, and/or cancer has spread to other organs, such as the liver or brain

surgery and/or radiation therapy, and/or chemotherapy depending on the size, type and location of the tumor, whether or not the cancer has spread, and the person's overall health. Biopsies are not commonly used as the primary method of diagnosis since, if cancer tissue is present, it may spread into the scrotum, leading to a more complicated treatment. Most often, the first approach is to remove the affected testicle through an incision in the groin in a procedure called a radical inguinal orchiectomy. Although this is a radical approach, it is important to remember that having only one testicle does not affect a man's ability to achieve a normal erection or have children and is therefore not quite as drastic as it might appear at first glance. One option that some men choose is to have an artificial testicle (saline filled prosthesis) implanted in the scrotum that both weighs and feels the same as a natural testicle.[12.3]

[12.3] Since this is typically a cancer found in younger men, they should consider storing frozen semen (sperm banking) before their treatment begins in case they decide later to start a family or have more children.

Open procedure   Laparoscopic procedure

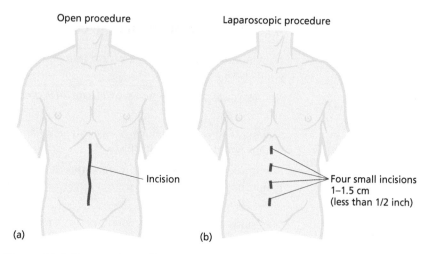

(a)   (b)

Figure 12.6 Retroperitoneal lymph node dissection. Lymph nodes from within the abdomen are removed in this procedure. (a) Typically the procedure involves a midline abdominal incision. (b) A minimally invasive approach is being studied in which doctors perform the surgery through several small abdominal incisions using a laparoscope.

Some lymph nodes within the abdomen may also be removed during surgery or in a subsequent operation called a retroperitoneal lymph node dissection (RPLND). There are two options for this procedure – the standard surgery which uses a large incision, or one that creates a small incision using a laparoscope (Figure 12.6). Obviously, the second is less invasive and the patient has a quicker recovery. One consequence of lymph node dissection is that damage can occur to the nearby nerves that control ejaculation. The man can still have an erection, but the sperm are diverted into the bladder rather than out of the body and this may result in infertility. This has led some surgeons to develop a nerve-sparing surgical technique, similar to the one developed for prostate cancer (Chapter 11).

Radiation and chemotherapy may be utilized

Patients typically receive approximately 15 treatments of external beam radiation therapy and may experience mild skin reactions, upset stomach, fatigue, and loose bowel movements. Sperm production may be temporarily affected, but this is usually not permanent. Advanced testicular cancer that has spread to the brain may also be treated by radiation therapy directed at the brain.

Chemotherapy is recommended for patients with metastatic non-seminoma testicular cancer. The most common regimen is a combination of three drugs – bleomycin (Blenoxane), etoposide (VePesid, VP-16), and

Box 12.2

A revolution in treatment

John Cleland was diagnosed with testicular cancer at the age of 22 in 1973, when the cure rate for this disease was about 5–10%. He was given three chemotherapy treatments, but they caused mouth sores so severe that he could not eat or drink and the cancer had metastasized to his lungs. His oncologist, Dr Lawrence Einhorn, currently a Distinguished Professor of Medicine at Indiana University, researched a then new platinum-based drug called cisplatin (Figure 12.7). *Ten days* after John's first treatment with cisplatin in 1974, the metastatic tumors were no longer visible in his lungs. The next year was spent watching and waiting, and for the

Figure 12.7 Lawrence Einhorn, MD. Dr Einhorn is credited with the development of a curative treatment for testicular cancer. He is also an authority on cancers of the lung, some types of urologic cancers, and certain other tumors and is an elected member of the National Academy of Sciences, one of the highest honors accorded to an American scientist. Photo courtesy of Dr Einhorn.

first time, the word *cure* was being used. Thirty plus years later, John Cleland is married, a father of three children, and a high school biology teacher. He spends much of his time talking to young men about their health and early detection of testicular cancer and to others who have been newly diagnosed. He has run a number of marathons but the race he is proudest of winning was against testicular cancer. Since the Food and Drug Administration approved cisplatin in 1978, it has been widely used against a number of other cancers, including ovarian, cervical, esophageal, head and neck, and gastric. While none of these has the same "cure" response as does testicular cancer, it has helped to extend the lives of many patients.

the platinum-based drug cisplatin (Platinol) – that is given intravenously. As always, the highest response rate is when the cancer is detected at an early stage. Once the chemotherapy regimen has been completed, X-rays or CT scans are utilized to see if there are any residual masses (they may be cancer or scar tissue), which may lead to further surgery. At this point, if any cancer is found (a rare occurrence), it is treated by some combination of the three traditional drugs or by using just etopaside and cisplatin. Chemotherapy is not without additional health considerations. The long-term effects can affect the brain resulting in memory loss, decreased speed of processing information, anxiety, depression, fatigue, and a lowered attention span. Scientists are working to find alternative treatments for the minority of men with advanced disease who do not respond as well to current chemotherapy regimens.

CAUSES AND TREATMENTS ARE BEING STUDIED

Much research is being conducted to determine the alleles of specific genes that are involved in the development of testicular cancer. The identification of particular genes associated with the disease is valuable to the development of diagnostic genetic screening tests as well as the determination of the most appropriate form of treatment. The majority of testicular tumors have some abnormality on chromosome number 12. Located on this chromosome is the *hiwi* gene which produces a protein in germ-line but not somatic cells of the testes. One study demonstrated that in nearly two-thirds of testicular seminomas that were analyzed, the *hiwi* gene was expressed at a level higher than normal. In 10 non-seminomas, the expression level of the gene was that found in a normal germ-line cell. The evidence indicates an association between the *hiwi* gene and testicular germ cell tumors (TGCTs).

This same gene is also being studied in those families in which men have a history of undescended testicles, or those with cancer in both testicles. Families containing several closely related men who have testicular cancer are being pursued in the hope of identifying other genes associated with the disease. Currently, there is evidence that there are at least three additional genes. Drugs used for the treatment of other cancers, such as paclitaxel (Taxol), are also being studied to determine if they lower the relapse rate. Clinical trials are underway to determine the effect of a combination of high-dose chemotherapy and stem cell transplants (Chapter 16) as a treatment for patients with advanced germ cell cancers or recurring testicular cancer. Hopefully, the knowledge gained will increase the cure rate of this young man's cancer even higher.

EXPAND YOUR KNOWLEDGE

1 Compare and contrast testicular cancer to prostate cancer in incidence, population, treatment, and cure rates.
2 At what age should TSE begin? When is the best time to perform it? Why is it not routinely recommended by cancer organizations?
3 Explore another chemotherapeutic drug that has been as successful as cisplatin in treating more than one type of cancer.
4 What other cancers are associated with HIV/AIDS?
5 Give some examples of other conditions that mimic the symptoms of testicular cancer.

ADDITIONAL READINGS

Marieb, E. N. and Hoehn, K. (2007) *Human Anatomy and Physiology*, 7th edn. Upper Saddle River, NJ: Prentice Hall.

Martini, R. and Bartholomew, E. (2007) *Essentials of Anatomy & Physiology*, 4th edn. Upper Saddle River, NJ: Prentice Hall.

National Comprehensive Cancer Network, Clinical Practice Guidelines in Oncology. Testicular Cancer. v. 2.2009 (2006) www.nccn.org/professionals/physician_gls/PDF/testicular.pdf

Purdue, M. P., Devesa, S. S., Sigurdson, A. J., and McGlynn, K. A. (2005) International patterns and trends in testis cancer incidence, *International Journal of Cancer*, **115**: 822–827.

US Preventive Services Task Force (2005) Screening for Testicular Cancer: Recommendation Statement. *American Family Physician*, **72**: 2069–2070.

13

Skin cancer

> If just one family, one person, were spared this series of events because they heard me speak, then it would be worth every scar, every tear, and every hour of not knowing [what would happen to me].
>
> Shonda Schilling, melanoma survivor, co-founder SHADE Foundation

CHAPTER CONTENTS

- Skin cancer statistics
- Structure and function of the skin
- Three types of skin cancer
- Risk factors for developing skin cancer
- Methods used to screen for skin cancer
- Surgery and chemotherapy are standard treatments for metastatic skin cancer
- What happens after skin cancer treatment
- Limited UV radiation exposure is the number one form of prevention
- Expand your knowledge
- Additional readings

Cancer: Basic Science and Clinical Aspects, 1st edition. By C. A. Almeida and S. A. Barry. Published 2010 by Blackwell Publishing, ISBN 978-1-4051-5606-6.

Skin cancer is the most common form of cancer, and fortunately one of the most curable when detected and treated early. Over the last several decades there has been a consistent and dramatic increase in the number of skin cancer diagnoses as well as a decrease in the average age at the time of diagnosis. This is thought to be attributable to two primary factors: early detection/self awareness, and an increase in outdoor, recreational lifestyles which give rise to sun exposure. It is estimated that >90% of skin cancers are caused from sun exposure yet <33% of people of all ages use proper sun protection.

SKIN CANCER STATISTICS

Currently one out of five Americans and one out of three Caucasians will develop some form of skin cancer in their lifetimes. Some estimates put the total number of skin cancer diagnoses at just over one million in the US – a truly staggering figure considering that it is equivalent to the total number of occurrences for all other forms of cancer *combined*! Actual numbers are not known because of a uniqueness in the reporting of skin cancer diagnoses. All other forms of cancer have established registries to which diagnoses are reported, allowing accurate statistics to be maintained. There are two types of skin cancer, the much more common nonmelanoma and the more dangerous melanoma. Nonmelanomas, which comprise approximately 96% of skin cancer diagnoses, are not reported to a single registry. As a result, the numbers of diagnoses reported are estimates only. While more than three times the number of people die of melanoma than nonmelanoma, skin cancer that has been detected and treated prior to it having metastasized has a cure rate of 90–95%. This is not to imply, though, that the disease is incapable of disfiguring and killing.

> **Most cancer sources do not list skin cancer's incidence rate. Why?**

Each type of skin cancer is named after the type of epithelial cell in the outer layer of the skin from which it develops. The first two, basal and squamous cell carcinomas, are referred to as nonmelanomas to distinguish them from the third and more serious type, melanoma, which derives from pigment-producing melanocytes. It was estimated that about 96% of the more than one million skin cancer diagnoses in the US in 2008 were nonmelanomas, and more than three quarters, approximately 800,000, were basal cell carcinomas. With only 62,000 diagnoses anticipated that year, melanoma is the rarest, but deadliest, form of the disease.

STRUCTURE AND FUNCTION OF THE SKIN

The outermost layer of our bodies is covered by skin that is composed of epithelial cells, connective tissue, muscles, and nerves. The variety of cell types and functions that the skin carries out classifies it as an organ, in fact, it is the largest organ in the body. Figure 13.1 provides a cross-sectional view of a typical segment of skin, which is composed of two major layers, the dermis and epidermis. The dermis is the lower of the two layers and consists of a somewhat dense layer of connective tissue that is rich in collagen, a type of protein that provides strength and flexibility.[13.1] Blood vessels that provide nourishment are found in this layer but do not extend into the epidermis. Hair follicles as well as sweat and sebaceous glands originate in the dermis and extend into the upper epidermal layer. The base of a hair follicle is lined with a specific type of epidermal cell that forms hair. As the epidermal cells divide,

> [13.1] Our ears and the tips of our noses possess a good deal of collagen and its resilience makes it possible to bend and flex them. Collagen is a substance in current vogue as people seeking a more youthful appearance have it injected into certain facial tissues.

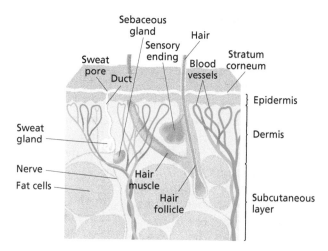

Figure 13.1 Cross-section of skin. The three principal layers of skin are the epidermis, dermis, and subcutaneous layer.

they move toward the center of the follicle and fuse together to form a hair that will continue to grow as long as the epidermal cells are dividing. Attached to each hair follicle is one or more sebaceous glands that produce sebum, a substance that contains a mixture of liquid fats and dead cells from the interior lining of the gland. Sebum coats and conditions the hair and top layer of skin to prevent drying. Areas of the body that do not have hair also possess sebaceous glands and the sebum is brought to the surface of the skin by ducts.[13.2] Sweat glands produce, as the name implies, sweat, which is essentially a salt water solution. The evaporation of the water from the skin has a cooling effect that helps to transfer body heat to the environment, thereby enabling the body to regulate its temperature. During extreme conditions, such as extreme exertion, the body is able to produce up to 1.5 liters (approximately 1.6 quarts) of sweat an hour. The dermis also possesses nerve endings, some of which extend into the epidermis. The nerves allow for the detection of pain, temperature changes, and various tactile sensations. Attached to each hair follicle is a nerve and a smooth muscle fiber. The nerve enables the detection of the slightest movement of the hair and contraction of the muscle causes the hair to "stand on end."

The outer layer of skin, the epidermis, is much less complex in its structure and composition yet plays a critically important

[13.2] Infections of sebaceous ducts and hair follicles can result when they become clogged with dirt or when there is an excess production of sebum, which often occurs during adolescence. Bacteria feed off of the fats and multiply triggering an immune response that causes each affected area to become swollen and red forming pimples or acne. The white discharge from a ruptured pimple is a collection of tissue fluid, dead bacterial cells, and white blood cells that gathered in the area to fight the infection.

role as the primary barrier between the environment and inner tissues and organs. When a cross-section of epidermis is viewed under a microscope, two main types of cells are revealed. The basal cells are round and form the lower layer of the epidermis. The thin external layer is composed of squamous cells that are flat in their appearance. The basal cells that are just above the dermis divide continuously, pushing the previously produced cells upward toward the outer layer of the epidermis. As the cells move farther and farther away from the dermis and become progressively flatter, they begin producing large amounts of the fibrous protein keratin and are therefore known as keratinocytes. As they continue their journey towards the surface, they compact tightly together and eventually die. The accumulated layers of dead cells create an external water tight cell layer because the large amounts of keratin act as an effective barrier to prevent not only the excessive loss of water from the body but also the penetration of harmful chemicals and microorganisms from the environment. Millions of dead cells in the outermost layer of epidermis are sloughed off every second of every day and replaced by the continual division of basal cells in the lowest level of the skin layer.[13.3]

The bottom most layer of the epidermis contains basal cells know as melanocytes that produce melanin, a collection of pigments that vary in color from black, brown, yellow, to red (Figure 13.2). Individuals differ in the number of melanocytes they possess as well as in the type and amount of melanin they produce, all of which is responsible for differences in skin color.

[13.3] A person's entire epidermis is replaced approximately every month to month and a half.

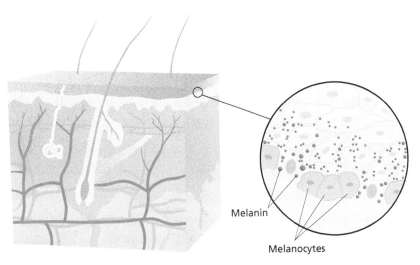

Melanin

Melanocytes

Figure 13.2 Melanocytes. The pigment melanin is produced by melanocyctes located in the lowest level of the epidermis. Adapted from: medpediamedia.com/

The skin is attached to the underlying muscle and bone by way of the hypodermis, the subcutaneous tissue that consists mostly of adipose (fat) cells, blood vessels, and nerves that are entwined in the connective tissue proteins collagen and elastin.[13.4] The layer of fat acts as an insulation barrier that helps maintain body temperature and as a shock absorber that protects the blood vessels and nerves that are nestled among the fat cells. This is the site where subcutaneous injections are administered as it is considered less susceptible to pain.

> [13.4] The skin can be thought of as being attached to the hypodermis in much the same way that a house is attached to its foundation.

> **Compare and contrast the major layers of the skin.**

THREE TYPES OF SKIN CANCER

Each of the three types of epithelial cells found in the epidermis are capable of becoming cancerous. The two most common forms of skin cancer arise from the malignant transformation of basal and squamous cells. These cancers are respectively termed basal cell carcinoma (BCC) and squamous cell carcinoma (SCC), or less frequently as keratinocyte carcinomas. The malignant transformation of melanocytes is termed a melanoma and, while it is the least common form of skin cancer, it is also the deadliest. Basal and squamous cell carcinomas are also classified as nonmelanomas.

Basal cell carcinoma

Approximately 80% of skin cancers are BCC, making it the most commonly diagnosed of all forms of cancer. Malignancies of the skin can develop on any area of the body but greater than 90% occur on those areas that are regularly exposed to the sun, such as the bald areas of the scalp, face, rim of the ear, tip or sides of the nose, neck, shoulders, chest, back, arms, hands, legs, and feet. Rarely does it occur on areas that are regularly protected from the sun. A characteristic of BCC is that it is slow growing and seldom metastasizes unless left untreated.

Squamous cell carcinoma

Approximately 16% of skin cancers are SCC and most often develop on sun-exposed areas but also have a tendency to form where there are scars, chronic inflammation, persistent sores, or areas that have been subjected to multiple doses of X-rays or harsh chemicals. SCC is more aggressive than BCC in that it grows faster and is more likely to metastasize. When metastasis does occur, it is most often to the subcutaneous layer and only rarely to the lymphatic system or distant tissues. Annually, there are more than 1000,000 people diagnosed with nonmelanomas in the US and of these, only 1,500 die of the disease.

Melanoma

The most aggressive and deadliest form of the disease is melanoma, which accounts for approximately 4% of the yearly diagnoses of skin cancer. Yearly estimates of diagnoses in the US are 62,000, while there are 8,000 deaths from the disease. Melanomas can develop anywhere on the body just as nonmelanomas do, but they traditionally tend to form on the trunk of men and on the lower legs of women. Since they arise from melanocytes, however, it is possible for a melanoma to develop wherever the pigment-producing cells are present in the body.

RISK FACTORS FOR DEVELOPING SKIN CANCER

Exposure to UV radiation is the number one risk factor

There are a number of risk factors that increase the likelihood of a person developing skin cancer, regardless of its type. Men develop skin cancer at a disproportionately higher rate than do women; a man's chance of developing basal and squamous cell carcinomas is two and three times that of a woman's, respectively. Caucasian men over the age of 50 make up the majority of people diagnosed with all forms of skin cancer. A significant reason for this is that, traditionally, men over the age of 40 have spent the most amount of time outdoors.

Besides gender and age, the primary risk factor is exposure to ultraviolet (UV) radiation from the sun, tanning beds or booths, or sun lamps. UV light is a form of electromagnetic radiation (Figure 13.3) that has more energy than visible light but less energy than X-rays. There are three forms of UV radiation. The damage to collagen fibers done by the penetration of UVA rays (320–400 nm) deep into the skin is believed to reduce the resiliency of skin over time and contribute to the development of wrinkles, a sign of aging. While it does not penetrate as deeply, the energy of UVB rays (280–320 nm) breaks the noncovalent hydrogen bonds between the nitrogenous bases of the DNA molecules in the skin cells (Chapter 2). Oftentimes, when the bonds are broken between two adjacent thymine-adenine basepairs, the two thymines will bond covalently with each other to form a thymine dimer (Figure 13.4). This causes a distortion in the DNA molecule that, if not repaired correctly, is very likely to cause an error when the DNA molecule is being replicated or result in an alteration in the DNA sequence (mutation) which, if located in a gene, could negatively affect the protein produced from it. UVC radiation

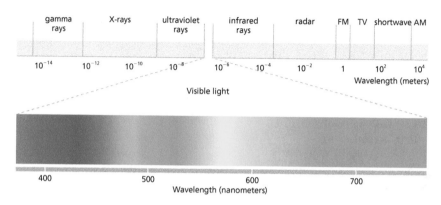

Figure 13.3 Electromagnetic spectrum of energy. Electromagnetic radiation spans from gamma rays that have the shortest wavelength and the most amount of energy to radio waves that have the longest wavelength and the least amount of energy. Source: http://www.dnr.sc.gov/ael/personals/pjpb/lecture

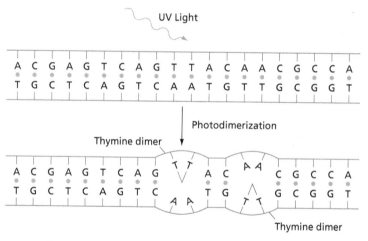

Figure 13.4 Thymine dimer formation. When UV radiation strikes a DNA molecule it breaks the hydrogen bonds between base pairs and adjacent thymines can become covalently bonded forming a thymine dimer.

(100–280 nm) is of less concern as the vast majority never reaches the Earth's surface because it is reflected or absorbed by the atmosphere, primarily the ozone layer.

Tanning and burning of the skin

The melanin pigments do not simply provide color to our skin. They actually absorb UV radiation, blocking the ability of the rays to penetrate deeply into the skin and thereby minimize the damage they cause. Melanocytes reside in the lowest level of the epidermis and have long extensions that

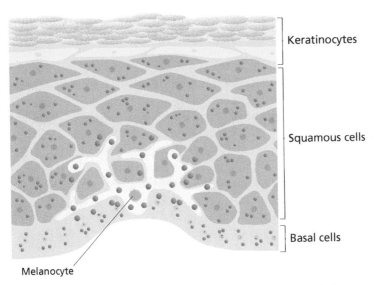

Keratinocytes

Squamous cells

Basal cells

Melanocyte

Figure 13.5 Melanin is packaged into vesicles that move through long cellular extensions of the melanocyte to be distributed to neighboring cells. The melanin accumulates on the outer surface of the nucleus that faces the external surface of the skin.

reach out among the surrounding keratinocytes (Figure 13.5). Once the melanin is produced it is packaged into granules that travel along the cellular extensions of a melanocyte and are then passed to the surrounding keratinocytes. In these cells, the melanin accumulates on the external or "sunny" side of each nucleus to create a UV shield to protect the DNA. The temporary increase in the production of melanin in response to increased UV ray exposure, commonly referred to as a tan, is the body's attempt to enhance its level of protection against the damaging radiation. A tan fades as the keratinocytes with the additional melanin progress upward to the outermost layer of the epidermis and are eventually sloughed off.

A typical sunburn is the result of extensive damage to the epidermis by UVB radiation. In response to the damage, the blood vessels in the dermis dilate in order to increase the amount of blood flow and bring cells of the body's immune system to the affected area. The damaged skin exhibits erythema, or redness, from the increased amount of blood in the area.[13.5] Chemicals in the blood irritate the local nerves, causing the sensitivity associated with a sunburn. Ironically, this pain is beneficial because it draws attention to the damaged area. This ensures that the area is tended to, thus minimizing the likelihood of further damage occurring. Peeling skin is the result of epidermal cells

[13.5] Pushing down on a sunburned area with a finger will cause it to turn white as the blood is forced from the area, but it will quickly become red again once the pressure of the finger is released and the blood returns to the capillaries.

undergoing apoptosis due to extensive DNA damage. It usually takes several days to complete the shedding and healing cycle.

People with fair skin have a higher risk of incurring UV-related skin damage and developing skin cancer than those with darker skin. For example, Caucasian Americans are 20 times more likely than African Americans to develop melanoma. In particular, those who have blonde or red hair, or blue, green, or gray eye color, have the highest risk because they tend to have skin that burns easily or freckles when exposed to the sun. A practical example of this is observed among the fair-skinned people of northern Australia. The area generally has temperate weather, yet the inhabitants who emigrated from the British Isles have the highest rates of skin cancer. Unfortunately, it is too often incorrectly thought that developing a tan helps protect against the damaging effects of the sun's rays. The truth is that damage to the connective tissue and the DNA of the skin cells caused by UV radiation accumulates over time.

The greater the amount of UV radiation exposure a person has, especially if it has resulted in moderate to severe sunburns, the greater the likelihood that skin cells will ultimately undergo malignant transformation. Long-term exposure to the sun's UV radiation may result in the formation of small, pink or red colored, scaly lesions called actinic keratosis. They usually appear on sun-exposed areas of the skin such as the face, ears, back of the hands, and arms of middle aged or older people with fair complexions. This is a precancerous condition that can, though not often, progress to squamous cell carcinoma.

> **Why do fair-skinned individuals have a higher risk for developing skin cancer than those with dark skin?**

Box 13.1

Treatment for psoriasis may increase skin cancer risk

Psoriasis is a chronic inflammatory condition that produces red, scaly patches on the skin (Figure 13.6). A form of treatment for the condition is therapeutic UV exposure or photodynamic therapy (Figure 13.7). The drug psoralen, which is either applied directly to the skin or taken orally, is known as a photosensitizer. It is in an inactive state when first absorbed by the skin cells. Penetrating UVA light converts it to its active form, which is toxic to the diseased cells. This form of therapy allows for the targeted destruction of skin cells while the UV light also suppresses the immune response that triggers the inflammation reaction that causes psoriasis. People who have undergone therapeutic UV exposure have increased rates of mutations in the *p53* tumor suppressor gene and the *Ha-Ras* oncogene and have an increased risk of developing skin cancer.

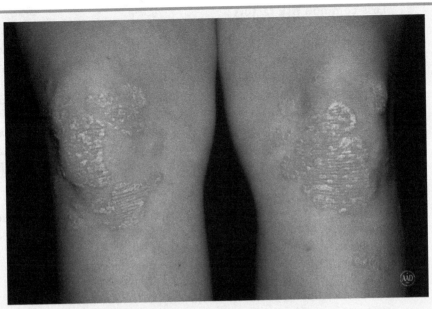

Figure 13.6 Psoriasis. Red, thick, and scaly skin are characteristics of the skin condition Psoriasis. Reproduced with permission from the American Academy of Dermatology, Copyright © 2009, All rights reserved.

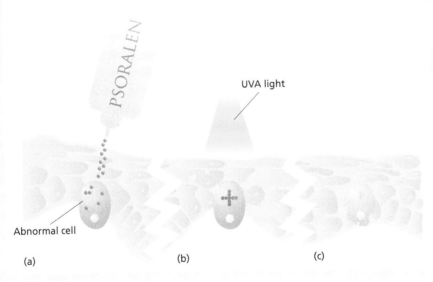

Figure 13.7 Photodynamic drug therapy. (a) A photosensitive drug is applied to the skin and accumulates in abnormal cells. (b) The drug is converted to its active form by exposure to UVA light. (c) The activated form of the drug causes damage that is lethal to the cell.

Less common risk factors

There is an increased likelihood that a person who was previously diagnosed with a form of skin cancer will be diagnosed with it again. The chance that a person will develop a second basal or squamous cell carcinoma is approximately 40% and 15%, respectively. For those diagnosed with melanoma, 11% will develop a second tumor and of those, 30% will develop a third lesion. A person's risk also increases if he or she has one or more first-degree relatives who have been diagnosed with skin cancer.

Approximately 10% of those diagnosed with melanoma have a family history of the disease. This does not automatically imply mutated genes that predispose a person to the development of the deadly cancer were inherited. In fact, research has demonstrated that while certain genes are often mutated in melanomas, the vast majority of these mutations are *induced* over a period of time rather than inherited. It is more likely that members of families with high incidences of melanoma share characteristics (e.g., pale coloring) or behaviors (e.g., excessive sun exposure) that are responsible for their increased risk.

The rare families that do demonstrate inherited or familial melanoma most often possess mutations in three tumor suppressor genes, *CDKN2A*, *CDK4*, and *MC1R*. Inheritance of mutations in these genes only predisposes a person to the development of the disease and does not always indicate causation. The determining factor will be whether or not a person acquires mutations in additional genes that are critical to the regulation

Box 13.2

Xeroderma pigmentosum

Skin cells produce enzymes that repair the mismatched DNA base pairing (i.e., thymine dimers) caused by UV radiation. While the repair system is not 100% accurate, it does an excellent job of minimizing the number of mutations introduced into the DNA. People with the inherited condition xeroderma pigmentosum (XP) have a dramatically reduced ability to repair DNA damage caused by UV light because they possess mutations in the genes that code for the repair enzymes. As a result, they have a very high susceptibility of acquiring and accumulating UV-induced mutations in their skin cells during each exposure to sunlight. Even short-term exposure to the sun causes severe burns and eye irritation.

Without stringent precautions to limit their exposure to UV light, many precancerous and cancerous lesions develop at a very young age. An ordinary clear glass window allows approximately 70% of the sun's UV rays to pass through it. The most common tinting or coating used to block the penetration of UV light still allows up to 25% of it through. People with XP not only have to limit their time outdoors until after sunset but they have to remain away from windows that do not have shades, drapes, or blinds blocking the sunlight. These extremely unfortunate circumstances have resulted in the formation of derogatory terms such as "children of the dark," "children of the night," and "vampire children." Full body suits that include a face shield that blocks UVA and B radiation are now available that give children with XP the ability to be outside during the day (Figure 13.8).

Those with XP have extremely dry skin and many freckles and nevi as a result of irregular distribution of melanin. A normal life expectancy is possible for someone with XP as long as they are diligent about limiting their exposure to the sun or artificial sources of UV light, have skin and eye examinations on a regular basis, and have suspicious lesions removed at their earliest stages.

Figure 13.8 Xeroderma pigmentosum. A child with xeroderma pigmentosum must cover all exposed skin when going outdoors during daylight hours. Courtesy of Xeroderma Pigmentosum Society.

of a cell's progression through the cell cycle. Only then will malignant transformation occur.

A person is at greater risk of developing skin cancer if he or she has a weakened immune system. A disease such as HIV attacks and destroys primarily T lymphocytes, a type of white blood cell that is a major factor in the body's defense system. A depleted supply of T lymphocytes prevents a person from carrying out immune responses against pathogens that have invaded the body or against body cells that are functioning in abnormal ways, such as cancer cells. People who have received organ transplants must take immunosuppressive drugs for the rest of their lives. The drugs are used to depress the immune system's ability to initiate a response against the foreign tissue so as to avoid its rejection. A consequence, though, that is also seen with HIV infection, is a reduced ability of the body's defense system to detect and destroy malignantly transformed cells. People who have received organ transplants have a much higher risk of developing skin cancer, and if it does develop it is often more aggressive.

There are many different types of the human papillomavirus (HPV; Chapter 10). Some cause the familiar small, rough, and elevated growths known as warts on the hands and feet. They are not precancerous growths and if left untreated may remain for a period of weeks or months until often receding on their own. Certain forms of HPV cause warts specifically in the areas around the genitals and anus and are associated with the development of skin cancer.

The uniform light pink to black pigmentation of nevi (singular nevus), more commonly known as moles, is caused by an unusual aggregation of melanocytes in the epidermis (Figure 13.9). All nevi are considered to be benign melanocytic tumors, and should therefore be monitored closely for changes in appearance since the melanocytes have the potential to become malignant. Nevi are usually no larger than 6 mm, can be round or oval, flat or elevated, and any shade from light pink to a dark

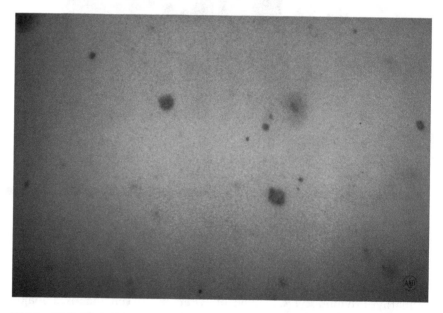

Figure 13.9 Nevi. Numerous nevi are present on an area of skin. Reproduced with permission from the American Academy of Dermatology, Copyright © 2009, All rights reserved.

black depending on the distribution of melanin.[13.6] At times they may
have several hairs growing from them.
Everyone has many nevi, some nearly
microscopic in size that may have been
present since birth or have become visible and darkened through childhood
until sometime during the teen years.

[13.6] 6 mm is approximately equal to
$^1/_4$ inch, or the diameter of a pencil
eraser.

METHODS USED TO SCREEN FOR SKIN CANCER

Skin examinations are the primary screening method

Any and all changes to the appearance of an area of skin, especially if
it persists for a significant period of time or involves a mole, should be
taken seriously. More than half of all skin lesions that are diagnosed as
malignant are first identified by the patient prior to seeing a healthcare

Table 13.1 Characteristics of nonmelanomas and melanomas

Malignancy	Characteristics
Basal cell carcinoma	Flat and pale, or raised with a light red or pink color with a shiny surface; may contain small capillaries that may rupture and bleed; may develop a crater in the center or blue, brown, or black pigmented areas
Squamous cell carcinoma	Red patches of skin or an elevated growth with a rough surface
Melanoma	Mole's surface becomes scaly, bleeds, or oozes fluid; edge becomes irregular; alteration in pigmentation; size increases

Table 13.2 Features of a mole that are assessed during an examination

Characteristic	Example
Asymmetry	
Border	
Color	
Diameter	

provider. Each of the three forms of skin cancer exhibits particular characteristics during its growth that are summarized in Table 13.1. In the vast majority of cases melanomas arise from existing moles and are pigmented. In rare instances they can arise in another location and may be pale. There are four characteristics, referred to as the **ABCDs of melanoma**, that are assessed to determine whether or not there is a malignant transformation of a mole:

Asymmetry: one half of the mole is not a mirror image of the other half
Border: the outer edge of the mole is irregular rather than smooth
Color: lack of uniformity in pigmentation across the mole
Diameter: the diameter is greater than 6 mm[13.6]

Table 13.2 contains pictures that illustrate alterations in each of the ABCD characteristics.

The principal means of detecting skin cancer is through routine visual inspections of the skin. The Skin Care Foundation, an international organization, has developed the slogan, "If you can spot it, you can stop it." Examinations of the skin can be a part of a person's monthly cancer self-examination along with a breast or testicular self-examination. It can also be part of a yearly physical examination by a healthcare professional, especially for those considered to be at high risk. A person should become very familiar with the moles, freckles, or birthmarks on all areas of his

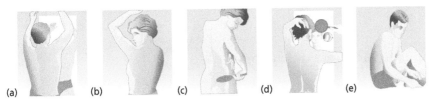

(a)   (b)   (c)   (d)   (e)

Figure 13.10 Skin self-examination. (a) Facing a full-length mirror and with arms raised the entire front of the upper body, including both underarms, is observed starting from the forehead and making the way down to the hips. Turning 90° to the left and right with the arms above the head enables the entire right and left sides of the torso to be viewed. (b) In turn each arm is raised with the forearm bent toward the head and then rotated to the right and left so that a close examination can be made of the entire arm as well as the back and palm of each hand. Examining each of the nail beds, the surfaces under each of the nails, for a brown or black streak is necessary since they are also locations where malignancies develop. (c) Facing away from a full-length mirror and using a handheld mirror the back of the neck, back, and buttocks are examined. (d) Using a handheld mirror in front of a full-length mirror helps in examining the entire scalp while using the fingers of the free hand, a brush, or blow dryer to part the hair. (e) The genitals, front and back of both legs, the tops and soles as well as the spaces between the toes of both feet are examined with the help of a handheld mirror.

or her body so that any changes in their appearances or the development of a new growth can be detected as early as possible. Performing the examination in a room that has good lighting will make it more likely that all areas of the body will be well illuminated, making the observation of small marks on the skin that much easier. Figure 13.10 illustrates the steps involved in a skin self examination.

Any new spots or lesions or suspicious changes to a mole, freckle, blemish, or birthmark should be brought to the attention of a dermatologist, a physician specializing in conditions that affect the skin, as soon as possible for a more thorough examination. It is sometimes beneficial to maintain a set of pictures of areas of particular concern so that they can serve as a baseline and be used to make comparisons with photos taken during subsequent examinations.

> **Define and describe the ABCDs of skin cancer detection.**

During an examination of a suspicious area a dermatologist may use a dermoscope, a handheld device that provides a magnified and more detailed view of a lesion than is possible with the unaided eye. The optics of the dermoscope actually permit visualization of the dermis below the external epidermal layer. This is extremely helpful in making a thorough assessment of a lesion by detecting the subtlest of changes that may be occurring that would otherwise go unnoticed. It also assesses whether a biopsy is necessary and, if so, whether additional tissue needs to be removed

Box 13.3

The Melanoma Education Foundation

Dan was 24 years old and living in California after graduating from the University of Miami in 1994 with a degree in environmental science and a minor in chemistry (Figure 13.11). He was academic captain of the UM Varsity Crew team, excelling in both rowing and academics, and was in excellent physical condition. Those who knew him described him as soft-spoken, considerate, sincere, conscientious, and bright with a good sense of humor.

On a visit home in 1996, one of his parents noticed a large dark brown thick mole on his lower back, nearly half an inch wide. He had had the mole as a child, though not at birth, and it had bled slightly about 4 or 5 months prior to his visit home. Since it was in a location that was difficult to view in a wall mirror, he was unable to see that it had undergone other changes. Upon his return to California he saw a dermatologist and had the mole removed, at which time it was determined that it was a late-stage melanoma requiring further treatment. After removing all 29 lymph nodes in the affected area, two of them tested positive for melanoma as well. He was treated for 1 year with interferon, an immune system

Figure 13.11 Daniel Fine died at the age of 26 from malignant melanoma. Melanoma Education Foundation (www.skincheck.org).

stimulant that causes flu-like side effects including exhaustion, aches, and depression. Within 4 months of the completion of his treatment, the melanoma had spread to his liver and lungs and was inoperable. He was given chemotherapy and experimental treatments but the cancer continued to increase in his abdomen and spread to his bones and brain.

Dan faced his illness with extraordinary courage and dignity, never complaining or expressing anger or asking "why me?" In his final days he was unable to walk or sit comfortably due to severe fluid accumulation in his abdomen and legs. Somehow though, one night in October of 1998, he mustered the strength to call his parents into his bedroom, sit up on the side of his bed, and put an arm around each of them. He said "I love you both" and died early the next morning.

In their son's honor, Stephen and Gail Fine established the Melanoma Education Foundation in the hopes of preventing other families from experiencing a similar tragedy. This nonprofit organization provides a comprehensive website for self-detection and prevention of melanoma and for educating about the disease. Their foundation increases awareness of melanoma in three ways by conducting workshops for high school and middle school health educators, providing information about early self-detection and the prevention of melanoma on their website, www.skincheck.org, and conducting talks and facial skin analyzer screenings for area organizations and businesses.

during the procedure and what an appropriate form of follow-up treatment would be, if necessary.

Several skin biopsy methods are available

A biopsy provides a definitive diagnosis of whether or not a lesion is malignant and if so, its type and extent of invasion as well as the likelihood of metastasis. The type of skin biopsy selected is based on whether the malignancy is suspected to be a nonmelanoma or melanoma, as well as on the size and location of the lesion. Excisional and incisional biopsies are the two main classes of procedures. During an excisional biopsy the lesion along with a surrounding margin of healthy tissue is removed (Table 13.3). If the lesion is large or in a location that requires the preservation of tissue, an incisional biopsy may be performed – only a portion of the affected area is removed for pathological examination. An incisional biopsy is typically done only when the lesion is suspected to be a nonmelanoma or when an excisional biopsy is not a viable option.

If a lesion is suspected to be a nonmelanoma, either a shave or punch biopsy is performed. A shave or curettage biopsy is the least invasive technique; first the area is cleansed and made numb with a local anesthetic and then a curette, a curved surgical blade, is used to slice across the area, removing all of the epidermis along with the very top of the

Table 13.3 American Academy of Dermatology recommendations for margin sizes

Tumor thickness (mm)	Excisional margin (cm)
in situ melanoma	0.5
<1.0	1.0
1.0–4.0	2.0
>4.0	≥2.0

dermis (Figure 13.12). A few passes across the area may be made to ensure that all of the cancer cells have been removed. A physician may choose to follow curettage with electrodessication, a procedure in which a very thin probe that produces a weak electric current is applied to the area where the skin layers have been shaved away. The electric current is used to destroy additional cells, including malignant ones, that remain in the area.

Mohs' micrographic surgery is a modification of a shave biopsy. After the first pass with the curette and the removal of all or nearly all of the cancerous cells, a subsequent shaving of close to a single layer of cells is performed. The specimen is immediately analyzed under the microscope for the presence of cancerous cells. Repeated shavings are taken and assessed until a clear margin in a specimen sample is viewed under the microscope.

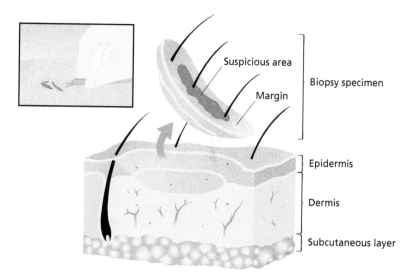

Figure 13.12 Shave biopsy. The epidermis and top layer of dermis are removed during a shave biopsy.

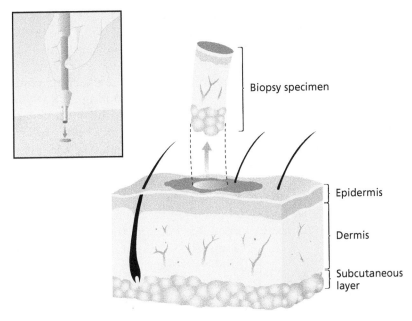

Biopsy specimen

Epidermis

Dermis

Subcutaneous
layer

Figure 13.13 Punch biopsy. The epidermis, dermis, and some subcutaneous tissue are excised during a punch biopsy.

The superficial nature of the procedure results in little, if any, bleeding and does not require stitches, making it a preferred technique for anywhere on the face to keep scarring to a minimum.

The more invasive punch biopsy is performed when there is evidence of deeper invasion requiring a full thickness skin specimen (Figure 13.13). Once the area is cleansed and numbed with a local anesthetic, a punch biopsy instrument that contains a round blade at the end is positioned vertically over the lesion. Pressure is applied downward as it is repeatedly rotated clockwise and counterclockwise until the blade has penetrated through to the subcutaneous fat. Once the blade is withdrawn, the sample of tissue is excised. There is a potential risk, although minor, of damaging nerves, arteries, or veins if the punch travels too deeply into the subcutaneous layer. When a lesion is beneath the nail of a finger or toe, a nail bed biopsy is performed. Either a portion or the entire nail is removed to gain access to the affected area and then an excisional or incisional biopsy is performed.

> **Which surgical procedure is preferred for facial skin cancer? Why?**

A physician may suspect metastasis based on the size and type of lesion or if one or more of the regional lymph nodes is enlarged. The determination of whether tissue below the subcutaneous layer possesses metastatic cells is made possible by performing a biopsy on a sample obtained using a needle inserted through the skin. A lymph node biopsy may be performed at the same time as one of the previously mentioned procedures if there is a concern that metastasis has occurred. Intact lymph nodes can be surgically removed or their contents can be sampled through a fine needle aspiration (FNA) biopsy. During a FNA biopsy, a needle is inserted directly through the skin into a lymph node and a small amount of its contents are aspirated. While a FNA biopsy is far less invasive than an excisional lymph node biopsy, it should be kept in mind that cancer cells present in a lymph node may not be taken up in the aspirant and therefore may not always be detected during analysis.

To increase the chances of sampling the lymph node most likely to have metastatic cancer cells, the physician may determine which of the regional nodes is the sentinel node, the one that the lymph first enters after draining from the affected area. A sentinel node biopsy (Chapter 6) begins by injecting a dye or radioactive substance into the area of the lesion (Figure 13.14). The sentinel node is identified as the first one to acquire the dye or become radioactive.

> **Name another cancer that would benefit from a sentinel node biopsy.**

Regardless of the method used to obtain the specimen, the tissue will be processed and stained to enable the detection of distinctive character-

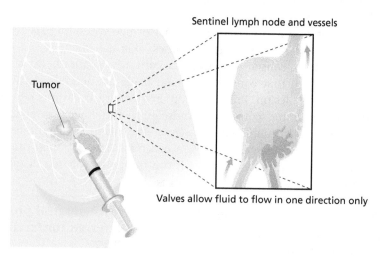

Sentinel lymph node and vessels

Tumor

Valves allow fluid to flow in one direction only

Figure 13.14 Sentinel lymph node biopsy. Either a dye or radioactive substance is injected into the area around the tumor. The first lymph node to take up the dye or radioactive substance is the sentinel lymph node. Adapted from: http://apps.uwhealth.org/health/adam

Table 13.4 Relationship between Breslow thicknesses and 5-year survival rates

Breslow thickness (mm)	5-year survival (%)
<0.76	97
0.76–1.50	92
1.51–2.50	76
2.51–4.0	62
4.1–8.0	52
>8.0	32

istics of cancer cells when it is viewed under the microscope. The analysis of skin biopsies is often performed by a dermapathologist, a physician whose specialty is the diagnosis of skin conditions. The biopsy report will indicate whether the specimen contains benign or malignant cells. If malignant cells are observed, the report would also state whether they possess nonmelanoma or melanoma characteristics. Also of importance is the estimation of how deep the cells have penetrated into the epidermal or dermal layers of the skin or subcutaneous fat layer. The vertical depth of the tumor, known as the Breslow thickness, is determined for its prognostic value (Table 13.4). The thicker the tumor, the deeper it has penetrated into the epidermal, dermal, and subcutaneous layers, the greater the possibility of metastasis as well as subsequent recurrence. It is very unlikely that a tumor with a Breslow measurement of less than 1 mm will have metastasized, but this likelihood increases significantly when a tumor has a depth greater than

> [13.7] 1 mm is approximately 1/25 of an inch or the width of a period (full stop).

1 mm.[13.7] The determination of the occurrence of metastasis can also be made from an examination of tissue from intact lymph nodes or from a needle biopsy. The verification of an intact margin of healthy cells surrounding the malignant cells is critical to ensuring that all of the cancer cells were removed. The removal of additional tissue would be warranted if a clear margin was not obtained or if the presence of affected cells was inconclusive. Part of the evaluation of a melanoma is whether or not there is a layer of epidermis over the top of the tumor. The tumor is classified as ulcerated if the epidermis is absent.

The information obtained during a biopsy of malignant tissue is also used to determine the stage of the cancer. The most commonly used TNM-based system (Chapter 6) for staging nonmelanoma and melanoma has been devised by the American Joint Committee on Cancer. Tables 13.5

Table 13.5 AJCC stage grouping for nonmelanoma cancer

Stage	Stage grouping	Explanation
0	Tis, N0, M0	Carcinoma *in situ*; aka Bowen disease; malignancy only present in epidermis (Tis); not present in regional lymph nodes (N0); has not spread to distant sites (M0)
I	T1, N0, M0	≤2 cm in diameter (T1); malignancy only present in epidermis; not present in regional lymph nodes and has not spread to distant sites (N0, M0)
II	T2, N0, M0 or T3, N0, M0	Diameter is 2–5 cm (T2) or >5 cm (T3); malignancy only present in epidermis; not present in regional lymph nodes and has not spread to distant sites (N0, M0)
III	T4, N0, M0 or Any T, N1, M0	Diameter can be any size and tumor has grown below the skin into muscle, cartilage or bone (T); malignant cells may (N1) or may not (N0) have metastasized to regional lymph nodes; malignancy has not spread to distant sites (M0)
IV	Any T, any N, M1	There can be any degree of invasion (any T) or lymph node involvement (any N); has spread to distant sites (M1)

and 13.6 provide the stage groupings for nonmelanoma and melanoma, respectively.

SURGERY AND CHEMOTHERAPY ARE STANDARD TREATMENTS FOR METASTATIC SKIN CANCER

By far the most common type of treatment for any form of skin cancer is its surgical removal. In fact, the majority of them are small enough in size to be removed in their entirety during the biopsy procedure, essentially affecting a cure. A second or third procedure may be necessary if the biopsy results indicate the possibility of cancer cells remaining at the excision site.

Treatment of a suspicious lesion may even occur without obtaining a specimen for biopsy. A dermatologist may determine from the size, depth, and appearance of a lesion that it is most likely precancerous or a very early stage nonmelanoma. While all of these conditions should be treated, they do not necessarily warrant a biopsy. A procedure may be chosen that destroys rather than excises the tissue. One option is cryosurgery, which entails the application of liquid nitrogen to kill the cells by freezing them at a temperature of −195.5°C (−320°F). A second option is laser surgery, which kills the cells by burning them with the high energy beam of light generated by a laser. A treatment option for superficial but not

Table 13.6 AJCC stage grouping for melanoma

Stage	Stage grouping	Explanation	5-year relative survival rate
0	Tis, N0, M0	Melanoma *in situ* (Tis); malignancy only present in epidermis; (N0, M0)	99.5–100%
IA	T1b, N0, M0 or T2a, N0, M0	Thickness is ≤1 cm (T1) or is ulcerated (Tb) or is 1.01–2 cm (T2) and is not ulcerated (Ta); not present in regional lymph nodes (N0); has not spread to distant sites (M0)	95%
IB	T1b, N0, M0 or T2a, N0, M0	Thickness is ≤1 cm (T1) and is ulcerated (Tb) or is 1–2 cm and is not ulcerated (Ta); not present in regional lymph nodes (N0); has not spread to distant sites (M0)	89–91%
IIA	T2b, N0, M0 or T3a, N0, M0	Thickness is 1.01–2 cm (T2) and is ulcerated (Tb) or is 2.01–4 cm (T3) and is not ulcerated (Ta); not present in regional lymph nodes (N0); has not spread to distant sites (M0)	77–79%
IIB	T3b, N0, M0 or T4a, N0, M0	Thickness is 2.01–4 cm (T3) and is ulcerated (Tb) or is >4 cm (T4) and is not ulcerated (Ta); not present in regional lymph nodes (N0); has not spread to distant sites (M0)	63–67%
IIC	T4b, N0, M0	Thickness is >4 cm (T4) and is ulcerated (Tb); not present in regional lymph nodes (N0); has not spread to distant sites (M0)	45%

Table 13.6 AJCC stage grouping for melanoma (*continued*)

Stage	Stage grouping	Explanation	5-year relative survival rate
IIIA	T1–4a, N1a, M0 or T1–4a N2a, M0	Malignancy can be of any thickness (T1–4) but is not ulcerated; metastasis to one (N1) or 2–3 (N2) regional lymph nodes and the metastases can be viewed only under a microscope (Na); has not spread to distant sites (M0)	63–69%
IIIB	Any T, N1a–N2c, M0	Malignancy can be of any thickness (T1–4) and may or may not be ulcerated (Ta or Tb); metastasis to one (N1) or 2–3 (N2) lymph nodes; metastases may only be viewed under a microscope (Na), may be visible to the naked eye (Nb) or is in the nearby lymphatic channels (Nc); has not spread to distant sites (M0)	30–53%
IIIC	Any T, any N, M0	Malignancy can be of any thickness (T1–4) and may or may not be ulcerated (Ta or Tb); metastasis to one (N1), 2–3 (N2), or 4 or more (T3) lymph nodes; metastases may only be viewed under a microscope (Na), may be visible to the naked eye (Nb) or is in the nearby lymphatic channels (Nc); has not spread to distant sites (M0)	24–29%
IV	Any T, any N, M1a, M1b, M1c	Malignancy can be of any thickness (T1–4) and may or may not be ulcerated (Ta or Tb); metastasis to one (N1), 2–3 (N2), or 4 or more (T3) lymph nodes; metastases may only be viewed under a microscope (Na), may be visible to the naked eye (Nb) or is in the nearby lymphatic channels (Nc); metastasis to distant skin, subcutaneous tissue, or lymph nodes (M1a); metastasis to the lung (M1b); metastasis to other organs (M1c)	7–19%

invasive SCC is photodynamic therapy that uses a photosensitizer drug and controlled UVA exposure to destroy the malignant epidermal cells (Figure 13.7).

> **What condition mentioned earlier in this chapter is treated using photodynamic therapy?**

A final option is the use of a topical chemotherapeutic, a cytotoxic drug applied directly to an affected area of the skin. Only the cells on the external surface of the skin that come in direct contact with the drug are affected because the drug does not penetrate very deeply. This is a viable option for the treatment of superficial precancerous or malignant squamous and basal cells. The most commonly used topical chemotherapeutic is 5-fluorouracil (5-FU), an analog of a nucleotide (Chapter 7) that interferes with the functionality of RNA or, when altered by the cell's metabolism, negatively affects the replication of DNA. The drug is applied by the patient twice a day, usually for several weeks. A benefit of topical chemotherapy is the minimization of side effects, which are typically limited to redness and tenderness in the affected area but can become severe. This can be controlled to some degree with the use of other drugs or cycle therapy, the application of the drug for an intermittent period of on and off days. The area that receives the drug is especially sensitive to the damage of UV radiation and so must be protected from the sunlight to prevent burning.

A regimen of systemic chemotherapy is typically planned when a biopsy has been performed and there is evidence of metastasis to one or more regional lymph nodes or to distant tissue, or if metastatic tumors have been detected. Systemic chemotherapy is usually treated on an outpatient basis and the drugs are administered through an intravenous line. Cisplatin, doxorubicin, 5-FU, mitomycin, and temozolomide are drugs commonly used in the treatment of metastatic skin cancer. The specific type of drug or combination of drugs used, their dosages, and the frequency between each administration are determined based on the stage of the tumor and its location. Also taken into consideration are the patient's overall health status and ability to tolerate the toxic side effects, which can be quite severe and debilitating due to the drugs' negative effects on healthy cells throughout the body.

Metastatic skin cancer, especially melanomas, are treated with various immunotherapies, drugs that enhance the body's own immune system. The two most widely used forms of immunotherapies are cytokines, proteins produced by immune system cells when carrying out an attack against an infective agent, such as a virus or bacterium, or a body cell that is no longer functioning properly, such as a cancer cell. The first is interferon-alpha, which stimulates the activity of natural killer (NK) cells that recognize, attack, and destroy cancer cells. The second is interleukin 2, which enhances the ability of lymphocytes (a type of white blood cell) to destroy cells that are functioning abnormally. To be effective, both of

Box 13.4

Isolated limb perfusion and infusion

Two procedures have been used to treat metastatic melanoma and soft tissue sarcomas in arms or legs. The first is isolated limb perfusion (ILP), a procedure first developed in the mid 1950s. ILP involves the reversible surgical manipulation of the vasculature of the limb to be treated so that its arterial and venous blood supplies form a loop and are isolated from that of the rest of the body. This enables the injection of a very high dose of one or more chemotherapeutics into the main artery of an affected limb. The drug(s) will remain within the closed-loop circulation and not enter the general circulation. The premise is that the high drug concentrations increase the lethality of the drugs on the faster growing cancer cells while being less toxic to the healthy cells of the limb's tissues. Since the drugs are not able to enter the general blood supply, the tumor(s) can be treated aggressively in isolation from healthy cells without the patient experiencing the severe toxic side effects normally caused by the drugs.

ILP is practiced only at a few cancer centers worldwide because of the highly complex nature of the surgical procedure and the potential lethality if the high concentration of drugs escapes from the limb's isolated circulatory system and enters the rest of the body. While there is success in achieving remission with ILP, there is also a high rate of recurrence, and ultimately, for patients with melanoma tumors that are 1.5 mm or greater in thickness, surgery with adjuvant ILP provides no increase in survival when compared with surgery alone. Therefore, the objective of ILP is to control rather than to cure the disease within a limb and is an alternative to amputation.

In the early 1990s a modified version of ILP was developed at the Sydney Cancer Centre in Sydney, Australia. The procedure, isolated limb infusion (ILI), requires far less complicated surgery and achieves results comparable to those of ILP. A catheter is inserted into the main artery of the limb to be treated and a pneumatic tourniquet is applied to prevent the outflow of blood, allowing for the administration and isolation of the high-dose chemotherapeutic drugs to the limb. Prior to the removal of the tourniquet, the limb's blood supply is flushed to remove remaining traces of the drug. The simplicity of the procedure makes it an available treatment option for those who would otherwise not be able to undergo the rigors of the surgery involved in ILP.

these drugs must be taken in high doses that cause side effects such as fever, chills, aches, depression, fatigue, fluid retention, and damage to the heart and liver that can be so severe that the patient must discontinue with the form of treatment. There are a large number of clinical trials currently underway investigating the effectiveness of various forms of immunotherapies in the treatment of metastatic skin cancers. Some of them involve the use of vaccines that entail the administration of specific components of the melanoma cells into the body to help stimulate an immune response against the intact cancer cells residing in the body. Unfortunately, this has proven to be a difficult goal to achieve and the results of trials to date have not been too promising.

Along with surgery and chemotherapy, radiation therapy is usually a mainstay of cancer treatment, but this is not the case with skin cancer. Radiation is rarely used to treat primary tumors located on the skin because of their superficial nature. It may be used, however, to treat metastatic tumors that are deep in the body or to reduce the size of tumors for palliative purposes.

WHAT HAPPENS AFTER SKIN CANCER TREATMENT

Skin self-examinations on a regular basis and follow-up examinations with a healthcare provider are important after the treatment for any form of skin cancer, especially melanoma. The frequency of follow-up examinations and the types of tests performed during them will depend on whether a patient had nonmelanoma or melanoma and, most importantly, on its stage.[13.8] Most often a recurrence of melanoma is in the skin at a site close to the location of the primary tumor. It is also possible that a tumor will develop in one of the local lymph nodes. In either case, the treatment would most likely be surgical excision. The distant locations where metastatic melanoma most often appear are the lung, bone, liver, and brain. Tumors arising in these locations may be treated with surgery, chemotherapy, and/or radiation depending on their locations, number, sizes, and stages at the time of detection.

> [13.8] A follow-up visit to a healthcare provider will usually include a skin examination, a check for swollen lymph nodes, blood work, and possibly imaging studies.

LIMITED UV RADIATION EXPOSURE IS THE NUMBER ONE FORM OF PREVENTION

Skin cancer is a highly preventable disease because the leading contributor to the development of mutations involved in the malignant transformation of squamous cells, basal cells, and melanocytes is exposure to the sun's UV radiation, something which an individual is capable of controlling. Protecting the skin from UV rays, whether they be from natural sunlight or artificial sources (e.g., tanning booths or beds, sun lamps), is the primary way of reducing the amount of DNA damage in skin cells. There are a number of things that a person can do with minimum effort that can

[13.9] Concrete, sand, water, and snow are surfaces capable of reflecting between 85% and 90% of the UV rays that strike them.

significantly reduce the amount of exposure to UV rays. The greatest danger from the sun is between 10 am and 4 pm when the rays are strongest. Therefore, outdoor activities during those hours should be limited. Altitude and surroundings can also affect the amount of UV exposure.[13.9] When going outside, it is advisable to wear pants (trousers) versus shorts, long sleeve versus short sleeve shirts, eyeglasses that do not allow UVA and UVB rays to penetrate, and a hat, preferably one with a wide brim to shield the face and ears. Any exposed skin should be protected by liberally applying sunscreen and lip balm that are broad spectrum (block both UVA and UVB rays) and have an SPF (sun protection factor) of at least 15. To maximize the effectiveness of sunscreens, they should be applied at least 30 minutes prior to going outside, reapplied every 2 hours, and unless waterproof, reapplied after swimming or vigorous exercise that causes excessive perspiration.

Sunscreen products have SPFs that range from 2 to 60. The SPF refers to the level of protection that the product provides by blocking the penetration of UV rays. The number is determined by observing the time it takes skin that has been applied with the product to burn in comparison to that of unprotected skin. For example, if it takes 15 minutes for unprotected skin to burn, it would take 30 minutes to burn with a product that has an SPF of 2, or 225 minutes, nearly 4 hours, with a product that has an SPF of 15. It is important to realize that a number of factors influence the effectiveness of a sunscreen, such as the amount applied to and absorbed by the skin, the time of day, the type of activities engaged in, and the degree of perspiration.

Explain the significance of SPF's.

Box 13.5

Teenagers are dying for a tan

Since 1975 the incidence of melanoma among US women ages 15–29 has doubled, and it is the second most commonly diagnosed cancer for young adults. While skin cancer, particularly melanoma, was once considered an "old" person's disease, dermatologists are reporting alarming changes in the cases they are currently seeing in younger patients. In 2006, the World Health Organization estimated that up to 60,000 deaths worldwide are caused annually by excessive UV exposure, particularly in teenagers with a compulsion for indoor tanning.

Tanning is increasing in its popularity, especially among young people, despite all of the warnings about its association with skin cancer and premature aging. There is no difference in terms of safety whether a tan is achieved by exposure to UV rays from the sun or a tanning salon. A much safer alternative is "sunless" tanning methods that use dyes applied either by a spray or cream.

Consumers Union, the publisher of *Consumer Reports*, investigated the use of tanning facilities in 2004 by means of a phone survey. They called approximately 300 sites in 12 cities and found widespread failures to inform customers about the risks of skin cancer and premature wrinkling, and learned that many ignored standard safety procedures such as recommending the use of eye goggles. Many respondents claimed that indoor tanning is safer than sunlight and is well-controlled, neither of which is true. Among other conclusions, the survey revealed that the salons extended customers the invitation to come as often as they desired, even if the person had never done indoor tanning, and about one-third denied that the practice of indoor tanning could cause skin cancer or that it could prematurely age the skin. A small percent said that protective eyewear was not required, even though exposure to intense UV light can damage the cornea or retina or eventually cause cataracts. About one in five said erroneously that minors did not need an adult's consent.

The results confirmed for the surveyors that the tanning industry is indeed minimally regulated, with poor compliance for what little rules do exist. Consumers Union is hoping that the FDA will call for the reduction of UV exposure limits and require additional consumer warnings, both in industry advertising and at the point of entry into salons. Further, all states would be mandated to have stringent licensing regulations, including training courses for employees, regular inspection of facilities, and parental consent required for minors.

A study from the *Archives of Pediatrics and Adolescent Medicine* reported that 40% of white adolescent females age 13–19 had used a tanning booth at least once and about 28% of them at least three or more times. The same study demonstrated that 12% of males used tanning booths at least once and 7% at least three or more times. Considering that about 89% of all sun exposure and subsequent damage occurs before the age of 18, this news is especially detrimental to women. The issue of vanity seems to be outweighing common sense. The National Cancer Institute says that 40–50% of Americans who use tanning beds more than once a month are 55% more likely to develop malignant melanoma. Despite the availability of this information to consumers, these salons flourish.

> **Are tanning booths safer than lengthy exposure to sun? Why or why not?**

EXPAND YOUR KNOWLEDGE

1 Why are people with dark skin more likely to be diagnosed with a more advanced stage of melanoma than those who have light skin?

2 Explain the risks and benefits of phototherapy for the treatment of psoriasis.

3 What imaging tests may be performed in an attempt to determine if malignant cells have metastasized and formed secondary tumors in distant organs?

4 Explore the use of sunless tanning products and whether or not there is a relationship with their use and skin cancer.

5 What is the minimal SPF factor recommended for use by the American Academy of Dermatology Association?

ADDITIONAL READINGS

Chiu, V., Won, E., Malik, M., and Weinstock, M. A. (2006) The use of mole-mapping diagrams to increase skin self-examination accuracy. *Journal of American Academy of Dermatology*, **55**: 245–250.

Indoor Tanning: Unexpected Dangers. *Consumer Reports* February 2005.

Marieb, E. N. and Hoehn, K. (2007) *Human Anatomy and Physiology*, 7th ed. Upper Saddle River, NJ: Prentice Hall.

Martini, R. and Bartholomew, E. (2007) *Essentials of Anatomy & Physiology*, 4th edn. Upper Saddle River, NJ: Prentice Hall.

Noschos, S. J., Edington, H. D., Land, S. R., Rao, U. N., Jukic, D., Shipe-Spotloe, J., and Kirkwood, J. M. (2006) Neoadjuvant treatment of regional stage IIIB melanoma with high-dose interferon alfa-2b induces objective tumor regression in association with modulation of tumor infiltrating host cellular immune responses. *Journal of Clinical Oncology*, **24**: 3164–3171.

Rawe, J. (2006) Why teens are obsessed with tanning. *Time*, August 7, 54–55.

Ron, I. G., Sarid, D., Ryvo, L., Sapir, E. E., Schneebaum, S., Metser, U., Asna, N., Inbar, M. J., and Safra, T. (2006) A biochemotherapy regimen with concurrent administration of cisplatin, vinblastine, temozolomide (Temodal), interferon-alfa and interleukin-2 for metastatic melanoma: a phase II study. *Melanoma Research*, **16**: 65–69.

Whiteman, D. C., Stickley, M., Watt, P., Hughes, M. C., Davis, M. B., and Green, A. C. (2006) Anatomic site, sun exposure, and risk of cutaneous melanoma. *Journal of Clinical Oncology*, **24**: 3172–3177.

14

Lung cancer

Four million unnecessary deaths per year, 11,000 every day. A cigarette is the only consumer product which, when used as directed, kills its consumer.
Dr Gro Harlem Brundtland, Director-General Emeritus,
World Health Organization

CHAPTER CONTENTS

- Lung cancer statistics
- Lungs are the site of the exchange of gases
- Risk factors associated with the development of lung cancer
- Lack of distinctive symptoms makes early diagnosis difficult
- Lung cancer is often diagnosed at an advanced stage
- There are two main categories of lung cancer
- Three traditional therapies are used in lung cancer treatment
- Is there discrimination in cancer research funding?
- Expand your knowledge
- Additional readings

Cancer: Basic Science and Clinical Aspects, 1st edition. By C. A. Almeida and S. A. Barry. Published 2010 by Blackwell Publishing, ISBN 978-1-4051-5606-6.

The majority of lung cancer cases occur in smokers. When a smoker is diagnosed with the disease, there is sadly often a sense of self-blame. Others tend to agree with that assumption, as if to intimate that it could not happen if someone does not smoke. If lung cancer does occur in a smoker, some look upon it as an unfortunate but self-induced punishment. Yet the idea that *only* smokers become afflicted is far from the truth. Only recently activists have begun wearing clear colored ribbons or bracelets as a way of referring to lung cancer as an "invisible" disease. This designation arises from the seeming lack of empathy and support that lung cancer patients receive in comparison with other cancer patients. There are a host of other contributing factors, many of them environmental, some genetic, and only now are they on the cusp of being identified or understood. This chapter will address causes, risk factors, diagnostic testing, and treatments for this type of cancer.

LUNG CANCER STATISTICS

Lung cancer has been the most common cancer worldwide since 1985, with the highest rates occurring in North American and European men, particularly those from Eastern Europe, yet it is also the most preventable of all cancers, since tobacco use is considered to be the most important factor in its development. It remains the leading cause of cancer death for both men and women worldwide, contributing to more deaths than any other type of malignancy. The disease is estimated to kill approximately 162,000 of the 214,000 Americans diagnosed with it in 2008, account-ing for approximately 29% of all cancer deaths during that time. Lung cancer claims more lives than breast, prostate, colon and pancreas cancer combined. At the present time, nearly 60% of affected patients will die within a year of their diagnosis and 85% within 5 years. In spite of the high numbers and serious prognosis associated with the disease, more than 400,000 people who have been diagnosed with lung cancer are alive today.

LUNGS ARE THE SITE OF THE EXCHANGE OF GASES

Lung cancer is particularly debilitating because it eventually robs the person of the ability to breathe normally. The lungs, located within the chest, are about 10–12 inches (25–30 cm) long and are two sponge-like organs that consist of a total of five sections or lobes (Figure 14.1). The right lung has three lobes, whereas the left lung is smaller with only two lobes because the heart takes up room on that side of the chest cavity. Air is taken in through the mouth and nose and then travels down the trachea to the lungs. In the lungs, the trachea divides into two smaller branches called bronchi which divide further, getting smaller until they become bronchioles. The ends of the bronchioles contain small sacs, or alveoli, where the air comes into contact with the circulatory system. The alveoli are surrounded by capillaries that allow for the efficient exchange of carbon dioxide for oxygen. In the lungs, the carbon dioxide gas that is given off as a meta-bolic waste product by our cells is released from the blood and exhaled and exchanged for the oxygen from the air we breathe. The oxygen is then carried by red blood cells from the lungs and delivered to all the cells of the body in order to fuel the production of the energy they need to func-tion and live.

RISK FACTORS ASSOCIATED WITH THE DEVELOPMENT OF LUNG CANCER

Any pollutants or toxic chemicals that are in the air and inhaled will enter the body through the lungs. Any of the tissues in the lungs can become

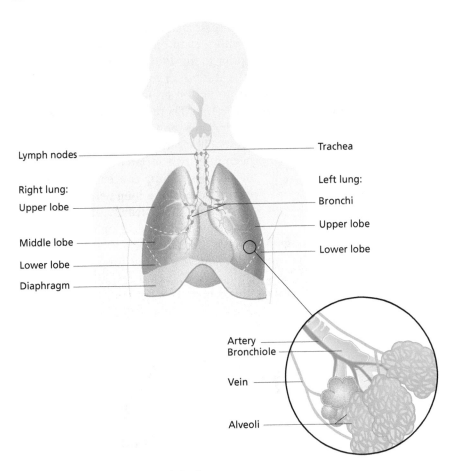

Lymph nodes

Right lung:

Upper lobe

Middle lobe

Lower lobe

Diaphragm

Trachea

Left lung:

Bronchi

Upper lobe

Lower lobe

Artery
Bronchiole

Vein

Alveoli

Figure 14.1 The anatomy of the lungs.

cancerous, but the cancer most often arises in the lining of the bronchi. There is an enormous amount of irritants and pollutants in an industrialized society that the body is exposed to on a daily basis. Excessive exposure to such pollutants can interfere with the body's natural defense system and inherently increase one's risk of developing cancer.

Smoking tobacco is the primary cause of lung cancer

Smoking harms nearly every organ of the body. The list of diseases associated with smoking includes abdominal aortic aneurysm, acute myeloid leukemia, cataracts, cervical cancer, kidney cancer, pancreatic cancer, pneumonia, and stomach cancer. These are in addition to the diseases previously known to be caused by smoking, which include lung, esophageal, laryngeal, oral, throat and bladder cancers, chronic lung diseases, coronary heart and cardiovascular diseases, and sudden infant death syndrome, as well as its association with certain reproductive effects.

Lung cancer was not common prior to the 1930s but increased dramatically over the next few decades as the use of tobacco products became more popular. Smoking is known to be the primary risk factor for any type of lung cancer (Figures 14.2, 14.3). There are at least 40 known carcinogens and about 4000 chemicals present in cigarette smoke, and they have been directly associated with nearly 87% of all cases of lung cancer.[14.1] Those who have smoked for a long period of time or who smoke more heavily than others are at a greater risk for developing the disease.

14.1 Some examples of chemicals in cigarette smoke are carbon monoxide, benzene, formaldehyde, hydrogen cyanide, and ammonia.

One piece of encouraging news is that if a person stops smoking early enough, before the cancer develops, his or her lung tissue returns to normal over a period of several years. Smoking cessation at any age lowers the risk and eliminates the primary contributing factor for this disease. The reduction in a person's risk after being smoke-free for

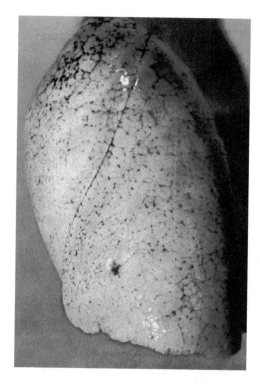

Figure 14.2 Healthy lung. This healthy lung was taken from a nonsmoker. Reprinted with permission © 2008 American Lung Association. www.lungusa.org

Figure 14.3 Cancerous lung. A lung that has been damaged by smoking. Reprinted with permission © 2008 American Lung Association. www.lungusa.org

5 years is evident by the substantial amount of repair that occurs in the lung tissue. After 10 years, the risk of developing the disease is one-third to one-half of the risk for people who continue to smoke. As the number of smoke-free years increases, the risk continues to decline, although former smokers will always remain at a somewhat higher risk than those who never smoked.

While many people believe that smoking cigars and pipes is less dangerous and that low tar cigarettes are safer, these are only myths and do not hold true. Tar is the term used to describe the toxic chemicals in cigarettes. Low tar cigarette smokers oftentimes take more or deeper puffs, thereby drawing larger amounts of smoke further into their lungs.

Smoking marijuana may also put a person at risk of developing lung cancer, since marijuana cigarettes have more tar than regular cigarettes while still containing many of the carcinogenic substances found in tobacco. It is also common for people who smoke "pot" to inhale it very deeply and hold the smoke in their lungs longer than is usually done with regular cigarettes. Due to the illegality of using marijuana, few studies have been performed to determine the impact of this practice. Theoretically, the number of years and amount of marijuana smoked will have an impact on a person's cancer risk.

Exposure to radon or asbestos puts people at risk

Radon is an invisible, odorless, radioactive gas that exists naturally in areas where there are concentrations of uranium in the ground. Radon is a form of radiation and radiation is known to damage DNA (Chapter 7). It can be located underneath many homes (depending on soil content) and is able to penetrate through the walls and foundations without any indication to the inhabitants. The remedy for a high level of the gas is to provide an escape avenue, such as an open basement window or a fan that exhausts to the outside. A simple kit purchased from a hardware store can be used to determine the level of radon gas in a house. People who have been exposed to radon have an increased risk of lung cancer over time, with approximately 12% of all lung cancer deaths linked to it.

> Are there any laws in your state banning the sale of homes with radon? Has your home been tested for radon?

Asbestos, a fire-retardant insulating material, was used for many years in numerous industrial and residential applications, but in 1989 the US government banned all new uses of the product. As long as it is contained in solid form and not released into the air, it is not believed to be harmful. People who work with or are exposed to asbestos have a higher risk of developing mesothelioma, a lung cancer which originates in the lining of the lung and has a high mortality rate. Their risk is further increased if they also smoke. Fortunately, this is now a well known fact and extreme precautions have been taken to protect today's workers.

Other possible cancer-causing agents in the workplace include uranium, coal products, metals such as chromium, cadmium or arsenic, gasoline and diesel exhaust, and a multitude of chemicals and solvents. In some cities, especially in those that are heavily industrialized, air pollution appears to put residents at a slightly higher risk of contracting lung cancer, but the risk is still far less than that caused by smoking.

The genetic influence on lung cancer development is being studied

Some people, especially those who develop the disease at a younger age, seem to have a genetic susceptibility for lung cancer. Research is currently underway to identify particular DNA alterations that are associated with its development. The Genetic Epidemiology of Lung Cancer Consortium studied 52 families with a strong history of lung, throat or laryngeal cancer to determine if there was a susceptibility to the disease. In 2004, they announced that those with a genetic abnormality on chromosome 6 will more readily develop lung cancer even if they do not smoke often. Markers on chromosomes 12, 14 and 20 also indicated a possible, albeit weaker, linkage to lung cancer susceptibility. Now that specific loci have been identified, expanded studies on more families will be performed. People who carry the trait would be especially vulnerable to lung cancer and could pass this trait on to their children.

Acquired DNA mutations in lung cells often result from tobacco smoke exposure. The *p53* tumor suppressor gene and the *ras* oncogene have been associated with lung cancer development.

> **Name some risk factors implicated in the development of lung cancer other than tobacco smoke.**

LACK OF DISTINCTIVE SYMPTOMS MAKES EARLY DIAGNOSIS DIFFICULT

Some examples of common symptoms

Many of the signs and symptoms of lung cancer are at first very subtle and are thus easily missed or mistaken for other conditions. Early lung cancer may have no specific symptoms but some of the more common signs and symptoms that a person could have include: a persistent cough that gets worse over time, constant chest, shoulder or back pain made worse by deep breathing, coughing up blood, shortness of breath, onset of wheezing or hoarseness, recurrent bronchitis or pneumonia, bloody or rust-colored sputum (phlegm) or any change in its color or volume, neck or face swelling, wheezing, hoarseness, difficult or labored breathing, or stridor (a harsh rasping sound with each breath). As always, unexplained weight loss, loss of appetite, or fatigue could be indicative of cancer or other serious conditions.

Lung cancer may metastasize before symptoms appear

Early stage lung cancer has been traditionally difficult to diagnose due to a lack of symptoms and as a result, may spread before being detected. Because the symptoms mentioned above might initially be muted or sporadic, an association with the disease might not be made. Lung cancer that has metastasized may result in enlarged lymph nodes in the neck or masses near the surface of the body such as above the collarbone. A tumor may lead to pain, numbness, or weakness of the arms or legs if it is compressing against nerves, a seizure if it is in the brain, or jaundice

Box 14.1

Smoking is a women's health issue

Although men have higher rates of lung cancer than do women (approximately 114,000 men and 100,000 women in 2008), there has been an alarming increase in the number of females afflicted while the male incidence rates have significantly decreased. Scientists have not yet identified why this has occurred, but some have theorized that men and women clear out toxins from tobacco smoke at different rates. In the late 1980s, lung cancer overtook breast cancer as the most common cause of cancer deaths in women, often theorized as a result of the advent of cigarettes and commercials aimed towards women in the decades before.

Smoking is a habit that frequently begins at a young age, often before the age of 16. Yet it was not always this way. In the early twentieth century, smoking was equated with poor character, low social status, and even prostitution. It was then that the tobacco companies realized that women could be a new group of persons to be converted into smokers – a marketer's dream population. Suffragettes had just won the right to vote and the earliest hints of feminism had begun to surface when the tobacco companies introduced advertising campaigns with such slogans as "Reach for a[cigarette]instead of a sweet." The implication, of course, was that the use of their products would satisfy a craving and yet still enable women to stay thin. An additional message demonstrated independence, as the manufacturers called cigarettes "torches of freedom." Over the next few years, more and more women took up the habit with the idea of throwing off the conventional shackles of society. By the 1960s, with the women's movement in full steam, one in three women were lighting up. In a stunning advertising move, Virginia Slims developed the ad "Tailored for the feminine hand, slimmer than the fat cigarettes men smoke" asserting that these cigarettes were designed just for women. This campaign triggered the biggest growth in history for tobacco advertising specifically targeting women.[14.2] Years later, as the dangers and effects of smoking were starting to be identified and increasing numbers of women began quitting, the tobacco companies responded by introducing new lowtar and light cigarettes. Although these had no demonstrated health benefits, there was a promotional implication that they did. Today, movies, television shows, and magazines regularly show young, attractive, thin women with a cigarette in hand, with the ultimate goal of maintaining the myth that smoking is still socially acceptable.

> [14.2] The US Public Health Cigarette Smoking Act of 1969 was signed into law by President Richard Nixon on April 1, 1970 putting an end to cigarette ads on TV and radio.

Box 14.2

Deena Soloway: a young woman's tragic story

Deena Soloway was only 28 when she died of lung cancer in 1998 (Figure 14.4). She began smoking at the age of 13 in the hopes of emulating the glamorous models and celebrities on magazine covers and in movies. Working in Los Angeles as a photographer on TV and movies sets in her twenties, she began to feel extreme fatigue and experienced violent coughing fits. After several medical examinations, a grapefruit-sized mass in her left lung was discovered, removed and determined to be cancerous. At that point, it was thought that the cancer was localized and contained, but 11 months later it re-emerged, this time in her brain. Deena lost all ability to speak about 3 weeks before she died, and was able to communicate only by notes and gestures. Her final note to her mother, Susan Levine, read: "Continue on, because if I need to go, I want somebody else to learn from my mistakes in life."

Figure 14.4 Deena Soloway.

(yellow coloring of the eyes and/or skin) if it has spread to the liver. Since these symptoms are often related to other conditions, it is easy to see why the association of lung cancer is not immediately made as it is with other cancers. If the cancer has already spread to other areas of the body by the time it is detected, a full recovery is an extreme challenge.

Lung cancer can occur among people who have never smoked

About 10% of men and 20% of women with lung cancer have *never* smoked. Lung cancer in nonsmokers may have specific physiological factors contributing to its cause. Lung tumors often possess estrogen receptors and there is evidence that they also express the *HER2* gene

---

Box 14.3

The Master Settlement Agreement: US vs. tobacco companies

In 1998, the US's leading cigarette manufacturers and the attorneys general of 46 states signed the Master Settlement Agreement (MSA). This settlement called for changes in the way tobacco products are advertised, promoted, and sold. It established the prohibition of tobacco advertising targeted toward young people less than 18 years of age, as well as guaranteed the payment of approximately $246 billion to the states over a period of 25 years. In addition, the tobacco companies agreed to not misrepresent the health consequences of smoking nor withhold or limit any research related to their products. Further, they agreed to release internal industry documents to the public, even though they were essentially self-incriminating. It was a protracted and hard won settlement, among the most extensive (and expensive) class action suits ever, but it was a stunning victory. The once thriving tobacco industry, although still intact and operational, was dealt a crippling blow.

   This victory, however, came with an unintended consequence; 43 out of the 46 states involved with the agreement have diverted the money away from its stated purposes of smoking and prevention awareness, cessation programs, education and research, and public service announcements, most likely the result of a lack of any legal requirement demanding specific uses of this money. Instead, cigarette tax and MSA revenues have been used for the building of roads, schools, debt service, or whatever the state legislatures deem necessary. In a time of shrinking tax bases, states have looked upon these settlements as a prime source of new revenue and structured their laws in such a way as to give themselves free reign over the spending.

(Chapter 8). Estrogen is being studied for a similar role to the one it plays in breast cancer, although the connection remains unproven. Adenocarcinoma, a type of lung cancer, is being diagnosed more often among nonsmokers and women.

While the causal association is not definitely known, one risk factor may be environmental tobacco smoke (ETS) or second-hand smoke. Nonsmoking spouses of smokers have a 30% greater risk of developing lung cancer than do spouses of nonsmokers. It has been estimated that 17% of all cases of lung cancer in nonsmokers are caused by second-hand smoke exposure during their childhood and adolescence. Early studies demonstrated that nonsmokers who are diagnosed with lung cancer are more likely to be women, possibly because more of their partners smoke, or because of genetic factors or hormonal influences. A 2006 study by the American Cancer Society, however, contradicted this long-held theory and research is ongoing to help establish or refute the reasons for the association of second-hand smoke and lung cancer.

LUNG CANCER IS OFTEN DIAGNOSED AT AN ADVANCED STAGE

Unfortunately, there are no early screening methods or tests for the detection of this disease. For many years, chest X-rays on smokers were the only available screening method. It has since been proven that although an X-ray may detect advanced lung cancer, it is not as effective in detecting it at an early stage when the tumor would be more operable. Lung tissue absorbs little radiation so there is not enough contrast between lung tumors and lung tissue to be observed on an X-ray. This is the main cause for the lower survival rates of those diagnosed with the disease.

Newer imaging tests provide a noninvasive means to visualize lung cancer

A newer and more advanced imaging technique called spiral computed tomography (Chapter 6) is currently being evaluated for possible benefits. The CT device is capable of scanning the entire chest with X-rays in about 15–25 seconds, and a computer creates a 3-D model of the lungs (Figure 14.5). Newer devices utilize a combination of PET (Chapter 6) and CT scans to more accurately determine the location of the tumor. Although the price of these scans is far more than that of a chest X-ray, the eventual savings for the

> **Why is lung cancer often diagnosed at an advanced stage?**

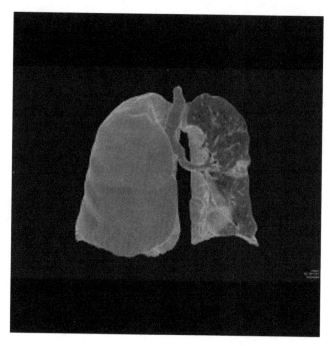

Figure 14.5 A CT image of the lungs. A 3-D image of both lungs with a portion of the left lung (shown on the right) digitally altered in order to view its interior. Image courtesy of David Gilmore at Beth Israel Deaconess Medical Center.

healthcare system would result from early detection techniques that will ultimately lead to more successful treatments as well as fewer and shorter hospitalization stays. Begun in 2002, the National Lung Screening Trial's (NLST) goals are to determine whether a standard chest X-ray or a spiral CT should be used as a screening tool for those at high risk for lung cancer and to evaluate which test will be more effective in ultimately reducing lung cancer deaths. Nearly 50,000 current or former smokers are now being studied at more than 30 sites across the country and will continue to be tested for 8 years.

Sputum cytology and lung biopsies are necessary for diagnosis

Patients with a phlegm (sputum) cough may be asked to give a specimen that can then be examined for cancerous cells. The physician may also perform a bronchoscopy, which is the insertion of a small, flexible, fiberoptic scope called a bronchoscope through the nose or mouth and down the throat to examine the bronchi, the major airways of the lungs,

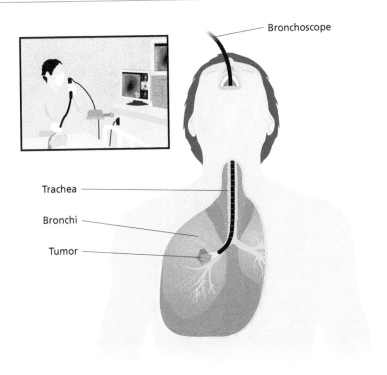

Figure 14.6 Bronchoscopy. While a patient is under anesthesia, a fiberoptic scope that allows the lungs to be viewed is inserted into the airways. Tissue samples can be removed at the same time for microscopic examination.

as well as to take a biopsy (Figure 14.6). The procedure may be performed under light sedation or a general anesthetic. A chest X-ray or CT scan of the chest may be utilized during the biopsy to assist in locating the tumor. Results of these procedures may be used for making a diagnosis, staging, or the selection of treatment.

Sometimes a biopsy, even when performed properly, may fail to diagnose some lung cancers, depending on the tumor's size and location. Therefore, even if such tests do not reveal cancer, it does not always mean that the lung mass is benign. Most lung cancers are located towards the edge of the lung rather than in a major bronchus, limiting the success rate of this procedure.

Lymph nodes may also be biopsied in an attempt to determine the stage of a cancer by examining the size of the tumor and the number and location of local lymph nodes that possess cancerous cells. A procedure that allows the surgeon to examine the nodes is a cervical mediastinoscopy, which is performed under general anesthesia. Through a one-inch incision

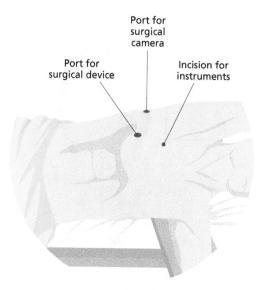

Figure 14.7 Thoracoscopy. The thoracoscopy procedure takes place while the patient is on his/her side. Incisions are made for the camera and other instruments necessary to examine the tissues and lymph nodes and, if necessary, to remove portions of the lung.

at the base of the neck just above the breastbone, the surgeon inserts a fiberoptic scope and follows the windpipe into the chest to locate and remove the lymph nodes for biopsy. One other option is the use of thoracoscopy, or video-assisted thoracic surgery (VATS), in which a fiberoptic camera is inserted into a one and a half-inch incision while surgical instruments are inserted into two other similar-sized incisions in order to remove specimens of the lung mass for examination (Figure 14.7). This technique may also be used to obtain lymph nodes or enable a visual examination for evidence of metastasis. Depending on the extent of the cancer that is observed, the surgeon may immediately decide to proceed with a more involved operation while the patient is still under general anesthesia.

THERE ARE TWO MAIN CATEGORIES OF LUNG CANCER

Non-small and small cell are two general types of lung cancer

Lung cancer is not really a single disease but rather a collection of diseases characterized by the cell type involved. There are two general types, each with different treatments and expected survival outcomes. They are non-small cell lung cancer (NSCLC) and small cell lung cancer (SCLC) and are classified as such by their cellular appearance under a microscope. Subclasses exist within each category.

Approximately 85% of lung cancers are NSCLCs, which typically grow and spread *more slowly* than do SCLCs. NSCLC has a 5-year survival rate of 60% if detected early and 40% if detected at a later stage. The three subcategories of NSCLC are named for the type of cells in which the cancer develops and reflect their differences in size, shape, and chemical make-up.

Squamous or epidermoid carcinoma generally arises near the bronchus and tends to grow slowly. Adenocarcinomas vary in size and grow near the outside surface of the lung. Finally, large cell undifferentiated carcinomas may appear in any part of the lung and grow and metastasize quickly.

Small cell lung cancer, also known as oat cell cancer, is less common than NSCLC (about 15% of lung cancers) but far more dangerous. It begins in the bronchi and can result in the spread of large tumors throughout the body. Therefore, it usually cannot be treated by surgery. It grows more quickly than NSCLC and is more likely to metastasize to other organs by the time it has been diagnosed.

THREE TRADITIONAL THERAPIES ARE USED IN LUNG CANCER TREATMENT

As with most cancers, treatment depends on the type of lung cancer and the extent of the metastasis and is often a combination of three standard therapies – surgery, chemotherapy, and radiation.

Surgery is an important part of treatment and staging

Surgery for lung cancer should be performed by a thoracic surgeon, a physician who specializes in surgical treatment of diseases or other conditions of the chest. If the disease is detected very early, the most common surgical procedure for the removal of lung cancer is a lobectomy, the excision of an entire lobe. Other options are a wedge resection, also known as a segmentectomy, the removal of only a portion of the lobe, or a pneumonectomy, the removal of the entire lung (Figure 14.8).

Staging of non-small cell lung cancer (NSCLC) relies on the TNM staging system, devised by the American Joint Committee on Cancer and described in Chapter 6. Staging of small cell lung cancer (SCLC) is unique in that it utilizes a two-stage system that classifies the tumor as either *Limited* or *Extensive*. Finding the cancer in only one lung and in lymph nodes on the same side of the chest puts it at the Limited Stage.

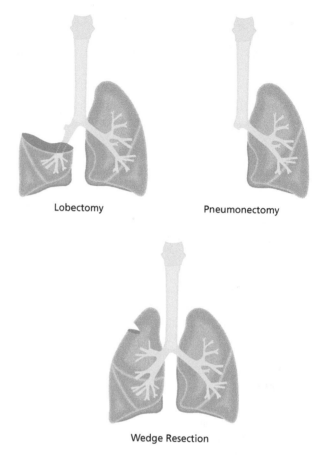

Lobectomy   Pneumonectomy

Wedge Resection

Figure 14.8 Common surgical procedures for lung cancer.

Cancer that is found in lungs and lymph nodes on both sides, in distant organs, or in fluid found around the lung would be considered an Extensive Stage. About two-thirds of the people with SCLC are at the extensive stage upon initial diagnosis. Patients diagnosed with SCLC are less likely to benefit from surgery since it usually is present in both lungs due to its tendency to grow rapidly and metastasize. Therefore, surgery is often performed for palliative (alleviate pain) reasons.

Radiation may be utilized alone or with other treatments

External radiation may be used as the principal form of treatment for lung cancer, particularly for those who are not strong enough for surgical intervention. It may be used concurrently with chemotherapy and/or surgery, to reduce the size of a tumor in an attempt to offer pain relief, to relieve bleeding, or to open up passageways that will enhance the

Box 14.4

The Bonnie J. Addario Lung Cancer Foundation

In 2003, Bonnie J. Addario had a recurring sharp pain in the left side of her chest. She was referred to a neurosurgeon who prescribed pain medication for herniated discs. The next month, she made her own decision to go to a facility that offered CT scans and screening, tests not always paid for by insurance companies. The results eventually led to a diagnosis of lung cancer for which she received surgery, intensive chemotherapy, and radiation. During her recovery, she decided to establish a foundation to educate people on lung cancer and the need for early detection.

The Bonnie J. Addario Lung Cancer Foundation, a San Francisco based non-profit organization, is the nation's largest philanthropic association devoted exclusively to eradicating lung cancer. The foundation is committed to changing this disease's horrific statistics and mortality rates. To achieve that goal, the foundation works collaboratively with a diverse group of physicians, researchers, corporations, and individuals to make meaningful change through research, early detection, prevention, and treatment.

Their 2008 subway advertisement campaign brought attention to the most under-funded and under-diagnosed but deadliest of cancers in a way that spread the word on the significance of early diagnosis. Their *One in a Million Campaign* asks everyone to donate $20 in the hopes that one million people will contribute. These creative ideas can go a long way towards raising money for a cause that has been largely ignored.

flow of air. Once chemotherapy and radiotherapy of the chest are completed, prophylactic cranial irradiation (focused on the brain) may be performed to prevent metastasis to the brain, which often occurs with SCLC.

Specific chemotherapeutic drugs are recommended

When the cancer has not spread beyond the chest cavity, it is treated primarily with two or more chemotherapeutic drugs *and* radiation. Studies have demonstrated that better results are obtained when chemotherapy and radiation are used in *combination* rather than in *succession* as may be done for other cancers. When metastasis has occurred beyond the chest, chemotherapy is often used alone. NSCLC is also treated with chemotherapy, although it is less responsive than SCLC. Surgery and radiation are more commonly used in the treatment of NSCLC because of its slow growth rate and tendency to not metastasize.

Typical chemotherapeutic drugs include cisplatin or carboplatin combined with paclitaxel for NSCLC patients who can tolerate a combination of drugs. For those who cannot tolerate this regimen, only one of the agents is used. Other drugs are available as second-line treatments if the first does not work or if the cancer continues to grow. The FDA has approved erlotinib (Tarceva), an oral medication that specifically targets cancer cells and interferes with their ability to grow. Other drug examples are Erbitux, an EGFR inhibitor (Chapter 7) and Iressa (gefitinib).

IS THERE DISCRIMINATION IN CANCER RESEARCH FUNDING?

Unfortunately, less research money has traditionally been allocated to the study of this disease in comparison to other cancers that may have a more sympathetic profile. In 2003, the US National Cancer Institute reported that lung cancer research received far less funding per death than other cancers. That year, the NCI spent approximately $1,740 per lung cancer death compared to $13,649 per breast cancer, $10,560 per prostate cancer, and $4,851 per colorectal cancer death. Although institutions and researchers have only finite amounts of money to work with, hopefully the number one cancer killer will become a major focus of attention.

EXPAND YOUR KNOWLEDGE

1 Which US states use the MSA revenues as intended?
2 Explore the latest findings on the clinical trials mentioned in this and other chapters associated with lung cancer.
3 Discuss the most recent findings on menthol and "light" cigarettes.
4 If smoking is associated with lung cancer, why isn't it banned?
5 Give some examples of smoking cessation products and programs. What are the success rates for these methods?

ADDITIONAL READINGS

American Lung Association's website: www.lungusa.org
The Bonnie J. Addario Lung Cancer Foundation's website:
   www.thelungcancerfoundation.org/
Carpenter, C., Wayne, M., Ferris, G., and Connolly, G., (2005) Designing cigarettes for women: new findings from the tobacco industry document. *Addiction*, **100**: 837–851.
Department of Health and Human Services, Centers for Disease Control and Prevention (2001) *2001 Surgeon General's Report – Women and Smoking.*
Lung Cancer Alliance website: www.lungcanceralliance.org

Marieb, E. N. & Hoehn, K. (2007) *Human Anatomy and Physiology*, 7th edn. Upper Saddle River, NJ: Prentice Hall.

Martini, R. & Bartholomew, E. (2007) *Essentials of Anatomy & Physiology*, 4th edn. Upper Saddle River, NJ: Prentice Hall.

Wainscott-Sargent, A. (2005) *A Breath Away: Daughters Remember Mothers Lost to Smoking.* Phoenix, AZ: Acacia Publishing.

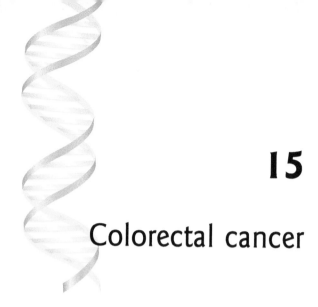

15

Colorectal cancer

Many people still are not getting screened because of fear and embarrassment. Some colorectal cancer organizations have the slogan, 'Don't die of embarrassment.' That's really true.

Bernard Levin, MD, Vice President, Division of Cancer Prevention & Population Sciences at The University of Texas MD Anderson Cancer Center

CHAPTER CONTENTS

- Colon and rectum are the last two sections of the gastrointestinal tract
- Risk factors for colorectal cancer
- Screening tests
- Treatment options
- Targeted therapies
- Screening tests performed after the course of treatment
- Expand your knowledge
- Additional readings

Cancer: Basic Science and Clinical Aspects, 1st edition. By C. A. Almeida and S. A. Barry. Published 2010 by Blackwell Publishing, ISBN 978-1-4051-5606-6.

The large intestine, also referred to as the colon, is the last segment of the digestive tract, and its primary function is the absorption of water to convert semisolid indigestible waste material into solid feces. The rectum, the last portion of the colon, stores the fecal matter until it is eliminated. Colorectal cancer, which is a disease of the colon and/or rectum, is more common in developed countries and, excluding most forms of skin cancer, is the third most common cancer among both men and women. Each year more than one million people worldwide are diagnosed with colorectal cancer and approximately 150,000 of those cases are recorded in the US and 194,000 in Europe. The disease affects nearly an equal number of men and women. It is the second most common cause of cancer deaths with just over 655,000 dying of it annually. The number of diagnosed cases has been declining for the better part of the past two decades largely due to awareness, successes with screening and early detection, and improvements in treatment methods. Nearly 40% of colorectal cancers are detected at an early enough stage before metastasis has occurred. The overall 5-year survival rate in the US is greater than 60% and, when detected at an early stage, approaches 90% with treatment.

COLON AND RECTUM ARE THE LAST TWO SECTIONS OF THE GASTROINTESTINAL TRACT

The energy our cells need in order to live is provided by the nutrients in the food we eat. Our bodies must first digest the food to make the nutrients accessible for absorption. Digestion occurs in the gastrointestinal (GI) or digestive tract, which is essentially a long tube that runs through the body from the mouth to the anus (Figure 15.1).[15.1] The GI tract is an organ system comprising the mouth, esophagus, stomach, small intestine, large intestine (a.k.a. colon and rectum), and anus. Digestion begins in the mouth when the teeth grind and tear the food into smaller pieces. The chewed food combines with saliva to form a ball known as a bolus. When swallowed, the bolus enters the esophagus, a muscular

15.1 The body can be thought of as a donut with a hole in the center. The donut represents the physical part of the body while the hole in the center is the digestive tract in which food enters and is digested, nutrients are absorbed, and wastes are stored and finally eliminated.

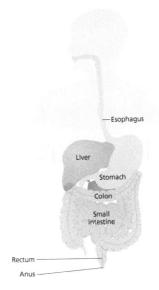

Esophagus

Liver

Stomach

Colon

Small Intestine

Rectum

Anus

Figure 15.1 The digestive system. The gastrointestinal tract includes the mouth, esophagus, stomach, small and large intestines, rectum, and anus.

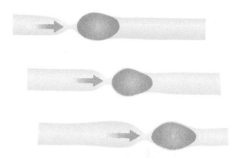

Figure 15.2 Peristalsis. Food is moved through the digestive tract by wave-like contractions of muscles.

---

¹⁵·² A person can swallow food while standing on his or her head because peristaltic contractions of the eso-phagus literally push the food toward the stomach.

---

¹⁵·³ The pH scale ranges from 1 to 14 with 7 being neutral. A pH below 7 is considered to be acidic while a pH above 7 is considered to be alkaline or basic. The stomach produces on aver-age 1–3 liters of gastric juice a day. The majority of this fluid is reabsorbed as it passes through the colon.

---

tube that connects the back of the throat with the stomach. The bolus is squeezed down the esophagus as a result of a series of wave-like muscle contractions called peristalsis (Figure 15.2).¹⁵·²

The majority of digestion occurs in the stomach and small intestine. The stomach is a muscular organ that con-tracts rhythmically to squeeze and churn its contents. It also produces gastric juice, a fluid that is very acidic (pH 1–3) and contains proteases, enzymes that degrade proteins, the principal type of food that is digested in the stomach.¹⁵·³ The com-bined activity of the acidic environment, which causes the proteins to denature (i.e., unfold), and the proteases, degrades proteins into short peptide fragments as well as individual amino acids. The inner surface of the stomach has a layer of mucus that protects it from the acidic gastric juice and digestive enzymes. After a period of time in the stomach, the food is transformed into an acidic liquid called chyme.

The majority of digestion and absorption of nutrients and water takes place in the small intestine, which in adults is approximately 5–6 m long and has a diameter of about 2.5 cm.¹⁵·⁴ Digested food passes from the stomach through the lumen of the small intestine, which has three portions. The first and shortest section, about 25 cm, is the duodenum where the majority of digestion takes place. The chyme from the stomach enters into the duodenum and mixes with secretions from the pancreas and gall bladder. These secretions contain

---

¹⁵·⁴ 1 inch = 2.54 cm; 1 m is equal to 3.28 feet.

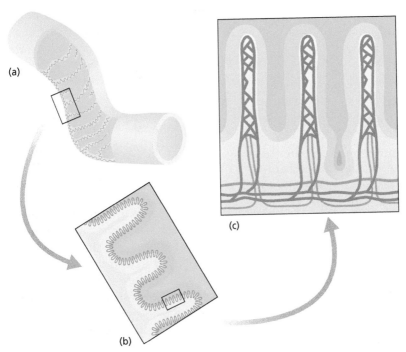

Figure 15.3 Inner lining of the small intestine. (a) The inner lining of the small intestine is highly convoluted forming villi. (b) The surface of each villus is highly convoluted forming microvilli. (c) Each microvillus has capillaries and lymphatic vessels that carry the absorbed nutrients to other areas of the body.

enzymes and bile salts that digest lipids, carbohydrates, and the remaining peptides. The pH of the duodenum is approximately 6, close to neutral but still slightly acidic.

Digested food passes from the duodenum into the jejunum, the central portion of the small intestine, which varies from 1 to 2 m in length and has a pH of approximately 6–7. The lining of the jejunum that faces the lumen is highly folded and each of the folds have many finger-like projections called villi (singular: villus) (Figure 15.3). Upon microscopic examination of the surface of a villus, very tiny projections called microvilli that resemble the villi are seen. The surface area of the lumen of the small intestine is dramatically increased by the presence of the villi and microvilli, making it highly efficient at absorbing the nutrients, water, and salts that are moved along through the intestine by peristalsis. The material that is absorbed enters the blood and lymphatic vessels within each villus. The final portion of the small intestine, the ileum, is the longest section with a length of 2–4 m. Similarly to the jejunum, its inner surface has villi and microvilli and its principal function is the absorption of digested fats and vitamin B_{12}, which is synthesized by bacteria in

15.5 The shift in pH from highly acidic in the stomach to slightly alkaline in the ileum provides a defense against pathogenic microbes that are swallowed. The vast majority of all organisms can only exist in an environment with a narrow pH range. Most organisms will not be able to survive the acidic environment of the stomach, but in the rare event one does, it almost certainly would not survive once it passed into the higher pH environment of the small intestine.

15.6 When considering all areas of the body that are populated with microorganisms, the colon has the largest concentration. In fact, microbes such as bacteria and fungi comprise approximately one-third of the mass of fecal material!

the intestine. The ileum has a pH of 7–8, neutral to slightly alkaline.[15.5]

The semisolid material that leaves the small intestine enters the large intestine or colon, which has an average length of 1.5–1.8 m and a diameter of 6.4 cm. Peristalsis moves material through the ascending, transverse, descending, and sigmoid portions of the colon (Figure 15.4). The inner surface, unlike in the jejunum and ileum, is smooth because there is less of a need for efficient absorption since the vast majority of nutrients and water were previously absorbed in the small intestine. What enters the colon is mostly indigestible carbohydrates known as roughage or dietary fiber. Most of the remaining water, salts, and vitamin K are extracted from the roughage as it is being mixed with mucus and bacteria that inhabit the colon, thus producing solid feces.[15.6] The feces are stored in the rectum, the last 15–16 cm of the colon. Defecation

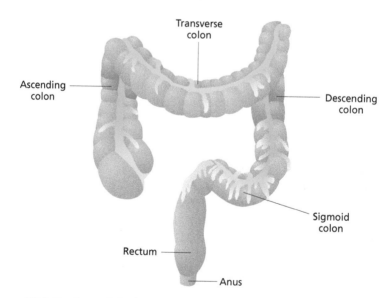

Figure 15.4 Portions of the large intestine. The segments of the large intestine are the ascending colon, transverse colon, descending colon, sigmoid colon, rectum, and anus. Adapted from: http://apps.uwhealth.org/health/adam

occurs when peristalsis pushes the feces through a ring of muscle that controls the opening and closing of the anus.

RISK FACTORS FOR COLORECTAL CANCER

Primary risk factors are age and family history

The principal risk factor for developing the disease is age; greater than 90% of people with colorectal cancer are older than 50 years of age. A person is at a much higher risk if he or she has one or more relatives, especially those who are first-degree, who were diagnosed with the disease. Approximately 20% of colorectal cancer patients have two or more closely related family members who have also been diagnosed with the disease. The only ethnic group with a higher risk is the Ashkenazi Jews, who are of Eastern European descent, for reasons that are not yet known.

Growths, chronic inflammatory bowel diseases, and lifestyle choices

Polyps are growths that extend from the surfaces of mucous membranes. Most polyps are pedunculated, meaning that they extend from the wall of the colon by a stalk, although some are classified as flat because the growth lacks a stem (Figure 15.5). If the polyp originates from the glandular tissue of the mucous membranes it is termed an adenomatous polyp.

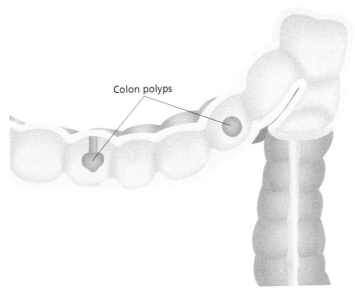

Colon polyps

Figure 15.5 Polyps. Growths along the inner lining of the colon are referred to as polyps.

While most colon polyps are benign, they are removed when detected during screening methods because of the likelihood of their becoming malignant, especially if they are adenomatous polyps. A person who has previously been diagnosed with an adenoma, had polyps, or was diagnosed and successfully treated for colorectal cancer, especially if before the age of 60, has a greater risk of developing cancer in another area of the large intestine.

Ulcerative colitis and Crohn's disease, two forms of chronic inflammatory bowel disease (IBD), increase a person's risk for colorectal cancer. Familial adenomatous polyposis (FAP) is an inherited condition that results in the formation of hundreds to thousands of polyps beginning as early as the age of five. If the affected area of the colon is not surgically removed, there is a 100% certainty that one or more of the polyps will become malignant by the time the person is in his or her early twenties. A similar condition is hereditary nonpolyposis colorectal cancer (HNPCC) in which only a few polyps develop at an early age. People with HNPCC have an 80% chance of developing colorectal cancer in their lifetimes and are also at an increased risk of developing numerous other cancers. Screening is vital for people with these conditions.

Certain lifestyle characteristics are associated with an increased risk of developing colorectal cancer. People who are obese, inactive, heavy alcohol drinkers, or who eat a diet high in red and processed meats, which is also likely to be high in fat (particularly saturated fat) and low in fruits and vegetables, are at higher risk for the disease. Conversely, fruits and vegetables, in addition to providing vitamins, minerals, and antioxidants, also provide dietary fiber. Fiber provides bulk that helps move the material through the colon. The slower the wastes travel through the colon, the longer the cells that line the interior of the colon are exposed to potentially harmful chemicals that may cause damage over a period of time.

Diabetics are 30–40% more likely to develop the disease. It is estimated that in comparison to nonsmokers, those who smoke and are diagnosed with colorectal cancer have a 30–40% greater risk of dying from the disease. Approximately 12% of the colorectal cancers that result in death are caused from the carcinogens that are swallowed or enter the blood while smoking.

A study that originated from the large Nurses Health Study and published in 2003 determined from an assessment of more than 78,500 female nurses that worked at least three nights per month for 15 or more years had a 35% increased risk of developing colorectal cancer. It is postulated that the nurses' melatonin production was suppressed by the exposure to light during nighttime hours. Melatonin is a hormone produced by the cells of the GI tract and its concentration varies according to a daily cycle. It is a powerful antioxidant that functions to protect the integrity of DNA and to maintain the proliferation of intestinal cells at a normal rate. It may be that having a decreased level of melatonin production put the nurses at an increased risk for the development of colorectal cancer.

SCREENING TESTS

There are a number of conditions besides cancer that affect the large colon, and many of them produce similar symptoms. Therefore, it is important to be alert for certain signs that should be reported as soon as possible to a physician for follow-up testing. Common conditions and symptoms include, but are not limited to, blood in the feces, frequent diarrhea or constipation, abdominal pain on a regular basis that may or may not be associated with a bowel movement, stool that is consistently thinner than normal, the frequent sensation that a bowel movement was not completed, weakness, tiredness, and sudden weight loss for an unknown reason. Symptoms are also likely to vary depending on the location of the cancer in the large intestine and severity of the condition. Tumors within the ascending colon typically extend from a single point of origin into the lumen of the intestine, unlike those in the remainder of the colon and rectum which tend to grow in the shape of a ring. The latter type is more likely to obstruct the movement of material through the intestine and cause a blockage.

A person exhibiting some of the signs mentioned above should undergo one or more of the same procedures used to screen for colorectal cancer as someone who is asymptomatic but at risk. Regular screening is essential to detecting the cancer at an early stage when it is most treatable. For the general population, screening should begin at the age of 50, but it may be started earlier for a person who has certain risk factors such as a family history of colorectal cancer. There are four tests most commonly used in the screening process. Three out of the four procedures enable the lining of the colon and rectum to be observed for the presence of a polyp. Polyps may range in size from small to large and, once detected, are removed and analyzed to determine if they are benign or malignant.

Fecal occult blood test

The fecal occult blood test (FOBT) detects the presence of minute amounts of blood in the stool that is undetectable (occult) by the human eye. A potential source of such blood is the blood vessels that are on the outer surface of a polyp. As fecal matter moves across the surface of a polyp the blood vessels may be damaged and release small amounts of blood. The FOBT is performed in two stages, the first at home and the second in a lab. A person uses a wooden stick to smear a test card with three small fecal specimens. The card is then returned to the doctor's office or is sent to a lab for chemical analysis to detect the presence of blood. A positive test result necessitates a follow-up procedure to determine if the blood originated from a polyp or another source such as a hemorrhoid or ulcer. Since not all tissue growths bleed, it is possible to obtain a false negative result when, in fact, a polyp is present.

> **Why do you think that FOBT is performed more often than FIT?**

The results of an FOBT can be affected by certain types of foods, such as those that contain blood (red meat) and high concentrations of vitamin C (fruits and vegetables, dietary supplements), or by nonsteroidal anti-inflammatory drugs (NSAIDs, such as aspirin, ibuprofen, and naproxen) that can cause bleeding. An alternate version of the FOBT is a fecal immunochemical test (FIT) that utilizes monoclonal antibodies which bind a specific protein present in human blood. This test is far less susceptible to the influences of foods and drugs that a patient may be eating or taking. However, it would still be unable to detect growths that do not bleed.

Sigmoidoscopy

> [15.7] An endoscopy is the use of a fiberoptic camera that is inserted into an orifice or small surgical incision to view internal areas of the body.

Two forms of endoscopy are used to observe the interior of the large intestine.[15.7] A sigmoidoscopy uses an endoscope that is approximately 60 cm (~2 ft) in length and has a diameter that

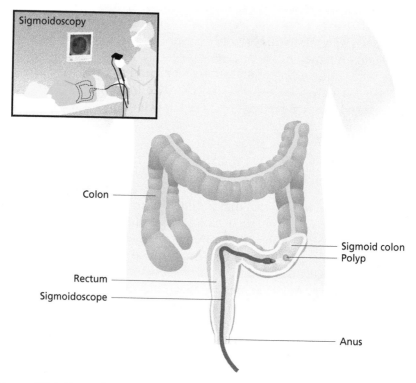

Figure 15.6 Sigmoidoscopy. A sigmoidoscope is used to view the inner lining of the rectum and sigmoid colon. Adapted from a figure of the NCI.

is approximately that of an average adult's index finger to view the interior lining of the rectum and sigmoid colon (Figure 15.6). A potent laxative is drunk the night before the procedure in order to cleanse the entire length of the colon. The removal of all fecal matter allows an unobstructed view of the lining of the last one-third of the colon and rectum. During the procedure, a flexible sigmoidoscope is inserted through the anus into the rectum and sigmoid colon, the two most common locations for intestinal polyps. The inner lining of these segments of the colon can be viewed on a monitor and if a polyp or suspicious area is observed, it can be removed for biopsy. During a sigmoidoscopy a patient may experience pressure and slight cramping, but rarely any significant pain.

Colonoscopy

A colonoscopy is a more extensive examination of the entire length of the colon (Figure 15.7). This procedure may be performed as an initial or routine screening method or as a follow-up procedure in order to examine the remainder of the colon when one or more polyps are found during a sigmoidoscopy. A colonoscope is similar to a sigmoidoscope except that it is a longer (135–165 cm; nearly 6 ft) instrument. The flexible

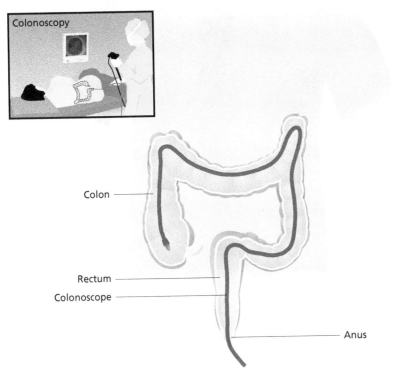

Figure 15.7 Colonoscopy. A colonoscope is used to view the inner lining of the entire colon. Adapted from a figure of the NCI.

Why is a colonoscopy more thorough than a sigmoidoscopy?

colonoscope is inserted through the anus and is slowly advanced through the rectum, sigmoid, descending, transverse, and ascending portions of the colon. It is also possible to excise tissue samples during the procedure for future examination. The procedure may take 15–30 minutes, depending on whether or not tissue samples are taken. Both sigmoidoscopy and colonoscopy carry the slight risk that the lining of the colon may be torn or punctured by the endoscope or during the process of excising a polyp. If this occurs, there is a risk that microbes could enter and travel through the bloodstream, which is normally sterile, causing an infection in one or more areas of the body. To minimize this risk, antibiotics may be taken prior to and after the procedure, depending on whether or not the patient has certain health conditions. Another minor risk is that a hemorrhage may occur during the removal of a polyp.

[15.8] Alternative terms for a virtual colonoscopy are colonography or CT colonography.

A newer, less sensitive and invasive procedure is a virtual colonoscopy (Figure 15.8).[15.8] A computed tomography (CT) scan (Chapter 7) takes numerous X-rays of the abdomen which can be

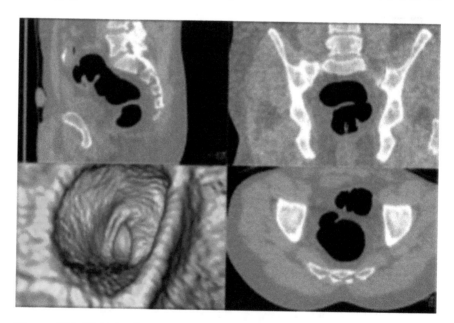

Figure 15.8 Virtual colonoscopy. Numerous 2-D CT scan images of the colon from different perspectives (upper left image: from left side of the body; upper right image: from the front of the body; lower right image: from the head down the length of the body) are used to produce a 3-D view of the inside lining of the colon that is artificially colored. Images courtesy of David Gilmore at Beth Israel Deaconess Medical Center.

viewed as a series of 2-D images or they can be assembled by a computer into a single 3-D image of the entire large intestine. The 3-D image can be rotated, allowing a physician to examine the colon from multiple angles. While the detail provided in the computer image does not compare to that obtained from a traditional colonoscopy, comparative clinical studies are being conducted to determine the value of each method. A virtual colonoscopy, however, would require a follow-up sigmoidoscopy or colonoscopy in the event that a suspicious area needed to be sampled.

Barium enema X-ray

Another means of observing the colon is a double contrast barium enema X-ray (Figure 15.9). Since the colon is composed of soft tissues and cannot be observed with much detail by X-ray, barium sulfate, a white metallic compound, is used as a contrast agent. A liquid laxative is drunk the night prior to the procedure to thoroughly cleanse the colon. The procedure begins with the administration of an enema, a liquid containing

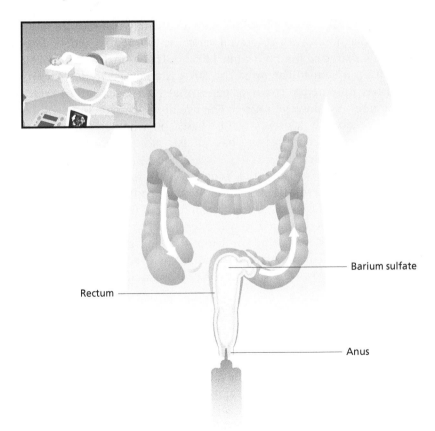

Barium sulfate

Rectum

Anus

Figure 15.9 Use of barium sulfate to view the lower portions of the GI tract. The use of a barium sulfate enema provides contrast to view the large intestine (colon) and rectum by X-ray.

barium sulfate that is injected into the colon through a tube inserted into the anus. Once the entire colon has been coated with a thin layer of the chalky substance, air is pumped through the same tube to inflate the colon. An outline of the interior of the entire colon can be seen on an X-ray of the abdomen, making it possible to detect protrusions into the lumen. Small or flat polyps, which are less common, may not be prominent enough to be detected by this method. A sigmoidoscopy is usually performed in addition to a barium enema as a safeguard since most polyps arise in the sigmoid colon and rectum.

The American Cancer Society recommends that, for the general population, screening should begin at the age of 50 and should consist of a FOBT or FIT on a yearly basis with a sigmoidoscopy every 5 years and a colonoscopy every 10 years. Recommendations for those who have an above average or high risk of developing colorectal cancer vary depending on the degree of the risk (Table 15.1).

Genetic testing

An option available to people who belong to a family that has had multiple members diagnosed with colorectal cancer is to undergo genetic testing. The DNA is extracted from the white blood cells in a small amount of blood drawn from a vein in the arm. The DNA is analyzed to determine if a person has inherited versions of genes that have been linked with the development of colorectal cancer. The age at which screening will begin and how frequently it occurs will be based on whether or not the testing has determined a person has or has not inherited detrimental versions of the genes.

Most cancers arise from the generation of spontaneous mutations in genes that play critical roles in maintaining control over the cell cycle and the integrity of DNA. Cells from the interior of the colon are constantly being shed and incorporated into the fecal matter. Consequently, there are DNA tests currently being studied to determine their effectiveness in detecting the presence of genetic mutations associated with the development of colorectal cancer within shed cells isolated from feces.

Biopsy of tissue removed during endoscopy

Malignant adenocarcinoma cells have distinctive characteristics that are associated with their role in producing mucus. A pathologist will look for these characteristics during a biopsy (from either a sigmoidoscopy or colonoscopy) in order to diagnose whether a growth is benign or malignant. Definitive staging of a malignancy is done following the surgical resection and pathological examination of the affected area of the intestine as well as the local lymph nodes. Not only does staging prove useful in deciding the most appropriate form of treatment, but it

Table 15.1 American Cancer Society guidelines on screening and surveillance for the early detection of colorectal adenomas and cancer in people at increased or high risk

Risk category	Age to begin	Recommended test(s)	Comment
INCREASED RISK – patients with a history of polyps on prior colonoscopy			
People with small rectal hyperplastic polyps	Same as those with average risk	Colonoscopy, or other screening options at regular intervals as for those at average risk	Those with hyperplastic polyposis syndrome are at increased risk for adenomatous polyps and cancer and should have more intensive follow-up
People with one or two small (less than 1 cm) tubular adenomas with low-grade dysplasia	5–10 years after the polyps are removed	Colonoscopy	Time between tests should be based on other factors such as prior colonoscopy findings, family history, and patient and doctor preferences
People with three to ten adenomas, or a large (1 cm +) adenoma, or any adenomas with high-grade dysplasia or villous features	3 years after the polyps are removed	Colonoscopy	Adenomas must have been completely removed. If colonoscopy is normal or shows only one or two small tubular adenomas with low-grade dysplasia, future colonoscopies can be done every 5 years
People with more than 10 adenomas on a single examination	Within 3 years after the polyps are removed	Colonoscopy	Doctor should consider possibility of genetic syndrome (such as FAP or HNPCC)
People with sessile adenomas that are removed in pieces	2–6 months after adenoma removal	Colonoscopy	If entire adenoma has been removed, further testing should be based on doctor's judgment
INCREASED RISK – patients with colorectal cancer			
People diagnosed with colon or rectal cancer	At time of colorectal surgery, or can be 3–6 months later if person doesn't have cancer spread that can't be removed	Colonoscopy to view entire colon and remove all polyps	If the tumor presses on the colon/rectum and prevents colonoscopy, CT colonoscopy (with IV contrast) or DCBE may be done to look at the rest of the colon
People who have had colon or rectal cancer removed by surgery	Within 1 year after cancer resection (or 1 year after colonoscopy to make sure the rest of the colon/rectum was clear)	Colonoscopy	If normal, repeat examination in 3 years. If normal then, repeat examination every 5 years. Time between tests may be shorter if polyps are found or there is reason to suspect HNPCC. After low anterior resection for rectal cancer, examinations of the rectum may be done every 3–6 months for the first 2–3 years to look for signs of recurrences

Table 15.1 American Cancer Society guidelines on screening and surveillance for the early detection of colorectal adenomas and cancer in people at increased or high risk (*continued*)

Risk category	Age to begin	Recommended test(s)	Comment
INCREASED RISK – patients with a family history			
Colorectal cancer or adenomatous polyps in any first-degree relative before age 60, or in two or more first-degree relatives at any age (if not a hereditary syndrome)	Age 40, or 10 years before the youngest case in the immediate family, whichever is earlier	Colonoscopy	Every 5 years.
Colorectal cancer or adenomatous polyps in any first-degree relative aged 60 or higher, or in at least two second-degree relatives at any age	Age 40	Same options as for those at average risk	Same intervals as for those at average risk
HIGH RISK			
Familial adenomatous polyposis (FAP) diagnosed by genetic testing, or suspected FAP without genetic testing	Age 10–12	Yearly flexible sigmoidoscopy to look for signs of FAP; counseling to consider genetic testing if it hasn't been done	If genetic test is positive, removal of colon (colectomy) should be considered
Hereditary non-polyposis colon cancer (HNPCC), or at increased risk of HNPCC based on family history without genetic testing	Age 20–25 years, or 10 years before the youngest case in the immediate family	Colonoscopy every 1–2 years; counseling to consider genetic testing if it hasn't been done	Genetic testing should be offered to first-degree relatives of people found to have HNPCC mutations by genetic tests. It should also be offered if one of the first three of the modified Bethesda criteria is met*
Inflammatory bowel disease: Chronic ulcerative colitis Crohn's disease	Cancer risk begins to be significant 8 years after the onset of pancolitis (involvement of entire large intestine), or 12–15 years after the onset of left-sided colitis	Colonoscopy every 1–2 years with biopsies for dysplasia	These people are best referred to a center with experience in the surveillance and management of inflammatory bowel disease

*The Bethesda criteria can be found in the "Can colorectal cancer be prevented?" section of our larger *Colorectal Cancer* document.
Reprinted by permission of the American Cancer Society, Inc. from www.cancer.org

Table 15.2 TNM staging for colorectal cancer

Category	Subcategory	Explanation
T		*Degree of invasion of the intestinal wall*
	T0	No evidence of tumor
	Tis	Cancer *in situ* (tumor present, but no invasion)
	T1	Tumor present but minimal invasion
	T2	Invasion into the submucosa
	T3	Invasion into the muscularis propria
	T4	Invasion completely through the wall of the colon
N		*Metastasis to regional lymph nodes*
	N0	No regional lymph node metastasis
	N1	1–3 lymph nodes involved
	N2	>4 lymph nodes involved
M		*Degree of metastasis beyond lymph nodes*
	M0	Metastasis beyond one or more regional lymph nodes has not been detected
	M1	Metastasis beyond one or more regional lymph nodes

also provides insight into a patient's prognosis. The TNM staging system most commonly uses the numerals 0, I–IV, to represent the least to most advanced stages of the cancer (Chapter 6). Table 15.2 provides a breakdown of the T, N, and M assessments, Table 15.3 provides the overall stage groupings and 5-year relative survival rates, and Figure 15.10 provides an illustration of the different stages.

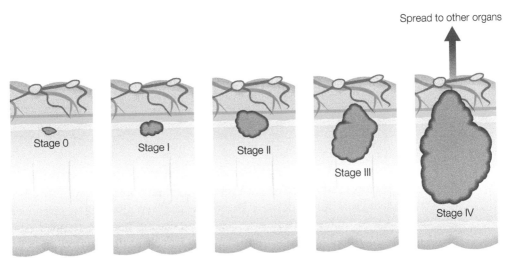

Figure 15.10 An illustration depicting the stages of colorectal cancer.

Table 15.3 AJCC stage grouping for colorectal cancer

Stage	Stage grouping	Explanation	5-year relative survival rate
0	Tis, N0, M0	Cancer is contained within the mucosa, the innermost lining of the colon; carcinoma *in situ*	
I	T1, N0, M0 or T2, N0, M0	Cancer has grown into the submucosa or muscularis propria; not present in regional lymph nodes and has not spread to distant sites	93%
IIA	T3, N0, M0	Cancer has grown into the muscularis propria or serosa; not present in regional lymph nodes and has not spread to distant sites	85%
IIB	T4, N0, M0	Cancer has grown beyond the wall of the colon into nearby tissues; not present in regional lymph nodes and has not spread to distant sites	72%
IIIA	T1, N1, M0 or T2, N1, M0	Cancer has grown into the submucosa or muscularis propria; has spread to 1–3 regional lymph nodes but not to distant sites	83%
IIIB	T3, N1, M0 or T4, N1, M0	Cancer has grown beyond the wall of the colon into nearby tissues; has spread to 1–3 regional lymph nodes but not to distant sites	64%
IIIC	Any T, N2, M0	There can be any degree of invasion; has spread to 4 or more regional lymph nodes but not to distant sites	44%
IV	Any T, any N, M1	There can be any degree of invasion or lymph node involvement; has spread to distant sites	8%

When colorectal cancer metastasizes, secondary tumors typically arise in the liver and lungs because those are the two sites where a cancer cell will most likely encounter a capillary bed and become trapped (see Chapter 5). Blood from the colon typically travels to the liver, which will metabolize the absorbed nutrients before they are utilized by the rest of the body. Since blood leaving the capillaries of the colon has been deoxygenated, it may also travel to the lungs to eliminate its carbon dioxide and become reoxygenated. The extent of metastasis can be determined with the aid of an ultrasound of the abdomen as well as CT and PET scans (Chapter 7).

TREATMENT OPTIONS

Surgery is the primary treatment method

A patient and his/her physician must make many decisions once a diagnosis of colorectal cancer is made. There are three standard forms of treatment: surgery, radiation, and chemotherapy, and there are various forms of each method. What must be decided after much discussion and the due diligence of a second opinion is the best course of action to be taken given the location, size, and potential stage of the cancer – remember that a definitive determination of the disease's stage is not made until an examination of the surgically excised tissue is performed. The patient's age and health status are also critical factors in the evaluation.

Surgery is the most common treatment method and it may be used with either a neoadjuvant or adjuvant form of treatment (Chapter 7). The type and extent of, discomfort from, and recovery time following surgery depends on several factors, including the size of the cancerous tissue, whether the cancer is in the proximal (i.e., beginning) or distal (i.e., end) portion of either the colon or rectum, and whether or not there are indications that the cancer has metastasized and, if so, to what tissues and organs.

The removal of the affected tissue may be done during a sigmoidoscopy or colonoscopy for stage 0 or I cancers. Cancers at this stage are small in size and have not invaded very deep into the tissues of the colon or rectum. Since the procedure is done without having to make an incision in the abdominal wall, it is called a local excision. If the cancer is in the form of a peduncle polyp then a polypectomy is performed by cutting through the base of the stalk (Figure 15.11). This may be all the surgery required if the pathologist's examination determines that the excised tissue has a clear margin, such as if the cancerous portion of the polyp was confined to the bulb and the stalk did not contain any malignant cells.

The surgical procedure most commonly performed to treat stage I–IV colorectal cancer is a segmental resection or colectomy, which involves the removal of the tumor and surrounding tissue and the reconnection of the remaining two portions of the intestine (Figure 15.12). The abdominal cavity is accessed through a 10–15 cm long incision made through the abdominal muscles just below the belly button. A visual inspection of the affected area is made to first evaluate the size of the tumor and whether or not it has protruded through the wall of the colon, and if so, to then

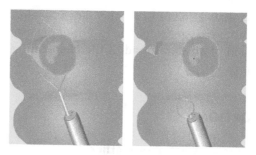

Figure 15.11 Polypectomy. A polyp can be removed by snaring and then cutting it free from the inner lining of the colon.

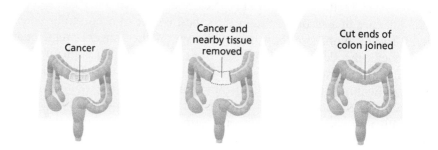

Figure 15.12 Colon resection. An area of the colon containing the cancerous tissue is removed and the two healthy ends are attached.

determine if there are signs that it has metastasized to nearby tissues. A decision is made as to how much of the colon and adjacent tissue is to be removed to ensure that all of the cancer and healthy tissue margins on either side of it are excised. In most cases 10–15 cm of intestine is resected or removed, along with regional lymph nodes. Next, the ends of the two remaining healthy portions of the colon are sewn together in a procedure known as an anastomosis. Despite the colon's smaller size, it typically functions close to normal.

There are occasions when the two ends of the intestine cannot be attached. This may be because the distance between them is too great or because the ends need to undergo a period of healing prior to being joined. In either case, the surgeon will perform a colostomy that will allow for the collection of fecal material that can no longer be eliminated through the anus (Figure 15.13). The procedure entails making an incision called a stoma in the abdominal wall through which the end of the large intestine that is closest to the small intestine will protrude. A colostomy bag is attached to the protruding end to collect the stool. Depending on the circumstances, the colostomy may be temporary or permanent. In

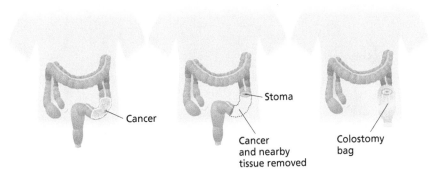

Cancer

Stoma

Cancer
and nearby
tissue removed

Colostomy
bag

Figure 15.13 Colostomy. If the two sections of the colon can not be reattached a colostomy is performed so that feces are collected into a bag.

either case, the patient will be trained on how to change and maintain it hygienically. Modern colostomy supplies are odor proof and designed to allow patients to continue their normal activities. During the first two days of recovery, nutrients are obtained through intravenous fluids, followed by a gradual reintroduction to foods beginning with those that are liquid or soft (broth, fruit juices, apple or cranberry sauce, and gelatin) and then those that are rich in protein yet tender, such as fish.

An option available to patients who do not have to have a large segment of the colon removed is to have the procedure performed laparoscopically. This is a much less intrusive method requiring only a few small incisions through which the surgical instruments are passed, and the recovery time is also much shorter. The level of discomfort following surgery is also much less because the procedure results in significantly less trauma to the abdominal muscles.

Surgery is also the most common form of treatment when the cancer is located in the rectum. Surgery in and of itself is often considered to be sufficient for stages 0 and I rectal cancers. The small size and superficial nature of these cancers can often be treated with a polypectomy or local excision during a sigmoidoscopy or colonoscopy. For stages II–IV rectal procedures, a segmental resection is typically performed. A tumor located in the proximal two-thirds of the rectum – the portion closest to the sigmoid colon – is removed by a low anterior resection (Figure 15.14). The procedure is very similar to what is done in the colon: an abdominal incision is made, the affected section of the rectum with clear margins on either side along with regional lymph nodes are removed, and the ends are joined in an anastomosis. If necessary, a temporary colostomy is performed.

When the cancer is in the distal or last third of the rectum the surgery is referred to as an abdominoperineal resection. In addition to an abdominal incision, one is also made in the perineum, the area around the anus. This allows for the removal of the anal sphincter, the ring of muscle that creates the anal opening, along with the affected portion of

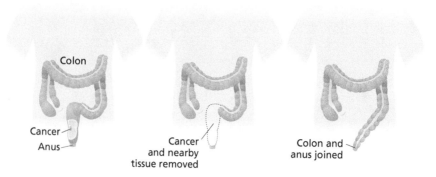

Colon

Cancer
Anus

Cancer
and nearby
tissue removed

Colon and
anus joined

Figure 15.14 Rectal resection. In a low anterior resection the distal portion of the rectum is removed and the sigmoid colon is attached to the remaining portion of the rectum.

> **Would it be better if a man who wants to have children is experiencing retrograde ejaculations or "dry" orgasms?**

the rectum. The loss of the anal sphincter will result in the permanent inability to regulate the passage of stool and will require the person to have a permanent colostomy. The cancer may have metastasized to the urinary bladder, prostate, or uterus which are adjacent to the rectum. If this has occurred, they may also be removed during the rectal cancer surgery, in which case the surgery is termed a pelvic exenteration.

Regardless of the form of surgery, a person typically remains in the hospital for 3–7 days and requires 3–6 weeks at home to recover from the abdominal incision. Beside the risks associated with any surgery involving anesthesia, there is a risk of complications, albeit rare, associated with colorectal cancer surgery. It is possible that a leak may occur at the suture site between the attached ends of the colon and/or rectum, allowing waste material and bacteria to enter the abdominal cavity and thereby cause an infection. If sufficient scar tissue forms at the suture site, it can create an adhesion that may interfere with the flow of material and cause a blockage. An abdominoperineal resection in a man may result in him experiencing a "dry" orgasm meaning that very little fluid is released upon orgasm. This may be caused by nerve damage during surgery that results in the failure to stimulate the release of seminal and prostatic fluids, or the fluids that are released into the urethra travel backward into the bladder rather than forward and out of the penis; this is referred to as retrograde ejaculation.

Surgery on stage IV tumors is typically palliative rather than curative because of the tumor's advanced nature and the presence of distant metastases. Surgery may be performed in an attempt to reduce the size of a blockage in the intestine created by the tumor or to remove secondary tumors if they are accessible and if their excision would not significantly compromise the functionality of the tissue or organ.

Chemotherapy and radiation

Chemotherapy and/or radiation are not curative forms of treatment with this disease. They may be used alone or in combination in an attempt to reduce the size of a tumor prior to surgery, to target a tumor that is surgically inoperable due to its position, to prevent the development of secondary tumors after surgery, or to provide palliative care when the cancer is very advanced or the patient is not healthy enough to undergo surgery. Surgery is considered sufficient for stage 0 and most of stage I tumors. Since stage II cancers have grown through the colon wall but show no evidence of cancer cells in any of the examined regional lymph nodes, any adjuvant therapy usually consists of only radiation to the affected area. The indication of metastasis by the presence of malignant cells in one or more lymph nodes in stage III cancers, or the presence of metastatic growths in stage IV tumors, requires chemotherapy and/or radiation.

Radiation therapy can be delivered from either an external or internal source. The traditional external beam treatment method (Chapter 7) requires that the patient lies on a table while multiple stationary beams, or a single beam of radiation moving in an arc around the patient, is aimed at the tumor site. This results in the radiation passing through healthy tissue, sometimes causing collateral damage, prior to reaching and destroying the target cancer cells. The treatment is typically administered 5 days a week for several weeks. An alternative method, endocavity radiation therapy, may be used to treat small tumors in the rectum. During this procedure, a probe that emits a high dose of radiation is inserted through the rectum so that the radiation passes through the least amount of healthy tissue before it reaches the affected tissue. The probe is in place for several minutes and the procedure is repeated once every 2 weeks over a 6-week period. It may also be used in combination with traditional external beam therapy, allowing a lower dosage to be supplied by that method. Brachytherapy (Chapter 7), the insertion of radioactive pellets or seeds directly into the tumor, is also used for the treatment of rectal tumors. Long-term consequences from radiation therapy can include rectal and bladder irritation, impotence in men, and vaginal irritation in women. These conditions can often be reduced with the use of various medications.

The use of chemotherapy is common following surgery and when the results of the biopsy indicate the likelihood that the cancer has metastasized. As an adjuvant form of therapy its goal is to kill any cancer cells that remain in the body undetected, thereby preventing the development of secondary tumors. The drug of choice for the treatment of colorectal cancer is 5-fluorouracil (5-FU), a nucleotide analog (Chapter 7). It is administered intravenously but there are numerous dosages and dosing schedules that are used. In some cases, it is given in a repeated series that consists of several daily injections followed by several weeks of recovery. In other cases, the administration may be continuous through an IV catheter. The drug's effectiveness is improved when combined with leu-covorin (LV), which is folinic acid, a form of a B vitamin. The addition of a third drug, oxaliplatin, is also common and the treatment strategy is referred to as FOLFOX (*Folinic* acid, *5-FU*, *ox*aliplatin). Patients taking oxaliplatin often experience temporary nerve damage that results in numbness and tingling, particularly in their arms and legs, as well as sensitivity to cold. Metastatic colorectal cancer is often treated with a combination of 5-FU, LV, and irinotecan. These drugs produce the side effects that are common with chemotherapy – nausea, diarrhea, mouth sores, loss of appetite, and hair loss.

TARGETED THERAPIES

There are two monoclonal antibodies (mAb) (Chapter 7) used in the treatment of colorectal cancer. During biopsy, the excised tissue can be examined to determine the amount of epidermal growth factor receptor (EGFR) on the surfaces of the malignant cells in comparison to healthy cells (Chapter 7). An elevated level of the receptor would allow the can-cer cells to bind a higher than normal amount of epidermal growth factor (EGF; Figure 15.15b). This results in the cells receiving an abnormally strong signal to divide. A patient whose tumor is overexpressing EGFR may be a candidate for treatment with cetuximab, a mAb that binds to the EGFR and does not allow EGF to bind (Fig.15.15c). This essentially prevents the cells from receiving external signals to divide, thereby slow-ing the growth of the tumor.

The second mAb, bevacizumab, is the first antiangiogenesis treatment approved for the treatment of colorectal cancer. It binds to and inactivates vascular endothelial growth factor (VEGF; Figure 15.16; Chapter 5).[15.9] The produc-tion of VEGF by the cancer cells stimu-lates the blood vessel growth necessary to form the vascularization that is feeding the tumor and removing its wastes. The binding of bevacizumab to VEGF, how-ever, prevents it from binding to the VEGF receptor, effectively removing the angiogenesis stimulus. In the absence of the stimulus, the angiogenesis inhibitors that the body produces on its own to control the growth of blood vessels become the predominant signal received by the endothelial cells that comprise the newly formed blood vessels and causes them to undergo apoptosis. Without sufficient vascularization, the environment becomes too toxic for the survival of the cancer cells. Adjuvant therapy that includes

[15.9] VEGF is a protein produced by the cancer cells that normally binds to the VEGF receptors present on the surfaces of endothelial cells of blood vessels, triggering a signaling cascade within those cells that ultimately causes them to divide and grow towards the malig-nant cells.

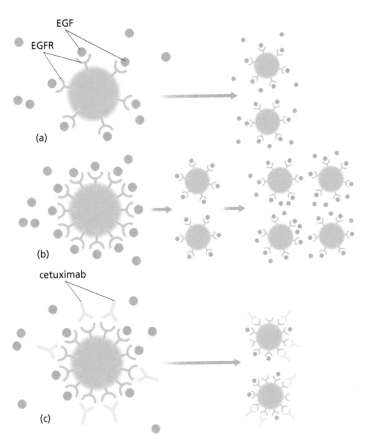

Figure 15.15 Mechanism of action of cetuximab. (a) A cell with a normal concentration of EGFR binds EGF producing an internal growth signal that stimulates the cell to divide for a certain time. (b) A cell with a higher than normal concentration of EGFR receptors will bind more EGF producing a stronger internal growth signal that will stimulate the cell to divide faster over the same time period. (c) The binding of the monoclonal antibody cetuximab to some of the EGFRs reduces the amount of EGF that can bind to the surface of the cell, thereby also reducing the growth signal and slowing the rate of division of the cell.

bevacizumab or cetuximab is currently being tested to determine whether either better reduces the recurrence of the cancer in comparison to the use of standard adjuvant chemotherapy.

Blood tests are performed throughout the course of treatment to monitor not only the effectiveness of the radiation and/or chemotherapy on the cancer but also their effect on various organs and the immune system. Certain types of cancers produce tumor markers, molecules that can be monitored in order to provide critically important information about the activity of a known tumor or to even detect the presence

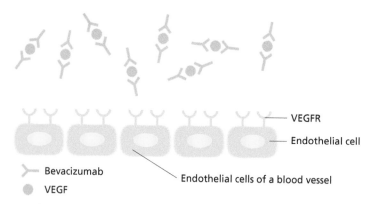

Figure 15.16 Mechanism of bevacizumab. The antibody bevacizumab binds to and neutralizes VEGF preventing it from binding to VEGF receptors on endothelial cells of blood vessels and thereby incapable of stimulating their growth.

of a previously unknown malignancy. Carcinoembryonic antigen (CEA) and carbohydrate antigen (CA 19-9) are two tumor markers that are synthesized at elevated levels by colorectal cancers. Therefore, determining their concentrations in the blood prior to treatment provides information about the cancer, such as size and aggressiveness, that is of value in determining a patient's prognosis. Monitoring before and periodically after surgery as well as during and after adjuvant therapy can provide critical information about the effectiveness of the treatment.

SCREENING TESTS PERFORMED AFTER THE COURSE OF TREATMENT

Most recurrences of colorectal cancers arise within 2–3 years of the initial diagnosis and surgical treatment. A colonoscopy is performed within the first year and then after 3 years and again after 5 years, assuming that a malignancy is not detected during each examination. It is likely that a patient will undergo CT or PET scans of the liver on a regular basis since that organ is a primary site of colorectal cancer metastases.

EXPAND YOUR KNOWLEDGE

1 The terms "small" and "large" in reference to the intestines seem to be misnomers since the small intestine is approximately four times the length of the large intestine. Explain why the terms are appropriate.

2 Provide a single sentence summation for each of the five stages of colorectal cancer.

3 Why would the first forms of solid food eaten following segmental resection surgery be rich in protein and low in fiber?

4 Why is it reasonable that endocavitary radiation therapy or brachytherapy are only used to treat rectal and not colon cancers?

5 The drug 5-FU is a nucleotide analog. What biomolecule does it become incorporated into and thereby disrupt its function?

ADDITIONAL READINGS

Chen, H. X., Mooney, M., Boron, M., Vena, D. Mosby, K., Grochow, L., Jaffe, C., Rubinstein, L., Zwiebel, J., and Kaplan, R. S. (2006) Phase II multicenter trial of bevacizumab plus fluorouracil and leucovorin in patients with advanced refractory colorectal cancer: an NCI Treatment Referral Center Trial TRC-0301. *Journal of Clinical Oncology*, **24**: 3354–3360.

Hendon, S. E. and DiPalma, J. A. (2005) U.S. practices for colon cancer screening. *The Keio Journal of Medicine*, **54**: 179–183.

Marieb, E. N. and Hoehn, K. (2007) *Human Anatomy and Physiology*, 7th edn. Upper Saddle River, NJ: Prentice Hall.

Martini, R. and Bartholomew, E. (2007) *Essentials of Anatomy & Physiology*, 4th edn. Upper Saddle River, NJ: Prentice Hall.

Nozoe, T., Rikimaru, T., Mori, E., Okuyama, T., and Takahashi, I. (2006) Increase in both CEA and CA19-9 in sera is an independent prognostic indicator in colorectal carcinoma. *Journal of Surgical Oncology*, **94**: 132–137.

Terstriep, S. and Grothey, A. (2006) First- and second-line therapy of metastatic colorectal cancer. *Expert Review of Anticancer Therapy*, **6**: 921–930.

16

Leukemia and lymphoma

Cancer is not a disease for reasonable people because there is no reason for it.
Lynn Sherr, television news reporter

CHAPTER CONTENTS

Cancer: Basic Science and Clinical Aspects, 1st edition. By C. A. Almeida and S. A. Barry. Published 2010 by Blackwell Publishing, ISBN 978-1-4051-5606-6.

Leukemia and lymphoma are both cancers of the blood and blood-forming system. The word leukemia is derived from the Greek word meaning "white blood," because the disease usually starts when one abnormal white blood cell (leukocyte) begins to continually replicate itself. A unique feature of leukemias is that they do not form discrete tumors. Lymphoma is a general term for a variety of cancers that begin in the lymphatic system. Unlike other cancers discussed, these cancers are diagnosed by blood and bone marrow tests. There are a wide variety of leukemias and lymphomas. This chapter will focus on the features of the most commonly known leukemias and will give some basic information on lymphomas.

LEUKEMIA STATISTICS

Many people mistakenly believe that leukemia is a childhood disease, yet it is ten times more common in adults. About 230,000 people of all ages are living in the US with the condition today. Approximately 44,000 new cases developed and there were 22,000 deaths from the disease in 2008 among Americans. It is the number one cause of cancer death in children and young adults. Although it can affect people of any age, it is more common in men and boys than in women and girls, and is more likely to occur in Caucasian than in African-Americans. The reasons for these differences are not yet known.

LEUKEMIA IS A CANCER OF THE BLOOD CELLS

Bone marrow is the soft, spongy, fatty inner part of large bones that possesses stem cells that divide to produce the mature erythrocytes (red blood cells) responsible for the oxygen/carbon dioxide transfer between the lungs and tissues, the leukocytes (white blood cells) that provide protection against infection and are involved in the body's immune system, and thrombocytes (platelets), cell fragments that play a role in blood clotting.

Leukemia is a cancer affecting leukocytes, cells that are grouped into two categories: lymphocytes and myelocytes. There are three types of lymphocytes: B lymphocytes make antibodies against infection, T lymphocytes regulate the body's immune responses, and natural killer cells attack cancer cells and viruses. Myelocytes, also called granulocytes since they possess tiny granules, are further divided into neutrophils (phagocytic), eosinophils and basophils (somewhat less phagocytic), and monocytes, (phagocytic, also process antigens) (Figure 16.1).

Although there are many different types and subtypes of leukemia, the four most common types are classified by how quickly the disease progresses – acute (fast-growing) or chronic (slow-growing) – and by the type of white blood cell involved (lymphocytic or myelocytic). Acute leukemia is the result of the production of abnormal immature white blood cells called blasts that are unable to carry out their typical functions. The disease worsens as the number of blasts increases, a process that occurs

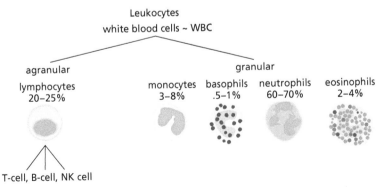

Figure 16.1 Types of leukocytes. There are five types of white blood cells.

16.1 Before the advent of treatment for leukemia, the term acute usually meant a disease with sudden onset that was ultimately fatal, while chronic meant a slow growing disease with longer-term survival.

rather quickly as the number of blasts produced exceeds the number of blasts that die off. The number of blasts increases more slowly in chronic leukemia, resulting in a gradual and slow progressing disease. Generally, these cells are more mature and able to carry out some degree of normal function.[16.1]

THE EXACT CAUSE OF LEUKEMIA IS UNKNOWN

As with most cancers, the exact cause of leukemia has not been identified, but certain risk factors have been associated with the disease. Some are associated with specific environments, while others may be inherited.

DNA alterations can be a cause of leukemia

Most leukemias are related to certain changes in the DNA of bone marrow cells that occur after birth rather than being inherited. People who have been exposed to very high levels of radiation have a higher incidence of the cancer, as was seen following the atomic bomb explosions in Japan during World War II or in nuclear plant accidents such as the one that occurred in Chernobyl, Russia in 1996. In years past, career professionals such as radiologists and radiologic technicians had an increased incidence of leukemia due to occupational exposure to radiation, eventually leading to stronger safety precautions to ensure a safer working environment. Cigarette smoking and working with certain chemicals, such as benzene and formaldehyde, have been demonstrated to put one at a slightly higher risk as well.

Certain conditions caused by abnormal chromosomes have also been implicated. One example of this is Down syndrome, which has a leukemia-linked chromosome abnormality called trisomy 21 (having an extra or third copy of chromosome number 21). Children with Li–Fraumeni syndrome, a rare genetic disorder, inherit a mutated version of the *p53* tumor suppressor gene and have an elevated risk of developing leukemia by the time they reach early adulthood.

EARLY SYMPTOMS OF LEUKEMIA

While each type of leukemia may have its own symptoms, commonalities do exist among them such as persistent fatigue or weakness, frequent infections, and easy bleeding or bruising. These are all due to the increased number of immature white blood cells. Healthy persons have a white blood cell count of approximately 4500–11,000/mm³ while those with leukemia will often have a highly elevated count, perhaps as high as 100,000/mm³, depending on the type of leukemia. Since the bone marrow is rapidly producing immature white blood cells (blasts), there is a shortage of normal white blood cells that results in a lower defense against infection. The crowding of the blasts causes a lower red blood cell count leading to anemia, a condition that causes shortness of breath during physical activities, that is accompanied by pale skin color. At the same time, there will be too few platelets, which may be reflected in frequent or severe nosebleeds, bruising, or bleeding from the gums. Additional flu-like symptoms may include fever or chills, loss of appetite, weight loss, swollen lymph nodes, enlarged liver or spleen, petechiae (tiny pinpoint red marks in the skin caused by bleeding), night sweats, and/or bone pain or tenderness. The reproduction of these compromised leukemic cells can lead to damage of the kidney, liver, or spleen. The spleen normally filters the blood and destroys old cells, but in a person with leukemia it may become enlarged and swollen as a result of having to deal with such large numbers of cells. Without intervention by treatment, the continued production of blasts can lead to their entering the central nervous system (CNS) resulting in headaches and seizures.

LABORATORY STUDIES ARE NECESSARY TO DETERMINE THE DIAGNOSIS

Physicians who specialize in blood-related disorders are called hematologists and may also specialize in oncology or cancer treatment. After an initial physical examination, testing of leukemia begins with a complete blood count (CBC), an analysis of the white and red blood cells and platelets. One component of the CBC is the White Blood Cell Differential, which determines the percentage of the five kinds of white blood cells (lymphocytes, neutrophils, eosinophils, basophils, and monocytes) since an increase, abnormality or immaturity in any of them may be significant.

To determine the type of leukemia, a bone marrow aspiration and biopsy will be performed. A special needle is injected into the ilium, more commonly known as the pelvic bone, after numbing the site (Figure 16.2). This area is chosen because, in most people, a portion of this bone is readily accessible from the lower back. First, a liquid bone marrow sample (aspirant) is removed, causing a feeling of pressure and sometimes a sharp but brief pain. A sample of the bone marrow is then removed by a biopsy needle, which may be slightly more painful. The aspirant and biopsy samples are sent to the pathology lab to be processed

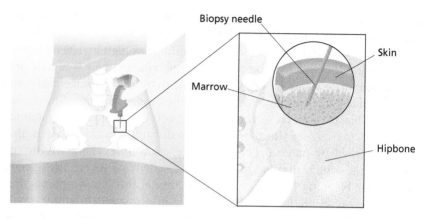

Figure 16.2 Bone marrow aspiration and biopsy. This procedure is performed on the patient's rear hip using local anesthesia. The aspiration is the removal of the soft, spongy material that lines the inside of most bones. The biopsy is the actual removal of a small piece of the bone marrow.

16.2 Normal human cells contain 23 pairs or 46 chromosomes. Twenty-two pairs are numbered from 1 to 22. The twenty-third pair consists of the sex chromosomes, called "X" or "Y." The letters "p" and "q" refer to the specific areas or "arms" of a chromosome.

How do you suppose the abnormal tyrosine kinase produced by the Philadelphia chromosome functions to create leukemic white blood cells?

and examined microscopically. Sometimes a lumbar puncture (spinal tap) is also done by placing a small needle into the spinal canal in the lower back. Cerebrospinal fluid is withdrawn to determine if leukemia cells are present.

Cytogenetic studies are performed to detect chromosomal abnormalities[16.2] (Figure 16.3). In certain types of leukemia, part of one chromosome may be attached to part of a different chromosome, an occurrence identified as a translocation. The Philadelphia chromosome (Ph[1]) is a translocation from chromosome 9 to chromosome 22 that creates an abnormal gene called *BCR-ABL*[16.3] (Figure 16.4).

16.3 The Philadelphia chromosome was first described in 1960 by Dr Peter C. Howell at the University of Pennsylvania School of Medicine and David Hungerford from the Fox Chase Cancer Center's Institute for Cancer Research, and was named after the city where it was discovered. It was the first chromosomal abnormality that was consistently found in any kind of malignancy and the first gene-based cause for cancer, giving rise to future research possibilities for the disease.

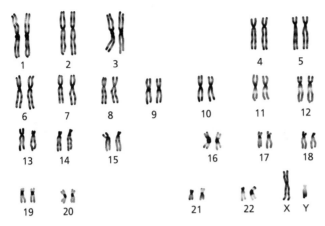

Figure 16.3 Karyotype. Most human cells have a nucleus that contains two sets of chromosomes, one set from each parent. Each set has 23 single chromosomes to include the sex chromosome. A normal female will have a pair of X-chromosomes and a male will have a pair that consists of an X and a Y.

This gene produces an abnormal tyrosine kinase, a mutation that causes many stem cells to develop into leukemic white blood cells. This finding is significant for determining treatment and prognosis in some forms of leukemia. The Philadelphia chromosome may be identified under a microscope or by chemical tests. Auer bodies (Auer rods), pink or red rod-shaped inclusions, are found in immature granulocytes in patients with certain leukemias. Additional sophisticated molecular tests that can examine very small changes in the DNA may also be performed.

THERE ARE MANY TYPES OF LEUKEMIA

There are over a dozen different types of leukemia, but some are more common than others. Treatment varies depending on the age and general health of the patient, the category of leukemia, and the presence or absence of malignant cells in the spinal fluid. More recently, researchers have discovered that the determination of the genetic make-up and numbers of specific cell types involved is beneficial in predicting the patient's prognosis. This section will address four commonly diagnosed leukemias in children and adults.

Acute lymphocytic leukemia

Acute lymphocytic leukemia (ALL), characterized by large numbers of immature lymphocytes, is most common in children, especially those between 3 and 6 years old. It can affect adults, usually those over 65 years

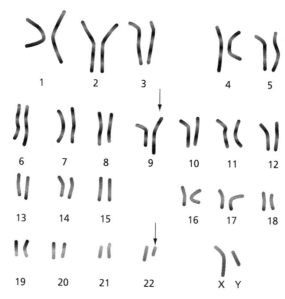

Figure 16.4 The Philadelphia chromosome. The Philadelphia chromosome (Ph¹) is a reciprocal translocation between one chromosome 9 and one chromosome 22, resulting in one chromosome 9 longer than normal and one chromosome 22 shorter than normal.

old, and it is uncommon in people between the ages of 16 and 50. If a child has leukemia, it is often, *but not always*, identified as ALL. The prognosis in children is very good as it is one of the cancers that can be considered "curable," but the results are not always as optimistic for adults (Box 16.1). The relative survival rate is 91% for children under 5 and 66% for adults.

ALL is typically treated using chemotherapy, which is oftentimes administered into the spinal canal, a common site of metastasis. Radiation therapy to the brain, another common site of metastasis, may also be utilized when necessary. Additional critical factors include the patient's age, white blood cell count, cytogenetic test results, and initial response to chemotherapy. Some positive ALL prognostic factors include a young age at diagnosis, an initial white blood cell count <50,000/mm³, and having malignant T rather than B cells. Having either the Philadelphia chromosome or a translocation between chromosomes 4 and 11 (which can occur in about 5% of ALL patients), and/or requiring more than 4–5 weeks of therapy to achieve complete remission are associated with a poor prognosis.[16.4]

16.4 According to a study published in the journal *Leukemia*, treatment with Gleevec® (imatinib) and methylprednisolone alternated with chemotherapy improves outcomes among elderly patients with Philadelphia chromosome-positive ALL.

Box 16.1

Dr Sidney Farber achieved the first clinical remission with chemotherapy ever reported for childhood leukemia

Sidney Farber (1903–1973) was born in Buffalo, NY and graduated from Harvard University Medical School in 1927 (Figure 16.5). Two years later, he became the first fulltime pathologist to be based at the Children's Hospital, Boston, Massachusetts. At that time, leukemia was similar to many other cancers in that patients had a very poor prognosis with no available treatments. Virtually no progress had been made since the disease was first identified in 1845, and painful death would occur usually within weeks of diagnosis. Dr Farber chose to specialize

Figure 16.5 Dr Sidney Farber. A specialist in children's cancers, Dr Farber developed the first successful chemotherapeutic treatment for childhood leukemia. Photo courtesy of Dana-Farber Cancer Institute.

in children's cancers and was the first to come up with "total care," a concept in which services such as clinical care, nutrition, social work, and counseling would be regularly provided for both the patient and his/her family.

Dr Farber was intrigued by the studies conducted during World War II, in which patients with certain types of anemia were cured with vitamin B_{12} or folic acid. Pernicious anemia and tropical anemia result in the bone marrow filling with immature "blast" cells, a condition similar to leukemia. Knowing that folic acid stimulates the growth and maturation of bone marrow, he reasoned that a drug that could chemically block folic acid would stop abnormal bone marrow production. In 1947, he attempted to prove this by using aminopterin, a new drug that was a derivative of folic acid (a natural chemical found in plants). Sixteen children seriously ill with leukemia were given the drug and 10 of them achieved temporary remissions, thus confirming his hypothesis. This was the first use of a drug that interferes with a cell's internal processes, and it ultimately led to an entire new category of drugs known as antimetabolites. Dr Farber reported his findings in the June 3, 1948 issue of the *New England Journal of Medicine*, but his results were not widely accepted due to the severe side effects that accompanied the treatment. The patients who did not die from the regimen were very close to death, and the success rates of the early trials were low.

There was much hostility surrounding Farber's use of chemotherapy, especially since it involved children whose suffering was all the more painful for physicians to watch. Throughout the 1950s and 1960s, Dr Farber continued his work with cancer treatment and began a fund-raising organization, the Jimmy Fund, which was named for a young cancer victim he called Jimmy. He eventually started a children's cancer research institution, now called the Dana Farber Cancer Institute, which today treats people of all ages with cancer. He used his accomplishments to appear before Congress to lobby for federal support for cancer research, making himself one of the first physicians to seek governmental involvement in the fight against cancer.

Acute myeloid leukemia

Acute myeloid leukemia (AML), seen rarely in children (10% of cases), is most often found in adults (the average age being 65). This cancer can be kept in remission for long periods or may even be ultimately cured. There are many different subtypes, so it is considered to be a group of diseases rather than a single one. The different subtypes (M0–M7) are based on the microscopic cell appearance. Auer rods are seen in blood cells of persons with AML, which sometimes helps to differentiate this leukemia from others (Figure 16.6). Survival rates depend on the subtype and the prognostic features as there is no standard staging system. Each has a different response rate to chemotherapy, the main form of treatment,

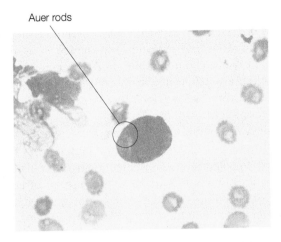

Figure 16.6 Auer rods. Clumps of reddish elongated material, seen in immature lymphocytes of patients with AML, are referred to as Auer rods. © Rector and Visitors of the University of Virginia. Charles E. Hess, MD and Lindsey Krstic, BA.

and several different types of treatment may be used (Table 16.1). Complete remission (CR) means that the blood and bone marrow counts are normal and the signs and symptoms are gone. Approximately 75% of younger adults and 50% of patients older than 50 achieve a CR after treatment for AML.

Table 16.1 French-American-British (FAB) classification of AML

AML subtype (originating cell type)	% Cases	Microscopic cellular appearance	Prognosis
M0	5%	Many blasts	Poor
M1	15%	Some maturing cells	Average
M2	25%	Mature cells, Auer rod	Excellent
M3	10%	Mature cells, Auer rods	Excellent, usually responds to certain drugs related to vitamin A
M4	25%	Mature cells	Average to excellent
M5	5%	Cells appear monocytic	Very poor
M6	5%	AKA erythroid or DiGuglielmo's syndrome, affects red blood cells	Very poor
M7	10%	Megakaryoblastic fragments	Poor

There are several chemotherapeutic drugs used in dealing with AML and the treatment is often divided into three phases: remission induction, post-remission consolidation, and maintenance, which is used infrequently. The most commonly used combination of drugs is cytarabine (Ara-C) and daunorubicin (Daunomycin) or idarubicin (Idamycin) given in one or two courses lasting from 3–5 weeks in the hopes of achieving a complete remission for the patient.

Of the various AML subtypes, M3 leukemia is unique in that it is treated by a nonchemotherapeutic drug known as ATRA (all-*trans*-retinoic acid), which is related to vitamin A, combined with other chemotherapeutic drugs. Since patients with this form of the disease often have serious blood clotting or bleeding problems, they may receive transfusions of platelets or other blood products, or are put on a blood thinner. Once the patients enter remission, they receive a minimum of two courses of chemotherapy and receive ATRA again for at least one year. This approach will cure 70–90% of patients with M3 leukemia.

One way to influence the choice of therapy of AML as well as to classify and predict the benefit of that treatment is to analyze the chromosomal changes (cytogenetics) in the leukemic cells. In adults, some favorable chromosome changes (60–80% survival rate) include abnormalities of chromosome 16, and/or translocations between chromosomes 8 and 21, or 16 and 17. Unfavorable changes (20% survival rate) include the presence of extra copies of chromosomes 8 or 13, a deletion of all or part of chromosomes 5 or 7, and/or abnormalities involving many chromosomes. Those with chromosomes that appear normal have a 40% survival rate. Generally, younger patients have more favorable chromosomal changes leading to successful AML treatment. Traditionally, the age of the patient and the amount of white blood cells influence the success of treatment but, as with all cancers, it is not possible to always precisely predict the outcome of each patient's disease. The overall survival rate for AML is 21% and for children under 15 it is 55%.

Chronic lymphocytic leukemia

Chronic lymphocytic leukemia (CLL), the most common type of leukemia in North America and Europe, is often a disease of middle aged and elderly people. Eighty to ninety percent of the cases are found in people over the age of 50, especially in Caucasians and males. CLL is more common in Jewish people of Russian or Eastern European descent while uncommon in Japanese and Southeast Asians. This disease is most often (95%) a result of the overproduction of B lymphocytes, causing suppression of the immune system. Often, patients are asymptomatic and the disease may be discovered by chance during routine blood tests, including a CBC, that are ordered during an annual physical examination. An abnormal blood smear would demonstrate an increased number of similar appearing mature

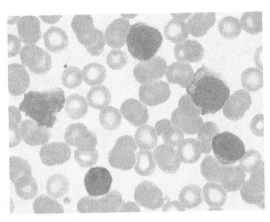

Figure 16.7 Blood smears from a patient with CLL. Note the presence of red blood cells and an abundance of lymphocytes (purple cells), and a lack of other white blood cells usually seen in a healthy person. © Rector and Visitors of the University of Virginia. Charles E. Hess, MD and Lindsey Krstic, BA.

lymphocytes, a condition known as lymphocytosis (Figure 16.7). A bone marrow biopsy is usually not needed to confirm the diagnosis. Other examples of CLL include hairy cell leukemia (a B-cell disease) and T-cell leukemia, both of which are far less common.

Many people with inactive CLL have no symptoms even when diagnosed, or have vague and nonspecific ones such as lymph node swelling (often in the neck), fatigue, weakness, weight loss, or repeated infections. Any of these symptoms could also be associated with many other conditions that are not cancer. Surprisingly, many people live for as long as 20 years following diagnosis without any treatment or additional symptoms. Since it is primarily a disease of the elderly, patients may even die of other illnesses or as a consequence of old age. If blood tests demonstrate a further increase in the total number of white blood cells or an increased number of lymphocytes, or if symptoms become more aggressive, the hematologist/oncologist may elect to begin treatment.

Staging is determined by information obtained from blood test results such as the number of lymphocytes in the blood and bone marrow, the presence or absence of anemia, the number of platelets, and clinical features such as the size of the liver or the size of the spleen, which acts as a filtering system for blood cells (Table 16.2). In chronic leukemia, the spleen becomes enlarged from storing so many cells, and it may compress other organs and/or interfere with chemotherapy. If this happens, a splenectomy (surgical removal of the spleen) is performed. Those who are anemic or do not have enough platelets have a poorer prognosis since the bone marrow does not produce enough healthy blood cells to fight infections, carry oxygen, or prevent bleeding. CLL has an overall survival rate of 76%.

Table 16.2 Staging of chronic lymphocytic leukemia

Stage	Risk	CBC changes	Other symptoms
0	Low, slow-growing	Lymphocytosis	None
I	Intermediate	Lymphocytosis	Enlarged lymph nodes
II	Intermediate	Lymphocytosis	Enlarged liver or spleen, may or may not have enlarged lymph nodes
III	High	Lymphocytosis, low red cell count, decreased hemoglobin	May or may not have enlarged liver or spleen or enlarged lymph nodes
IV	High	Lymphocytosis Decreased platelets, possible decreased red cell count	May or may not have enlarged liver or spleen or enlarged lymph nodes

Once treatment is necessary, chemotherapy is usually given not as a cure but as a means to help relieve symptoms such as the enlarged lymph nodes and spleen. Traditionally, a mild drug called chlorambucil (Leukeran), which kills cancer cells by interacting with their DNA, has been given orally with few side effects. A newer drug called fludarabine (Fludara) is given intravenously and has demonstrated such positive results in controlling the disease that it is now being used more often, even in patients whose leukemia has reoccurred. Combinations of fludarabine with other chemotherapeutic agents sometimes result in a complete remission. For patients who do not respond to Fludara, alemtuzumab, a monoclonal antibody specific for a protein on the surface of CLL cells, has been utilized. Rituximab (Rituxan), another antibody drug, may be used on its own or in combination with other chemotherapy. Depending on the stage and blood results, transfusions of blood or platelets may be necessary.

Chronic myelogenous leukemia

Chronic myelogenous leukemia (CML), also known as chronic granulocytic leukemia (CGL), is almost always found in older adults while rare (2%) in children. This less common leukemia is almost always associated with the presence of the Philadelphia chromosome (Ph[1]), which leads to an overproduction of myelocytic (granulocytic) white blood cells and the presence of both mature and immature white blood cells in the blood stream. The presence of Ph[1] is a useful tool in the diagnosis of CML and in determining an appropriate form of treatment. CML is staged

Table 16.3 CML phase classification

Phase	% Blast cells found in blood and bone marrow	Additional symptoms	Traditional prognosis (excludes Gleevec)
Chronic	<10%	Mild classic leukemia symptoms	Responds to standard treatment
Accelerated	10–30%	Fever, poor appetite, weight loss	Less responsive to standard treatment, presence of Ph[1]
Blast crisis	30%+	Fever, poor appetite, weight loss. Blast cells have spread to tissues and organs	Poor response to traditional treatment

according to the three main phases of the disease, by the number of blast cells in the blood and bone marrow, and by the severity of symptoms (Table 16.3). CML has often spread to the liver and spleen by the time it is diagnosed, so additional factors and information are taken into account to determine the prognosis of the patient as well as the course of treatment.

Traditionally, approximately half of the people who are diagnosed with CML survive this disease. Since its approval by the FDA in 2001 imatinib mesylate (Gleevec) has become known as a wonder drug in the treatment of CML. Its effectiveness has been so remarkable that it is prescribed more and more frequently. Thanks to the success of this drug, in 2005 more than 90% of CML patients were alive 4 years after diagnosis and only 3% of the deaths were actually caused by CML. Gleevec is a new kind of therapy that interferes with the activity of the abnormal tyrosine kinase produced by the Philadelphia chromosome and, in turn, inhibits the production of abnormal white blood cells. Almost all patients respond to this oral treatment, making it the newest form of therapy for this disease.

NONCHEMOTHERAPEUTIC TREATMENT OPTIONS HAVE SIGNIFICANTLY IMPROVED SURVIVAL RATES

Bone marrow transplants may be a treatment option

One option for certain leukemias is a bone marrow transplant (BMT). In this procedure, the leukemic bone marrow cells are destroyed by using very high doses of chemotherapy and/or radiation therapy. Once the treated bone marrow stops producing diseased cells, healthy bone marrow or stem cells are given intravenously to stimulate new bone marrow production. There are three types of these transplants: autologous BMT, the removal and re-infusion of the patient's own bone marrow cells (usually not performed for leukemia since the diseased cells cannot be separated from the healthy ones); syngeneic BMT, when a patient receives a transplant from an identical twin; and allogeneic BMT, the infusion of cells from a donor that is compatible to the patient and his or her tissue type. The donor could be a family member, friend, or stranger.

Tissue type is based on certain proteins present on the surfaces of body cells that identify "self" versus "non-self." The human leukocyte antigens (HLA) or major histocompatibility complexes (MHC) are found on white blood cells and nearly all other cells of the body. The HLAs routinely typed for are HLA-A, B, and DR for a total of six (one is inherited from each parent, resulting in two A's, two B's, and two DRs). The best case scenario is when all six HLA types are identical between the donor and recipient, as they would be with identical twins. If there is not an exact match, the closest one possible will be used. Usually five out of the six HLA types must be the same. The donor could be a brother or a sister, or an unrelated donor could be selected from a national registry of volunteers who have had their HLA types identified. The patient has a 1 in 4 chance of being a match with a family member and a 1 in 20,000 chance of being a match with anyone else. The closer the HLA tissue match is between the donor and the recipient, the better the chance that the transplanted cells will not be rejected and that new blood cells will begin reproducing on their own. A newer form of transplant uses peripheral (circulating) blood stem cells (PBSC) which are removed by apheresis from the donor's blood.[16.5]

> [16.5] Apheresis is a technique in which blood is drawn from a person's vein into a separation instrument that works by centrifugation. A selected portion of the blood (i.e., stem cells, white blood cells or platelets) is removed and stored, while the rest of the blood is returned to the person either into the same vein or into a vein in the other arm.

Umbilical cord blood is an excellent source of stem cells

Sometimes, umbilical cord stem cells, isolated from the umbilical cord that has been cut and removed from the mother after a baby is born, are used for transplants (Figure 16.8). These unique cells have the potential to develop into healthy white and red blood cells and platelets and are being studied for their use in the treatment of certain diseases. To make this option available, public and private cord blood registries have been established. One example of a *public* registry is the National

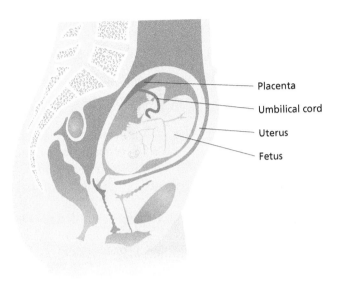

Placenta

Umbilical cord

Uterus

Fetus

Figure 16.8 Umbilical cord blood. Fetal blood taken directly from the umbilical cord after birth is a good source of stem cells.

Box 16.2

How does one join a marrow registry?

The procedure to become a member of a national registry is simple. Anyone between the ages of 18 and 60 who meets appropriate health guidelines and is willing to donate to any patient in need can be tested. Cotton swabs are used to obtain samples from the inside of the mouth. This specimen is then sent to a laboratory that performs HLA testing. Most insurance companies pay for this analysis. You will be notified if you are a match for someone who is on the waiting list. On any given day, more than 6000 men, women and children need a bone marrow transplant and consult a national registry. Contact www.marrow.org for specific information.

Marrow Donor Program Registry which has more than 90,000 cord blood units in storage. Families can contact this organization to inquire about the free, painless, donation process, but it is not available in all states at this time. If a woman chooses to donate her baby's cord blood with the hope of helping someone with leukemia or other blood disease, arrangements must be made prior to delivery in order to save the cells properly. Guidelines have been established to protect the health of the recipients, so a medical history form must be completed. There is no charge for this

^{16.6} As storing cord blood becomes more common, it is important for all to be familiar with the concept of public and private registries. The American Academy of Pediatrics studied the question of whether or not parents should store their child's cord blood and ultimately advised against it since they felt that there was little evidence that a child would need his or her own blood stem cells in the future.

service. Another option for expectant parents is a *private* cord blood bank that will store their baby's cord blood in case the child or other family member needs a stem cell transplant in the future. There is a charge for the process of obtaining the cord blood cells as well as an annual charge for storing them.[16.6] Pregnant women will probably receive mailings, see flyers in their obstetrician's office, and read ads in parenting magazines about the potential use of umbilical cord stem cells in case they are needed in the future.

Undergoing BMT or PBSC is a difficult procedure for the patient

Bone marrow or PBSC transplants are complex procedures and should only be performed in hospitals that have experience and success with them. Some programs may perform these procedures only on patients who have related donors, while other programs have strong success rates with those who have unrelated donors.

Prior to receiving *any* transplant cells, the patient must undergo a number of rounds of high-dose chemotherapy and sometimes radiation therapy to eliminate all cells in the bone marrow and the immune system including those that are cancerous. The donated cells from the bone marrow, peripheral blood, or umbilical cord blood will be thawed and given intravenously. It may take a few days or up to 3 weeks for the transplanted cells to take up residence in the patient's bone marrow and begin to grow and produce new white blood cells, then platelets, and finally red blood cells in a process called engraftment. During this period, the patient is given immunosuppressive drugs that intentionally weaken the immune system in an attempt to prevent the body from rejecting the transplant as well as antibiotics to minimize the risk of infection. He or she must remain in protective isolation to guard against exposure to microbes and pathogens until the white blood cell count rises above 500/mm^3, and may be allowed to leave the hospital when the count is over 1000/mm^3. The recipient must be carefully monitored for rejection of the transplant as well as for a post-transplant complication called graft-versus-host disease (GVHD). This occurs when the immune cells from the donated blood or marrow (graft) attack the recipient (host). These and other complications can be controlled often, but not always, and the success rate for these transplants is dependent on many variables.

A transplant option for older persons has demonstrated some success

For many years, transplants were performed only on young or middle-aged patients because it was felt that persons over the age of 55 could not tolerate the harsh procedure, leaving this population with limited treatment options. A newer form of transplant, sometimes referred to as a "mini" or nonmyeloablative transplant, has been used with some success in this population. They receive lower doses of chemotherapy and radiation, often on an outpatient basis, and consequently still possess some of their own blood cells when they receive transfusions of marrow. These procedures are not common in all hospitals and criteria are being established regarding what kinds of illnesses can be treated and who can expect to have a good prognosis.

LYMPHOMA IS A MALIGNANCY OF THE LYMPHATIC SYSTEM

The lymphatic system is a series of lymph vessels that run alongside the blood vessels throughout the entire body (Figure 16.9). As blood circulates through the tissues, fluid exits the blood vessels, bathes the surrounding tissues, and the majority of it reenters the capillaries. If the excess fluid was left to build up over time, the tissues would swell. This does not happen because the remaining fluid that did not reenter the capillaries is taken up by the lymphatic vessels. Multiple lymphatic vessels merge and eventually deliver the lymph to lymph nodes. As the fluid passes through the lymph nodes, material such as cellular debris, viruses, and bacteria that were obtained from the tissues is filtered out. Various lymphocytes also reside in the lymph nodes and can become activated and produce an immune response when presented with material that should not be in the body.

Any lymphoma is a result of the uncontrollable multiplication of lymphocytes that eventually crowds out healthy cells in the lymph nodes or other lymphoid tissues. The lymph nodes, located in the neck, underarm, groin, pelvis, and abdomen, are small bean-shaped structures that filter substances in the lymph fluid traveling throughout the body. Since lymphatic tissue is present throughout the body, this disease can begin in areas such as the stomach or intestines and spread to other organs.

HODGKIN'S VS. NON-HODGKIN'S LYMPHOMA

Lymphomas are divided into two main categories: Hodgkin's lymphoma or disease and the more common non-Hodgkin's lymphoma, which is usually referred to as simply lymphoma. The two types are

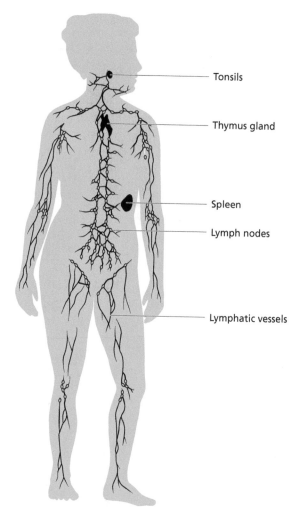

Figure 16.9 Lymphatic system. The lymphatic system consists of lymphatic vessels, lymph nodes, lymph fluid, tonsils, thymus gland, spleen and Peyer's patches (tissue on the surface of the small intestine).

different from each other microscopically as well as in clinical symptoms and risk factors. About 75,000 Americans, both adults and children, were diagnosed with a form of lymphoma in 2008. Hodgkin's lymphoma occurs less frequently than non-Hodgkin's lymphoma, accounts for about 11% of all lymphomas or about 8,000 new cases per year, and is characterized by the growth of Reed–Sternberg cells.[16.7] The Reed–Sternberg (R-S) cell is a large,

> [16.7] For about 170 years, the disease was called Hodgkin's disease but changed to Hodgkin's lymphoma when it was scientifically established that the cancer originated in a lymphocyte.

malignant white blood cell with two or more nuclei, whose appearance resembles "owls eyes."[16.8] Hodgkin's lymphoma is more often found in adolescents and young adults, while lymphoma is generally found in those over 60 years old.

There are about 30 different types of non-Hodgkin's lymphoma affecting about 66,000 people in 2008. The incidence of this disease has nearly doubled over the last 50 years for reasons not yet known. These lymphomas are currently classified by the clinical status of the patient, imaging studies, and the appearance and genetic characteristics of the affected cells.

> [16.8] The Reed–Sternberg (R-S) cell is named for scientists Dorothy Reed and Carl Sternberg who discovered it more than a century ago. While these cells are essential in the diagnosis of Hodgkin's lymphoma, they are not sufficient since they may also be identifiable in other conditions.

CERTAIN RISK FACTORS ARE ASSOCIATED WITH AN INCREASED INCIDENCE OF LYMPHOMA

Having a weak immune system, caused either by infection with a virus or bacterium or by an inherited condition, is considered to be a risk factor for lymphoma. Cases of the disease in those infected with Human Immunodeficiency Virus (HIV), the virus that causes AIDS, occur about 50–100 times more often than in those who are uninfected. In Africa, an infection with Epstein–Barr Virus (EBV), a member of the Herpesvirus family, is associated with Burkitt's lymphoma, a very rare form of cancer. EBV may also be found in those who have had organ transplants and have weakened immune systems because of the immunosuppressive drugs taken to prevent rejection. Human T-cell leukemia/lymphoma virus (HTLV-I) has been identified as a risk factor in certain geographic regions in Southern Japan, the Caribbean, South America, and Africa. The hepatitis C virus is now being studied as one of the risk factors as well. Infection with *Helicobacter pylori*, the bacterium that causes stomach ulcers, can increase the possibility of lymphoma in the stomach lining. There is an increased incidence of the disease in some farming communities, perhaps from the use of herbicides and pesticides. Other rare inherited conditions as well as the use of hair dyes before 1980, which contained certain chemicals, are being studied to determine if there is a possible linkage to lymphoma. Siblings of patients with Hodgkin's lymphoma also have an increased incidence of the disease.

The most common symptom of either form of lymphoma is a painless swelling of the lymph nodes. Other symptoms include weight loss, fever, night sweating, itchy skin, and fatigue. One distinctive but uncommon feature of Hodgkin's lymphoma is lymph node pain after drinking alcohol.

DIAGNOSING LYMPHOMA INVOLVES BIOPSIES AND IMAGING TESTS

Diagnosing lymphoma requires the biopsy of an involved or enlarged lymph node and subsequent microscopic examination of the specimen by a pathologist. Additional tests include a Complete Blood Count and blood chemistries to determine how well the other organs are functioning. CT scans, MRIs and PET scans may also be ordered to determine the extent of additional organ involvement. Cytogenetic studies will determine chromosomal abnormalities which may help in choosing which drug should be utilized for treatment. Since there are so many types of lymphoma, many factors will influence treatment.

LYMPHOMAS MUST BE CLASSIFIED TO DETERMINE APPROPRIATE TREATMENT

There are more than 30 subtypes of lymphomas, each of which has its own prognosis and treatment regimen. One method of classification is the determination of whether the lymphoma is slow-growing (low-grade) or fast-growing (aggressive) along with its stage (Table 16.4). In addition to staging, it must be determined whether the affected cells are T, B, or NK lymphocytes. This is done by distinguishing certain cellular features using molecular diagnostic techniques. Physicians will also take into consideration the age of the patient and if he or she has any other medical conditions, as is often the case with older and/or immunocompromised people. Finally, the types of symptoms, classified as A or B, is considered. Fever, excessive sweating, and loss of over 10% of body weight are referred to as B symptoms and are significant in treatment decisions. The designation A refers to the absence of these three symptoms.

Table 16.4 Lymphoma staging (Hodgkin's and non-Hodgkin)

Stage	Location
I	Detected in one lymph node or one organ
II	Detected in two or more lymph nodes near to each other (examples: neck and chest, abdomen)
III	Detected in several lymph nodes in the neck, chest and abdomen, or on both sides of the diaphragm
IV	Detected in several lymph node areas and in multiple organs (examples: lungs, liver, intestines, bones, brain) besides the lymph system

TREATMENT OPTIONS DEPEND ON THE TYPE OF LYMPHOMA, STAGE, AND EXTENT OF METASTASIS

Early stage lymphoma, especially when localized, may be treated with radiation, while widespread disease will require chemotherapy or a combination of both. Stem cell transplants may be considered for those who have relapsed. Hodgkin's lymphoma is currently considered to be one of the most curable forms of cancer by chemotherapy, especially after the first treatment. For those who do not respond initially, the second treatment is often very successful. For some patients with mild symptoms, the practice of "watchful waiting" may be the first and only choice of treatment.

EXPAND YOUR KNOWLEDGE

1 Explain how leukemia and lymphoma are different from other cancers in diagnosis and treatment.
2 Give examples of other forms of leukemia not discussed.
3 Identify some marrow donor registry drives in your city or state.
4 How is graft vs. host disease treated?
5 Why would a parent choose a public cord blood registry rather than a private one, or a private over a public registry?

ADDITIONAL READINGS

American Academy of Pediatrics' website: www.aap.org/
Caitlin Raymond International Registry's website: www.crir.org
Delannoy, A., Delabesse, E., Lheritier, V., Castaigne, S., Rigal-Huguet, F., Raffoux, E., Garban, F., Legrando, O., Bologna, S., Dubruille, V., Turlure, P., Reman, O., Delain, M., Isnard, F., Coso, D., Raby, P., Buzyn, A., Cailleres, S., Darre, S., Fohrer, C., Sonet, A., Bilhou-Nabera, C., Bene, M.-C., Dombret, H., Berthaud, P., and Thomas, X. (2006) Imatinib and methylprednisolone alternated with chemotherapy improve the outcome of elderly patients with Philadelphia-positive acute lymphoblastic leukemia: results of the GRAALL AFR09 study. *Leukemia*, **20**:1626–1632.
Marieb, E. N. and Hoehn, K. (2007) *Human Anatomy and Physiology*, 7th edn. Upper Saddle River, NJ: Prentice Hall.
Martini, R. & Bartholomew, E. (2007) *Essentials of Anatomy & Physiology*, 4th edn. Upper Saddle River, NJ: Prentice Hall.
National Marrow Donor Program's website: www.marrow.org (Be the Match Registry)

We cannot change the cards we are dealt, just how we play the hand.

Randy Pausch, Professor, Carnegie Mellon Institute, pancreatic cancer patient, and author of *The Last Lecture*, with Jeffrey Zaslow

Glossary

ABCDs of melanoma Four characteristics used to classify melanoma: asymmetry, border, color and diameter.

abdominoperineal resection The removal of a tumor in the distal portion of the rectum along with the anal sphincter followed by an anastomosis.

actin A protein found along the periphery of the cell just inside the plasma membrane associated with the cytoskeleton.

actinic keratosis A precancerous condition of small, pink or red colored, scaly lesions on the skin due to long-term exposure to the sun's UV radiation that can progress to squamous cell carcinoma.

acute lymphoblastic leukemia (ALL) A type of leukemia characterized by large numbers of immature lymphocytes and most commonly found in children.

acute myeloid leukemia (AML) A type of leukemia most often found in adults.

adenocarcinoma Cancer of the glandular cells.

adenosquamous Also known as mixed carcinomas.

adipose Fat tissue.

adjuvant Therapy that is given after surgery.

agonist Chemical that is structurally and functionally similar to another molecule.

alkylating Attaching alkyl groups.

allele Alternative versions of genes.

allogeneic bone marrow transplant The infusion of stem cells from any donor that is compatible with the patient's tissue type.

alpha-fetoprotein (AFP) A protein marker for testicular cancer.

alveoli Milk-producing sacs of the breast.

amino acids The building blocks (monomers) of proteins.

analog A compound that is structurally similar to the naturally occurring molecule.

anaplasia The appearance of undifferentiated cells that bear no resemblance to the cells normally found in that tissue.

anastomosis The procedure in which the two healthy portions of the colon are sewn together during a segmental resection.

anchorage dependence The tendency for cells to adhere themselves to neighboring cells and the extracellular matrix.

androgen deprivation therapy A treatment of prostate cancer which seeks to either prevent the synthesis of testosterone or block target cells from responding to testosterone in order to slow the growth or reduce the size of the tumor.

androgen independent A prostate tumor that is no longer responsive to hormone therapy.

androgens Male sex steroidal hormones.

androgen suppression therapy *see* androgen deprivation therapy.

anemia Reduction in the number of red blood cells.

angiogenesis The formation of new blood vessels.

angiogenesis inhibitor Molecules that inhibit blood vessel growth.

antiandrogens Antagonists that bind in place of natural hormone to the receptors in the cells of the testes, effectively blocking their ability to respond to sex hormones.

antibiotics Compound produced by a microorganism, or a chemically modified form synthesized in a lab, that inhibits the growth of bacteria.

antibody Protein that has a very high degree of specificity for another molecule; naturally produced by the body's immune system during infection to target foreign molecules for destruction.

antiemetic Antinausea medications.

antiestrogens Antagonists that bind in place of natural hormone to the receptors in the cells of the breast, effectively block their ability to respond to sex hormones.

antimetabolites An analog introduced to the cell that will disrupt the cell's ability to function.

anus The opening at the lower end of the alimentary canal that functions to expel undigested solid waste from the body.

apheresis A technique in which blood is drawn from a person's vein, the desired components are separated out, and the remaining portion is returned to the person's bloodstream.

apoptosis Programmed cell death.

aromatase enzymes Convert androgens produced by the adrenal cortex into estrogens.

aromatase inhibitors Molecules that inhibit the activity of aromatase enzymes.

arteriole Small blood vessel between arteries and capillaries.

artery Blood vessel which carries blood away from the heart.

asbestos A fire-retardant insulating material found to be carcinogenic when released into the air.

ascending colon The portion of the large intestine between the small intestine and the transverse colon.

asparaginase therapy The process of injecting large doses of the enzyme asparaginase into the circulatory system eliminating asparagines in the blood, starving the cancer cells of their sole source of the amino acid.

asymptomatic Not exhibiting any signs of illness.

atom The building blocks of matter that are composed of negatively charged particles called electrons that spin in shells or orbitals around a nucleus that possesses protons, positively charged particles, and neutrons, uncharged particles.

atypical glandular cells Abnormal glandular cells.

atypical nevi Pigmented spots or moles on the skin of the back, chest, abdomen, or face.

atypical squamous cells of undetermined significance (ASC-US) Most common abnormal Pap test result reported when the microscopic identification does not differentiate between an infection, irritation, or a precancerous condition.

autologous bone marrow transplant Removal and re-infusion of patient's own bone marrow cells.

B lymphocytes White blood cells that produce antibodies during infection.

barium enema X-ray A procedure that allows the outline of the soft tissues of the colon and rectum to be observed.

basal cell carcinoma (BCC) One of the most common forms of skin cancer; arises from the malignant transformation of basal cells.

basal cell A round cell that forms the lower layer of the epidermis.

basal level The activity level exhibited by a protein in order to satisfy the minimal needs of a cell.

basement membrane A supporting layer of extracellular material composed of a variety of glycoproteins and carbohydrates and provides a defining boundary between the epithelium and the underlying layer of connective tissue.

basophil A somewhat phagocytic myelocyte that plays a role in allergic responses.

Bcl2 A protein that functions to inhibit apoptosis.

benign Tumors are not cancerous.

benign prostatic hyperplasia The portion of the prostate surrounding the urethra exhibits a noncancerous growth by increasing the number of cells.

benign prostatic hypertrophy The portion of the prostate surrounding the urethra exhibits a noncancerous growth.

Bethesda System, The (TBS) Means of classifying Pap test results.

biological therapy Also known as biotherapy, immunotherapy or biological response modifier therapy. Use of drugs or methods to enhance the body's ability to detect and destroy cancer cells and repair or replace cells damaged from the use of other treatment methods.

biopsy Removal of tissue in order to determine whether or not it is malignant.

biopsy gun A hand-held device that uses a spring-loaded needle to obtain a sample of tissue.

blasts Abnormal, immature white blood cells unable to carry out their typical function.

bolus The ball of chewed food and saliva that is swallowed.

bone marrow The inner part of bones possessing stem cells that produce red and white blood cells and platelets.

bone marrow aspiration and biopsy Bone marrow is removed, typically from the ilium (the pelvic bone), using a special needle.

bone marrow transplant (BMT) Leukemic bone marrow is destroyed by using high doses of chemotherapy and sometimes radiation therapy, stopping production of diseased cells so that healthy bone marrow or stem cells can be given intravenously to stimulate new bone marrow production; the three types are autologous, syngeneic, and allogeneic.

bortezomib A proteosome inhibitor that has been approved for the treatment of multiple myeloma.

brachytherapy Also known as internal radiation therapy. Treatment of a smaller area with a higher dose of radiation.

breast self examination An examination that patients perform on themselves on a monthly basis in order to detect cancerous conditions.

Breslow thickness The vertical depth of the tumor used for its prognostic value.

bronchoscopy A test in which a small flexible scope is inserted through the nose or mouth to examine the bronchi and to take a biopsy.

CA-125 Blood test currently used for women at high risk for ovarian cancer, with suspicious symptoms, or to determine reoccurrence.

cancer vaccine Particular proteins from specific types of cancer cells that are used to jumpstart an immune response against those cells.

capillary Smallest vessel in which a single blood cell can pass through to tissues.

carbohydrate antigen (CA 19-9) A tumor marker that is synthesized at elevated levels in colorectal cancer.

carbohydrates Biomacromolecule more generally known as sugars.

carcinoembryonic antigen (CEA) A tumor marker that is synthesized at elevated levels in colorectal cancer.

carcinogen A cancer-causing agent.

carcinoma *in situ* Benign tumors which are contained in the original site and have not invaded surrounding tissues.

carcinoma The most prevalent type of cancer that originates in the skin or epithelium of the internal organs and glands.

caretaker A tumor suppressor gene that repairs structural damage to the chromosomes, corrects mutations in the DNA sequence that may occur during DNA replication, and sorts the chromosomes into daughter cells during division.

catheter A thin tube placed into a large vein to administer drugs.

cell cycle The process of cellular growth and development which consists of five phases: G_0, G_1, S, G_2, and M.

cell A collection of atoms, molecules, macromolecules and macromolecular structures creating the simplest unit of life.

Central Dogma of Biology The processes that lead to the expression of genetic information.

centromere The point at which two sister chromatids remain attached to one another.

centrosome A region adjacent to the nucleus from which microtubules originate.

cervical canal The spindle-shaped canal extending from the uterus to the vagina.

cervical punch biopsy The removal of a sample of tissue from the cervix for examination.

cervix The lower portion of the uterus which opens to the vagina.

checkpoints A point in each phase of the cell cycle where progression temporarily halts until a set of conditions is met.

chemistry panel A series of tests performed on a sample of blood that are usually grouped according to the functional aspect being analyzed.

chemotherapy The treatment of cancer with drugs that kill or inhibit the growth of cells.

choriocarcinoma A type of nonseminoma.

chromosomes Individual DNA molecules complexed with protein.

chronic lymphocytic leukemia (CLL) A type of leukemia resulting from an overproduction of B lymphocytes; the most common type of leukemia in North America and Europe.

chronic myelogenous leukemia (CML) Also known as chronic granulocytic leukemia; a rarer form of leukemia that is almost always associated with the presence of the Philadelphia chromosome and is characterized by an overproduction of myelocytic white blood cells.

chyme The acidic liquid that forms in the stomach as a result of the mixing of nutrients and gastric juice.

classical seminoma The most common type of seminoma.

clinical breast examination An examination performed by a healthcare provider to look for unusual growths in breast tissue.

clinical target volume All of the area within the gross tumor volume, as well as areas where there is a high probability that cancer cells reside but are presently undetectable.

colectomy see segmental resection.

colon Also known as the large intestine. Responsible for absorption of water and some vitamins as well as mixing indigestible material and roughage with mucus and bacteria inhabiting the colon to form feces.

colonoscope A long, flexible tube with a camera at one end used in a colonoscopy to view the inner lining of the large intestine.

colonoscopy Examination of the inner lining of the intestine using a small video camera on the end of the flexible tube.

colony-stimulating factors (CSF) Proteins that promote the bone marrow's production of red and white blood cells as well as platelets.

colostomy A procedure in which a bag is attached to one end of the intestine to collect fecal matter after a segmental resection in which the two ends of the intestine could not be reattached.

colposcopy A procedure in which an acetic acid solution is applied to the cervix, and the vagina and cervix are observed using a colposcope.

combination hormone therapy The use of various types of hormone therapies in the treatment of hormone-responsive tumors.

combined androgen blockage *see* combination hormone therapy.

complementary base pairing The particular way in which the nitrogenous bases pair within nucleic acid molecules.

complete blood count (CBC) Procedure that examines the numbers and kinds of white blood cells as well as the number of red blood cells and platelets.

computerized axial tomography (CAT) Test that creates digital sectioning or cutting of the body using X-rays to produce detailed images of inner body structures.

computerized tomography (CT) *see* computerized axial tomography.

conization The surgical removal of a larger, cone-shaped sample of tissue from the cervix and cervical canal using a knife, laser, or electricity.

consolidation therapies Chemotherapy treatments given after the initial therapy to help prevent recurrences.

core biopsy *see* incisional biopsy.

core needle biopsy A procedure that uses a biopsy gun to sample tissue from the prostate.

Crohn's disease A form of chronic inflammatory bowel disease that increases one's risk for colorectal cancer.

cryosurgery Freezing of malignant cells using a spray of liquid nitrogen or an ultracold probe.

cryosurgical ablation *see* cryosurgery.

cryosurgical therapy *see* cryosurgery.

cryptorchidism Undescended testicle.

curettage biopsy A procedure in which a curette is used to remove the affected epidermis along with the very top of the dermis.

curette A small spoon shaped instrument that has a sharp edge or cytobrush and is used during a curettage biopsy or endocervical curettage.

cycle therapy The application of a chemotherapeutic drug for specific periods of time.

cyclin A protein encoded by a proto-oncogene whose concentration rises and falls cyclically with progression through the cell cycle.

cyclin dependent kinase (cdk) A type of proto-oncogenic protein.

cyst Fluid-filled sac that can create a suspicious lump.

cytokines Proteins that the immune system cells use to communicate with one another.

cytokinesis The final step in cell division in which the furrow becomes deeper and deeper until the membrane pinches off forming two daughter cells.

cytology test Testing of fluids for the presence of cancer cells.

cytoreduction surgery *see* debulking.

cytoskeleton Internal framework of proteins that contributes to a cell's shape and stability and also acts as a rail system that transports vesicles around the cell.

cytosol The region of the cell between the nuclear and plasma membranes.

D and C Procedure in which the cervical opening is stretched (dilatation) and a curette is inserted to scrape and clean out cells from the lining of the uterus and cervical canal (curettage).

debulking surgery Performed to remove only a portion of the tumor, with the hopes that by reducing its size there will be an increase in the effectiveness of radiation or chemotherapy.

deletion Genetic mutation in which one or more nucleotides are removed from the DNA sequence.

deoxyribonucleic acid (DNA) A type of nucleic acid.

deoxyribose The sugar component found in deoxyribonucleic acid.

Dermapathologist A physician who specializes in the diagnosis and interpretation of skin conditions.

dermatologist A physician who specializes in conditions that affect the skin.

dermis The lower of the two skin layers and consists of a somewhat dense layer of connective tissue that is rich in collagen.

dermoscope A handheld device that provides a magnified and more detailed view of a lesion on the skin than is possible with the unaided eye.

descending colon The portion of the large intestine between the transverse colon and the sigmoid colon.

diagnostic surgery Also known as staging surgery. Performed as a means of obtaining tissue samples for biopsy to make a definitive diagnosis of cancer.

diagnostic tests Array of tests used to confirm that a person has a particular condition or disease.

differential gene expression The expression of genes only at certain times during development, under specific conditions, or in particular tissues.

digestive tract A long tube that runs through the body from the mouth to the anus that functions in the digestion and absorption of nutrients.

digital rectal exam A procedure in which the prostate is palpated through the rectum using the index finger.

digital X-ray imaging An alternative form of X-ray imaging that utilizes a computer sensor, rather than film, to detect the radiation and view internal tissues.

diploid A cell in which there are two copies of every chromosome with each chromosome in each pair coming from each parent.

disaccharide Two monosaccharides linked together.

double contrast barium enema X-ray The use of barium sulfate as a contrast agent in order to better view the lining of the colon by X-ray imaging.

ductal carcinoma *in situ* (DCIS) An early form of breast cancer in which the tumor is located in the ductal cells of the breast but has not invaded surrounding tissues.

ductal lavage A nonsurgical approach to detect abnormal cells in the breast by the collection of cells from the ducts.

duodenum The first and shortest section of the small intestine where the majority of digestion occurs.

dysplasia A disorganized arrangement of cells.

E2F Transcription factor that regulates the expression of genes that code for proteins needed in the S phase of the cell cycle.

edema Tissue swelling due to a buildup of interstitial fluid.

electrodessication A procedure in which a very thin probe that produces a weak electric current is applied to the affected area on the skin to destroy malignant cells.

electrosurgery High frequency electrical currents are used to kill cells.

embryonal carcinoma A type of nonseminoma.

endocavity radiation therapy Insertion of a probe that emits a high dose of radiation into the rectum, allowing the radiation to pass through the least amount of healthy tissue before reaching the affected tissue; repeated once every 2 weeks over a 6-week period.

endocervical curettage A scraping of cells from the inside of the endocervical canal for microscopic analysis.

endocrine system A collective term for all of the hormone producing organs in the body.

endometriosis A painful condition in which the tissue of the uterine lining grows outside the uterus and may adhere to the ovaries.

endoplasmic reticulum (ER) Organelle composed of two forms, the smooth ER involved in synthesis of various lipids and steroid hormones and the rough ER that is involved in protein modification.

endoscope A long flexible tube with a video camera on the end as well as an instrument that is able to remove tissue samples for examination.

endoscopic biopsies Biopsy performed within respiratory, gastrointestinal, or urogenital tracts using an endoscope.

endoscopy Procedures used to observe internal body structures with the use of a fibreoptic camera.

engraftment The process in which transplanted cells take up residence in the patient's bone marrow and begin to independently grow and reproduce.

environmental tobacco smoke (ETS) Also known as second hand smoke.

enzymes Proteins that catalyze reactions that would otherwise occur too slowly to sustain life.

eosinophils Somewhat phagocytic myelocytes.

epidermal growth factor A molecule that binds to a particular receptor on specific cells, thus triggering their growth.

epidermal growth factor receptor (EGFR) A specific protein found on the membrane of many different cells in the body which when bound to the protein, epidermal growth factor growth or reproduction is triggered.

epidermis The outer layer of skin that serves as the primary barrier between the environment and inner tissues and organs.

epidermoid carcinoma *see* squamous carcinoma.

epithelial cell abnormalities A Pap test result that indicates precancerous or cancerous changes within the cells lining the cervix.

epithelial ovarian cancer The most common type of low malignant potential tumor.

epithelial ovarian tumors Benign tumors of the ovary.

erythrocytes Commonly known as red blood cells; cells of the blood that contain hemoglobin and function to transport oxygen from the lungs to tissues throughout the body.

esophagus A muscular tube that connects the back of the throat with the stomach.

estrogen Female sex steroidal hormones.

estrogen receptor (ER) A nuclear receptor that triggers transcriptional changes upon estrogen binding.

estrogen receptor status The determination of the presence or absence of estrogen receptors during biopsy or surgery of a breast tumor.

estrogen replacement therapy (ERT) Used to alleviate the symptoms experienced by postmenopausal women.

excisional biopsy The removal of an entire tumor.

external beam radiation therapy A form of radiation therapy that delivers an X-ray beam directly at the tumor.

extracellular matrix (EM) A network of material that cells attach to, to ensure the structural integrity of the tissue.

false negative A result is reported as negative when in fact it should have been positive.

false positive A result is reported as positive when in fact it should have been negative.

familial adenomatous polyposis (FAP) An inherited condition that results in the formation of hundreds to thousands of polyps that, if not removed, will become malignant.

fatty acid A long chain of carbons and hydrogens that has a carboxylic acid at one end; a component of most cellular lipids.

fecal immunochemical test (FIT) An alternative form of the FOBT that utilizes monoclonal antibodies specific for a protein found in human blood.

fecal occult blood test (FOBT) Test that detects the presence of blood in the stool.

fibroadenoma A nonmalignant moveable lump of fibrous or glandular tissue.

fibrocystic breast A noncancerous condition in which cysts or scar-like connective tissue create areas of lumpiness in the breast.

fine needle aspiration Also known as needle core biopsy. A needle is used to extract a tissue sample.

flare reaction Temporary growth spurt seen in cancer initially after treatment with GnRH agonists because of the high concentration of LH and FSH that is released at the beginning of treatment.

5-fluorouracil (5-FU) A nucleotide analog administered intravenously for the treatment of colorectal cancer.

folic acid A type of B vitamin required for the synthesis of DNA bases A, T, and G.

follicle stimulating hormone (FSH) Released by the pituitary gland to stimulate cells within the gonads to produce androgens and estrogens.

free radical A molecule that has lost an electron and is consequently unstable.

frozen section exam A procedure used to determine whether cancerous cells are present in biopsied lymph nodes.

G_0 The quiescent phase of the cell cycle in which there is no immediate likelihood of division.

G_1 The time spent by cells carrying out their normal functions or those tasks necessary to prepare for division; occurs before chromosomal replication.

G_2 The time spent by cells carrying out their normal functions or those tasks necessary to prepare for division; occurs after chromosomal replication.

gain-of-function mutation A mutation converting a proto-oncogene into an oncogene by altering the structure of a protein such that its function is enhanced, or that increases the expression of a gene to yield more functional protein.

gastric juice A highly acidic fluid produced by the stomach that contains proteases to degrade proteins into short peptide fragments.

gastrointestinal (GI) tract *see* digestive tract.

gatekeeper A suppressor gene that prevents a cell from proceeding further along the cell cycle unless certain conditions have been met.

gene A specific stretch of nucleotide sequence within a DNA molecule that codes for a particular molecule.

gene therapy Introduction of DNA into cells to express a protein that will carry out a specific function.

genital warts Soft, moist, pink or flesh-colored swellings caused by certain types of HPV found on the inside or outside of the genitals.

genome A complete set of genetic information on the chromosomes.

germ cells Reproductive cells. The testes in males produce sperm and ovaries in women produce eggs.

germ-line The tissue that produces gametes (germ cells).

germ cell tumors Tumors of the germ cells in the ovary; make up about 5% of ovarian cancers.

glandular cells Cells that make up glands and secrete various types of substances.

Gleason grading system The most commonly used grading system for prostate cancer that utilizes a five point scale based on the appearance and differentiation of prostatic gland cells.

Gleason score A measure of the aggressiveness and likelihood of metastasis of prostate cancer obtained from the addition of the two most common grades found in a biopsy sample.

glycolipids Proteins bound to lipids and located in the outer membrane of cells.

glycoprotein Proteins bound to oligosaccharides and located in the outer membrane of cells.

glycosylation Process of modifying carbohydrates and lipids by adding carbohydrates to them.

GnRH agonists Synthetic chemicals that bind to and stimulate the GnRH receptors on pituitary cells.

GnRH antagonists Synthetic chemicals that bind to and block the GnRH receptors on pituitary cells.

Golgi apparatus (complex) An organelle composed of a series of flattened membranous sacs positioned between the rough ER and the plasma membrane.

gonadotropin Hormones released from the pituitary gland that stimulate the gonads to produce and release the sex hormones and to make gametes.

gonadotropin-releasing hormone (GnRH) Influences the pituitary hormone to produce and secrete a variety of hormones into the circulatory system, including luteinizing hormone and follicle stimulating hormone.

gonads The organs that produce gametes; ovaries in females and testes in males.

grade The degree of cellular abnormality during microscopic pathological examination.

graft-versus-host disease (GVHD) Occurs when the immune cells from donated blood or marrow (the graft) attack the recipient (the host).

granulocyte colony stimulating factor (GCSF) Growth factor that increases the number of white blood cells, reducing the risk of infection.

granulocyte *see* myelocyte.

gross tumor volume Consists of all areas within a region that are known to contain cancer cells.

Gynecological Cancer Education and Awareness Act A law implemented in 2007 that provides programs which increase the awareness and knowledge of both women and healthcare providers with respect to gynecologic cancers.

haploid Containing only half of the genome of the germ-line parent cell.

hemoglobin A protein that transports oxygen from the lungs to the tissues and carbon dioxide from the tissues to the lungs.

Her2/*neu* A type of EGFR found in some breast malignancies and is associated with aggressive tumor growth.

hereditary nonpolyposis colorectal cancer (HNPCC) A condition in which only a few polyps develop that give a person an 80% chance of developing colorectal cancer.

heterozygous Having different copies of a gene at a specific locus.

high grade SIL A category of squamous intraepithelial lesion that is likely to develop into cancer if not treated.

high-dose rate (HDR) brachytherapy Brachytherapy that implants pellets of high radioactivity for 5–15 minutes; repeated up to three times over a series of consecutive days.

Hodgkin's lymphoma A type of lymphoma that is characterized by the presence of Reed–Sternberg cells; 11% of all lymphomas.

homeostasis A state of internal balance.

homologous Chromosome pairs that contain the same information.

homozygous Having two identical copies of a gene at a specific locus.

hormone ablation *see* hormone deprivation.

hormone depletion *see* hormone deprivation.

hormone dependent Cancers that require the presence of androgens or estrogens in order to proliferate.

hormone deprivation Also known as hormone ablation or depletion. Treatment of hormone dependent cancers in which either the production of androgens and estrogens is eliminated or they are prevented from binding to their receptors.

hormone replacement therapy (HRT) An alternative to estrogen replacement therapy that uses estrogen in combination with progestins.

hormones Chemical signaling molecules in the body produced by a variety of cells that enable communication to occur between cells over short or long distances.

housekeeping genes Genes that are expressed in every cell which regulate the structures and functions that every cell must have and perform.

HPV DNA test A type of HPV test that can identify 13 of the high-risk types of HPV that are most likely to lead to cervical cancer.

HPV PCR test A type of HPV test that is able to identify other high-risk HPV types in addition to those that can be identified by the HPV DNA test.

HPV test The collection of cervical cells during a pelvic examination to detect the presence of the HPV virus.

human chorionic gonadotropin (hCG) A protein marker for testicular cancer.

human immunodeficiency virus (HIV) A virus affecting T cells of the body's immune system that predisposes one to the development of AIDS.

human leukocyte antigens (HLA) Also known as major histocompatibility complexes. Proteins present on the outside of cells of the body which identify "self" versus "non-self."

human papillomavirus (HPV) A virus transmitted sexually or via skin-to-skin contact that causes more than 90% of cervical cancers.

hydrocele Enlargement of testicle due to abnormal fluid collection.

hydrophilic Water loving polar (positively or negatively charged) molecules.

hydrophobic Water fearing nonpolar (uncharged) molecules.

hyperplasia An increase in the number of cells.

hypertrophy An increase in the size of individual cells.

hypodermis Subcutaneous tissue consisting mostly of adipose cells, blood vessels, and nerves entwined in collagen and elastin.

hypothalamus A glandular structure at the base of the brain responsible for regulating the endocrine system.

hysterectomy Surgical removal of the cervix and uterus.

ileum The final and longest portion of the small intestine.

immune system The collection of cells and tissues in the body that is specialized to fight infections by invading pathogens and organisms.

immunotherapy *see* biological therapy.

in situ Tumors in their original location.

incisional biopsy The removal of a portion of a tumor.

inflammatory breast carcinoma (IBC) A rarer form of breast cancer typically found in younger women and often not detected in routine screening tests.

informed consent Process of providing all details of a clinical study to participants both orally and in writing.

inguinal hernia A hernia in the groin area.

inherited (familial) melanoma Skin cancer that results from mutations in the tumor suppressor genes CDKN2A, CDK4, and MC1R.

inherited mutations Mutations in the DNA contributed by the sperm and/or egg at the moment of conception.

integral proteins Proteins embedded in both halves of the membrane.

interferon alpha An immunotherapy used to stimulate the activity of natural killer (NK) cells which recognize, attack, and destroy cancer cells.

interferon Proteins secreted by a cell infected with a virus warning neighboring cells of the imminent threat of viral infection.

interleukin 2 An immunotherapy used to enhance the ability of lymphocytes to destroy abnormally functioning cells.

interleukins Molecules that the immune system cells use to communicate with one another.

intermediate filaments Proteins scattered throughout the interior of the cell but not extending into the organelles associated with the cytoskeleton.

internal radiation therapy *see* brachytherapy.

interphase Stage of cell life cycle in which the cell grows in size, carries out its activities, replicates its chromosomes, and prepares to divide.

interstitial fluid Liquid that bathes tissues.

intraperitoneal (IP) A catheter placed through the skin and into the abdomen.

invasive ductal carcinoma (IDC) An advanced stage of ductal carcinoma *in situ* in which malignant cells have broken through the initial epithelial cell layer and have invaded the surrounding connective tissue.

invasive lobular carcinoma (ILC) A more invasive form of lobular carcinoma *in situ* in which malignant cells have invaded surrounding tissues.

ionizing radiation Radiation that consists of high-energy waves that are capable of penetrating cells and wreaking internal havoc such as denaturing and/or mutating proteins and DNA. Examples of this radiation include gamma and X-rays.

isolated limb infusion (ILI) A procedure in which a catheter is inserted into the main artery of a limb and a pneumatic tourniquet is applied to prevent the outflow of blood, allowing for the administration and isolation of high-dose chemotherapeutic drugs to the limb.

isolated limb perfusion (ILP) The reversible surgical manipulation of the vasculature of a limb so that its arterial and venous blood supplies form a loop that is isolated from those of the rest of the body, enabling the administration of a very high dose of chemotherapy.

jejunum The central portion of the small intestine between the duodenum and ileum.

karyotype A lab analysis in which the chromosomes isolated from a single cell are paired and ordered based on their sizes and other characteristics.

keratinocyte carcinoma A collective term for basal and squamous cell carcinomas.

keratinocytes Flattened squamous cells of the skin that have been pushed upward towards the surface and have begun producing large amounts of the fibrous protein keratin, creating a water tight cell layer.

Klinefelter's syndrome Rare genetic condition seen in men that have an extra X chromosome, resulting in low levels of male hormones, sterility, enlarged breasts, and small testes.

lactate dehydrogenase (LDH) An enzyme released into the blood when cellular damage occurs.

lactation The time a woman breast feeds.

laparoscope A slender tube with a video camera at one end used during laparoscopy.

laparoscopic surgery/laparoscopy Procedure that uses a laparoscope that can be inserted through a small incision in the abdominal wall, and used to guide several other instruments that are inserted through other small incisions.

laparotomy Procedure performed when there is a need to explore a large portion or the entire area of the abdomen.

large cell undifferentiated carcinoma A form of NSCLC that may appear in any part of the lung and grow and metastasize quickly.

large intestine *see* colon.

laser surgery Utilizes high powered beams of light to vaporize cancerous cells, enabling more precise incisions and targeting the destruction of tumor cells while sparing nearby healthy ones.

leukemia Cancer of the bone marrow.

leukocoria The appearance of a white pupil caused by the abnormal reflection of light off of the retina of the eye; sometimes referred to as "cat's eye."

leukocytes Commonly known as white blood cells; cells of the blood that are involved in the body's defense system and help to fight infection. The two subgroups are lymphocytes and myelocytes.

leukocytopenia A decreased white blood cell count.

Leydig cells Cells of the testes that produce male sex hormones.

lipids A biomacromolecule more commonly known as fats.

lobectomy The excision of an entire lobe of a lung.

lobular carcinoma *in situ* (LCIS) A form of breast cancer that develops and is contained within the epithelial cells of the lobules of the breast.

loop electrosurgical excision procedure (LEEP) The procedure in which a thin wire heated by electricity is used to perform a conization.

loss-of-function mutation This mutation alters the structure of a protein such that its function is weakened, or decreases the expression of a gene to yield less functional protein.

low anterior resection A procedure used to remove a tumor and regional lymph nodes in the proximal two-thirds of the rectum followed by an anastomosis.

low grade SIL A category of squamous intraepithelial lesion that may go away without further treatment.

low-dose rate (LDR) brachytherapy Brachytherapy that uses 80–120 pellets that are permanently implanted and decay to undetectable levels within 3–6 months.

lumbar puncture Also known as a spinal tap. A sample of cerebrospinal fluid is removed from the spinal canal of the lower back using a small needle.

lumpectomy Breast conserving surgery in which only the tumor and surrounding tissue is removed.

luteinizing hormone (LH) Release by the pituitary to stimulate cells of the gonad to produce androgens and estrogens.

lymph Fluid of the lymphatic system.

lymph node biopsy Dissection that is performed to determine whether or not cancer cells have broken free from the tumor, migrated into the lymph system, and been trapped in a local lymph node.

lymph nodes Point of convergence of lymphatic vessels.

lymphatic system A collection of vessels that drains excess fluid from tissues and returns it to the circulatory system.

lymphedema An accumulation of lymphatic fluid due to inadequate draining of the area.

lymphocyte A type of white blood cell; the three types are B lymphocytes, T lymphocytes, and natural killer cells.

lymphoma Cancer of the lymphatic system.

lysosomes Sacs that contain enzymes that break down large molecules that the cell has taken in or are no longer being used by the cell.

macromolecules Large cellular structures composed of smaller monomer units; the four major types are carbohydrates, lipids, nucleic acids, and proteins.

magnetic resonance imaging (MRI) An imaging technique that uses a magnetic field and radio waves to produce pictures of organs and structures inside the body.

major histocompatibility complexes (MHC) *see* human leukocyte antigens.

malignant Cancerous neoplasms.

mammogram A standard X-ray film or digital image produced by mammography that is examined for the presence of abnormal breast tissue that may or may not be felt during a manual examination.

mammography A low dose X-ray of the breast tissue.

mastectomy The surgical removal of the breasts and possibly underarm lymph nodes and/or chest wall muscles under the breast.

matrix metalloproteinases (MMPs) Enzymes that degrade the extracellular matrix proteins that will enable the growing vessel to migrate between the tissue cells toward the cancer cell.

mediastinoscopy A procedure which allows a physician to locate and remove the chest lymph nodes for biopsy.

meiosis A form of cell division that creates haploid daughter cells.

melanin A pigment produced by melanocytes that is responsible for differences in skin color.

melanocytes Cells found in the bottom-most layer of the epidermis that produce the pigment melanin.

melanoma The malignant transformation of melanocytes.

mesothelioma Cancer associated with asbestos exposure that originates in the lining of the lung and has a high mortality rate.

metaplasia The presence of a normal cell type in an inappropriate location.

metastasis Process by which cancerous cells separate from a tumor and spread to other areas of the body.

microtubules Proteins that radiate toward the plasma membrane in a starburst fashion from the centrosome associated with the cytoskeleton.

microvilli Microscopic projections on the surface of the villi in the small intestine that enhance the absorption of nutrients.

mini transplant *see* nonmyeloablative transplant.

mitochondrion An organelle that is involved in the release of energy from carbohydrates and lipids.

mitosis A somatic cell process in which the cell divides to form two diploid daughter cells.

mitotic index The ratio between the number of cells undergoing mitosis and the total number of cells within the field of view.

mitotic inhibitor Compounds which interfere with the assembly of tubulin into spindle fibers, result in cell death since they are unable to properly segregate their chromosomes during cell division.

MMP inhibitors Chemicals capable of inhibiting the activity of matrix metalloproteinases.

Mohs' micrographic surgery Performed on skin cancers, particularly around the eye, and involves the shaving of one layer of skin and examining it under a microscope and repeating the process until what is removed does not show any signs of cancer.

molecule Two or more atoms bonded together.

monoclonal antibodies Proteins produced by cells of the immune system that bind specifically to a particular area of an infectious microbe, foreign molecule, or cancer cell, targeting it for destruction.

monocyte A phagocytic myelocyte that also functions in antigen processing.

monomer Small molecules that link together to form large macromolecular polymers.

monosaccharides Simple sugars; the building blocks of more complex carbohydrates.

M-phase The phase in the cell cycle in which a somatic cell undergoes mitosis or a germ cell undergoes meiosis.

mRNA A type of RNA that is synthesized during transcription that contains the genetic information that will be transcribed into a protein.

myelocyte A type of white blood cell that possesses granules; the four types are neutrophils, eosinophils, monocytes, and basophils.

nail bed biopsy A procedure in which a portion of a nail or the entire nail is removed to gain access to the affected area in order to perform a biopsy.

natural killer cell A type of lymphocyte that will release perforin proteins that insert into the membrane of the targeted cell.

needle core biopsy *see* fine needle aspiration.

negative for intraepithelial lesion or malignancy A Pap test result that indicates no sign of cancer, precancerous changes, or significant abnormalities.

neoadjuvant Therapy that is given prior to surgery in an attempt to increase its success.

neoplasia An abnormal growth that lacks normal organization of the cells and possesses an increase in the number capable of dividing as compared to normal tissue.

neoplasm A tumor.

neutropenia A lowering of white blood cells which could increase the risk of infections.

neutrophil A phagocytic myelocyte.

nevus (pl. nevi) An unusual aggregation of melanocytes in the epidermis considered to be a benign melanocytic tumor. More commonly known as moles.

non-Hodgkin lymphoma A type of lymphoma with 30 different subtypes; usually referred to simply as lymphoma.

nonmelanoma A collective term for basal and squamous cell carcinomas.

nonmyeloablative transplant Also called a "mini" transplant. A newer form of bone marrow transplant that utilizes lower doses of chemotherapy and radiation so that the patient still possesses some of his or her own blood cells when receiving the transfusion; more commonly used in older patients.

nonpolar Neutral or uncharged.

nonseminomas Type of testicular cancer characterized by earlier onset, faster growth rate, and faster spreading than seminomas.

non-small cell lung cancer (NSCLC) A type of lung cancer characterized by slower growth and metastasis; 85% of lung cancers.

nuclear imaging Imaging that analyzes the physiology or biochemical activity rather than the anatomy of the body.

nucleic acids Biomacromolecules that store and convey the genetic information.

nucleotide The monomer unit of nucleic acids.

nucleus The dense region in the center of the atom that contains positively-charged protons and uncharged neutrons.

oat cell cancer *see* small cell lung cancer.

oligosaccharide Three to ten monosaccharides linked together.

oncogene A cancer producing agent.

oncology The medical subspecialty dealing with the study and treatment of cancer.

oophorectomy The surgical removal of the healthy ovaries and fallopian tubes.

orchiectomy The surgical removal of the testicles.

orchitis The painful inflammation of the testicle, often the result of a bacterial or viral infection.

organ Two or more tissues working in a coordinated fashion to carry out one or more specific functions.

organ system Multiple organs working together to perform a particular function.

organelles Membrane-bound structures that allow for the compartmentalization of cellular functions.

organism A single living being consisting of anywhere from one cell to quadrillions.

other malignant neoplasms A Pap test result indicating the presence of malignancies such as malignant melanoma, sarcomas, or lymphoma that rarely occur in the cervix.

ovaries Glands of the female reproductive system that produce the female gamete (the egg) and are the main source of female sex hormones.

p21 Protein that encodes for a protein that arrests the cell cycle binding to and inhibiting cdks.

p53 Protein that is a gatekeeper that functions at the checkpoints late in G_1, S and late in G_2 regulating the expression of many genes that promote cell cycle arrest, DNA repair, and apoptosis.

palliative surgery Surgery done to remove or debulk cancerous masses that may be compressing nerves causing pain or interfering with the functionality of nearby organs.

Pap test A microscopic examination of the cells taken from the inner surface of the cervix.

paracentesis An aspiration of abdominal fluid.

pathologist A physician who specializes in interpreting and diagnosing changes in bodily fluids and tissues that occur in response to disease.

pelvic examination An internal examination of the uterus, vagina, ovaries, fallopian tubes, bladder, and rectum.

pelvic exenteration The removal of the urinary bladder, prostate, or uterus during rectal cancer surgery.

pelvic lymph node dissection A removal of the lymph nodes adjacent to the female reproductive organs to examine for the presence of cancerous cells.

penetrance The proportion of individuals who possess a certain genotype and express its particular trait.

percent-free PSA The amount of free PSA relative to the total amount of PSA.

perineum Tissue between the anus and scrotum.

peripheral blood stem cells (PBSC) Stem cells present in the circulatory system.

peripheral proteins Proteins that are embedded in only half of the membrane's bilayer.

petechiae Tiny pinpoint red marks in the skin caused by bleeding.

Philadelphia chromosome (Ph¹) The chromosome resulting from a translocation from chromosome 9 to chromosome 22; creates a gene that produces an abnormal tyrosine kinase found mostly in leukemic white blood cells.

phospholipid bilayer Cell membrane in which phospholipids are aligned side-by-side in two inverted layers such that the polar heads are on the outside surfaces and the nonpolar tails are in the interior of the membrane.

phospholipid A molecule that consists of two fatty acids and a phosphate group bound to a glycerol molecule; the principle component of cellular membranes.

phosphorylation The addition of a phosphate group (PO_4^{2-}) to one of the amino acids on the surface of a target protein.

photodynamic therapy *see* therapeutic UV exposure.

pituitary gland A pea-sized tissue located at the base of the brain and is a site of hormone production and secretion.

placebo Inactive molecule that has no treatment value.

planning tumor volume Includes a margin beyond the clinical target volume to compensate for the possibility of a slight shift in the position of the diseased tissue that may occur as the result of a patient's movements between or even during treatment.

plasma Liquid portion of blood with clotting factors.

plasma membrane Outer membrane that acts as a safety barrier and provides a critical layer of protection that allows cells to properly function.

platelets Cell fragments in the blood involved in blood clotting.

pneumonectomy The removal of an entire lung.

polar Having an uneven charge distribution, creating partially charged regions within the molecule.

polymer A large molecule formed from the linking of smaller molecules called monomers.

polyp A growth that extends from the surface of a mucous membrane.

polypectomy Removal of the peduncle polyp from the colon.

polysaccharide A carbohydrate containing more than ten monosaccharides linked together.

positron emission tomography (PET) A diagnostic examination that assesses the activity levels of tissues in an attempt to pinpoint the location of cancerous tissue, which is identifiable by being more metabolically active than normal tissue.

previvors Those who do not have cancer but are at a higher risk for its development based on results from genetic testing or the presence of abnormal microscopic cells in an examined tissue.

primary peritoneal carcinoma A rare type of cancer that originates outside the ovaries in the peritoneum and mimics the symptoms of ovarian cancer.

private cord blood registries Store units of cord blood donated by families who have just had a baby; the cord blood is used only for the child or its family members.

progesterone A female hormone that plays a role in pregnancy and breast development.

prognosis Outlook for recovery.

prophylactic cranial irradiation Radiation focused on the brain; often used on lung cancer patients to prevent metastasis to the brain.

prophylactic mastectomy The removal of breast tissue prior to the development of cancer as a means of prevention in high-risk women.

prophylactic surgery Also known as preventative surgery. Surgery performed to reduce the risk of developing cancer.

ProstaScint scan A procedure which uses a radioactive antibody specific for PSMG to determine the location of cancerous cells within the prostate and surrounding tissues.

prostate A gland of the male reproductive system that produces a fluid to protect the sperm.

prostate-specific antigen (PSA) A protein component of prostate fluid that can be detected in a blood sample.

prostate-specific membrane antigen (PSMA) *see* prostate-specific membrane glycoprotein.

prostate-specific membrane glycoprotein (PSMG) A glycoprotein found only in the plasma membranes of benign hypertrophic and cancerous prostate cells.

prostatic intraepithelial neoplasia (PIN) A condition of the prostate in which the epithelial cells have altered characteristics but are not considered cancerous; theorized to be the initial stages in the development of prostate cancer.

protein Type of biomacromolecule composed of amino acids that is capable of a wide array of functions.

proteomics A method of analyzing proteins in the blood to detect a characteristic "signature" that may signal a particular cancer at the earliest stages.

proteosome Subcellular structure that degrades proteins that are damaged or no longer needed by the cell; often referred to as the garbage disposal of the cell.

proteosome inhibitors Treatment that results in cell cycle arrest and death by apoptosis.

proto-oncogene A precursor of an oncogene, which could be converted into an oncogene with a gain-of-function mutation.

PSA density One measure used to determine the likelihood of prostate cancer; determined by dividing the PSA value by the volume of the prostate.

PSA velocity The rate at which the PSA level rises.

public cord blood registries Store units of cord blood donated by families who have just had a baby; the cord blood can be given to anyone who needs it and is a match.

public health The practice of maintaining and improving community health policies, standards, and programs.

punch biopsy An instrument with a round blade at the end is positioned vertically over the lesion and pressure is applied downward as it is repeatedly rotated removing the skin lesion.

quiescent Phase of cell cycle entitled G_0 in which cells are arrested in their growth.

radiation therapy A form of treatment that uses X-rays to destroy tumors.

radical hysterectomy The reomoval of the uterus, cervix, the upper parts of the vagina, and the connective tissues that keep the uterus in place.

radical inguinal orchiectomy The removal of an affected testicle though an incision in the groin.

radical laproscopic prostatectomy A radical prostatectomy performed through several small incisions just above the pubic bone through which a laparoscope and surgical instruments are inserted.

radical perineal prostatectomy A radical prostatectomy performed through an incision in the perineum.

radical prostatectomy The surgical removal of the entire prostate.

radical retropubic prostatectomy A radical prostatectomy performed using a T-shaped incision above the pubic bone.

radiologist Physician that specializes in the use of X-rays or radioactive substances in the diagnosis or treatment of medical conditions.

radionuclide bone scan A procedure that utilizes the injection of radioactive material into a patient's bloodstream to locate areas in the skeleton that contain highly metabolically active cells.

radiotracer A radioactive molecule that accumulates in abnormal tissue after injection into the body.

radon An invisible, odorless, radioactive gas that exists naturally in areas where there are concentrations of uranium in the ground.

Rb Protein that binds to E2F and inhibits the expression of S phase genes.

reconstructive surgery Surgery that attempts to restore a natural appearance to a person's body following the removal of cancerous tissue.

rectovaginal exam A procedure used to feel the location and size of the pelvic organs.

rectum The last 15–16 cm of the large intestine where the feces are stored.

Reed–Sternberg cell A large, malignant while blood cell with two or more nuclei; characterizes Hodgkin's lymphoma.

resectoscope An instrument that is used during a TURP to cauterize excess prostatic tissue.

restriction point A major regulatory checkpoint late in G_1 of the cell cycle that is largely under the control of external growth factors.

retroperitoneal lymph node dissection (RPLND) The removal of lymph nodes within the abdomen through either a standard surgery or laparoscopy.

retroperitoneal Behind the abdominal cavity.

ribonucleic acid (RNA) A type of nucleic acid.

ribose The sugar component found in RNA nucleotides.

ribosomes Macromolecular complexes where protein synthesis (translation) occurs.

robotic surgery A form of laparoscopic surgery in which the surgeon controls all of the instruments indirectly through a set of hand controls.

rough endoplasmic reticulum The portion of the ER that possesses ribosomes on its cytoplasmic surface, producing a rough, knobby appearance under a microscope.

Rous sarcoma virus (RSV) An oncogenic virus that causes connective tissue tumors in chickens.

S phase The phase in the cell cycle in which DNA synthesis takes place and chromosomes are replicated.

sarcoma Cancer that originates in the connective tissue such as cartilage with bone, muscle, or fat.

second-look (look-back) surgery Allows the surgeon to determine the success or failure of treatment for ovarian cancer.

segmental resection Removal of the affected portion of the colon and surrounding tissue and the reconnection of the remaining two portions of the intestine; most commonly performed on patients with stages i–iv colorectal cancer.

segmentectomy *see* wedge resection.

seminomas Type of testicular cancer believed to arise from the epithelium of the sperm-producing/transporting tubules in the gland.

sensitivity The probability that a test will be positive for a person who has the disease.

sentinel node The first lymph node that receives lymph collected from the area of interest.

sentinel node biopsy/dissection A revised lymph node biopsy in which only the sentinel node is removed for assessment. Subsequent nodes are removed only if cancerous cells were present in the sentinel node.

Sertoli cells Cells of the testes that provide nourishment to the germ cells and that produce sperm.

sextant biopsy A biopsy of the prostate in which up to 24 core samples, from the top, middle, and bottom of both the right and left sides, are taken to obtain a representative sample of the prostate.

shave biopsy A procedure in which a blade is used to remove the outer layer of skin from an area suspected of containing a cancerous lesion; also known as a curettage biopsy.

sigmoid colon The small portion of the large intestine between the descending colon and the rectum.

sigmoidoscope A thin flexible tube with a camera at one end used in a sigmoidoscopy to observe the inner lining of the rectum and sigmoid colon.

sigmoidoscopy An examination of the interior lining of the rectum and sigmoid portion of the large intestine using a sigmoidoscope.

signal transduction cascade pathway The series of reactions that takes place in a cell in response to the binding of a signaling molecule to its receptor.

simple hysterectomy The removal of the uterus and cervix.

sister chromatid Two identical chromosomes bound to a single centromere.

skin self examination An examination recommended to be performed on a regular basis in order to detect changes, abnormalities, or cancerous lesions.

small intestine The portion of the digestive tract where the majority of digestion and absorption of nutrients takes place.

small cell lung cancer (SCLC) A type of lung cancer that arises in the bronchi and is characterized by faster growth and metastasis; 15% of lung cancers.

smooth endoplasmic reticulum The portion of the ER involved in lipid synthesis and modification and the degradation of toxic substances.

somatic tissue Any body cell that is not a germ cell.

sonogram A 2-D black and white image of the area being examined generated by the ultrasound.

specificity The probability that a test result will be negative for a person without the disease.

speculum A metal or plastic device used to examine the vagina and cervix during a pelvic examination.

spermatogenesis Production of sperm.

spermocytic seminoma A rarer form of seminoma.

spindle fibers Protein structures which pull chromosomes to the opposite poles of the cell and deliver them into the daughter cells during cell division.

spiral computed tomography (CT) scan An advanced form of CT in which the X-ray beam proceeds through the body in a spiral or helical fashion while the table moves forward at a constant rate.

splenectomy Surgical removal of the spleen.

sporadic mutations Mutations that occur spontaneously during the lifespan of a cell.

squamous carcinoma A type of NSCLC that arises near the bronchus and tends to grow slowly.

squamous cell carcinoma A category of squamous intraepithelial lesion composed of flat, thin squamous cells covering the surface of the cervix and that may become invasive. A common form of skin cancer arising from the malignant transformation of squamous cells.

squamous cells Flat cells that form the thin outer layer of tissue or organ.

squamous intraepithelial lesions (SILs) A Pap test result indicating changes in cells of the cervix. The three subcategories are low grade SIL, high grade SIL, and squamous cell carcinoma.

stage grouping The staging of prostate cancer using both the TNM system and Gleason score.

staging The process of determining the degree to which cancer has developed and potentially spread.

staging surgery A procedure in which tissues from the affected area are removed for biopsy in order to determine the stage of the tumor.

stem cells Undifferentiated cells that have the ability to divide continuously.

stoma The incision made in the abdomen during a colostomy through which a portion of the large intestine protrudes.

stomach A muscular organ that contracts rhythmically to squeeze and churn its contents and is the site of some protein digestion.

stromal cell carcinoma Tumors found in the stromal cells of the ovary.

stromal cells Cells of the gonads that produce the sex hormones.

stromal tumor Rare type of testicular cancer that develops in the supportive tissues of the testicles where hormones are produced.

subcutaneous tissue Tissue located just beneath the skin.

surgery A procedure used to repair or remove a damaged or diseased portion of the body.

surgical biopsy A biopsy in which one or more regional lymph nodes are removed with the tumor.

syngeneic bone marrow transplant The infusion of stem cells from an identical twin.

systemic chemotherapy Chemotherapeutic drugs that are administered intravenously so that they travel through the circulatory system to combat any malignant cells that have metastasized.

systemic radiation therapy Radiation therapy used to treat metastatic cancer found in many locations.

T lymphocyte A type of white blood cell that regulates the body's immune responses.

tamoxifen Estrogen antagonist used in the treatment of metastatic breast cancer.

telomerase An enzyme that prevents the shortening that typically occurs when the chromosomes are replicated.

telomere DNA sequence at the ends of each chromosome.

teratoma A type of nonseminoma.

testicular self-examination A recommended examination for males to perform on a regular basis so that changes or abnormalities can be detected.

therapeutic UV exposure A treatment for psoriasis that involves the application of a drug that, when exposed to UVA light, becomes toxic to diseased skin cells.

thoracic surgeon A physician who specializes in surgical treatment of diseases/conditions of the chest.

thoracoscopy Laparoscopic procedure of the chest cavity.

thrombocytes *see* platelets.

thrombocytopenia A decrease in the number of platelets, which could cause excessive bleeding.

thymine dimer Two thymines adjacent to one another in a single DNA strand that have covalently bonded to one another.

tissue A collection of cells that work together to perform a particular function.

TNM system The most frequently used staging system for many cancers.

topical chemotherapeutic A cytotoxic drug applied directly to the affected area of the skin.

topoisomerase An enzyme that unwinds the DNA helix during DNA replication.

totipotent Potential to produce all of the different types of cells found in the adult.

tracers Radioactive molecules used by nuclear scans that accumulate in specific areas of the body and can be detected by a computer.

transcription A copy of a gene's sequence is made in the form of mRNA.

transcription factor Protein that binds to genes and regulates whether or not they will be expressed.

translation The mRNA message synthesized during transcription is used to synthesize a protein.

transrectal ultrasound (TRUS) A procedure used to obtain an internal image of the prostate through the rectum.

transurethral resection of the prostate (TURP) A procedure in which excess tissue of the prostate is cauterized.

transvaginal ultrasound (TVUS) A procedure used to visualize the organs of the pelvic area.

transverse colon The portion of the large intestine between the ascending and descending colons.

tubal ligation Surgical tying of the fallopian tubes.

tubulin A protein that aggregates in a chain-like fashion to form the polymers of microtubules.

tumor Also known as neoplasm, an abnormal mass of cells with no useful function that can impact the functionality of the tissue in which it is present or surrounding areas.

tumor dormancy The appearance of secondary tumors only after the surgical removal of a primary tumor.

tumor markers Molecules that can be monitored in order to provide critically important information about the activity of a known tumor or to even detect the presence of previously unknown malignancy.

tumor progression A process whereby cells mutate independently of one another as they grow, resulting in a collection of genetically different subpopulations with unique growth and metastatic potentials.

tumor suppressor gene A gene that inhibits a cell's proliferation and promotes its death.

ulcerated A melanoma that does not have a layer of epidermis over the top of the tumor.

ulcerative colitis A form of chronic inflammatory bowel disease that increases one's risk for colorectal cancer.

ultrasound Also known as ultrasonography or sonography. An imaging technique that uses frequency sound waves to view tissues and organs inside the body.

ultraviolet radiation A type of radiation emitted from the sun, tanning booths and sunlamps that can damage DNA and alter the functionality of a cell.

umbilical cord stem cells Stem cells isolated from the umbilical cord that has been cut and removed from the mother after a baby is born and have the potential to develop into healthy red and white blood cells and platelets.

urinalysis A variety of chemical and microscopic tests performed on urine samples.

UVA 320–400 nm ultraviolet radiation that can damage collagen fibers.

UVB 280–320 nm ultraviolet radiation that breaks noncovalent hydrogen bonds between nitrogenous bases of DNA.

UVC 100–280 nm ultraviolet radiation that never reaches the Earth's surface because it is reflected or absorbed by the atmosphere.

vaccination Stimulation of the immune system to provide protection in the form of immunity against an infectious agent.

varicocele Enlargement and lumpiness in the testicle caused by the dilation of the veins.

vascular endothelial growth factor (VEGF) A potent angiogenic activator produced by tumor cells.

VEGF receptor A protein on the surfaces of endothelial cells of blood vessels that bind VEGF.

vein Blood vessel that carries blood to the heart.

venule Smaller blood vessels between the veins and capillaries.

video-assisted thoracic surgery see thoracoscopy.

villi Finger-like projections in the small intestine that enhance the absorption of nutrients.

virtual colonoscopy A CT scan of the abdomen that creates 2-D and 3-D images of the large intestine.

virus Infectious agents incapable of reproducing independently.

watchful waiting No active form of treatment is administered but the patient is tested frequently and observed closely; also known as active surveillance.

wedge resection The removal of only a portion of the lobe of a lung.

xeroderma pigmentosum (XP) A condition in which genes for DNA repair enzymes possess mutations so that there is a dramatically reduced ability to repair DNA damage caused by UV light.

X-ray A higher energy electroradiation that can be used for diagnostic imaging.

X-ray imaging A diagnostic technique that utilizes X-rays to provide an internal view of the body.

yolk sac carcinoma A type of nonseminoma.

zygote The initial cell of a human being created by the fusion of the sperm and egg.

Index

Page numbers in *italics* refer to figures, those in **bold** refer to tables or boxes.